Hazardous Waste Siting
and
Democratic Choice

American Governance and Public Policy

A SERIES EDITED BY

Barry Rabe and John Tierney

This series examines a broad range of public policy issues and their relationship to all levels of government in the United States. The editors welcome serious scholarly studies and seek to publish books that appeal to both academic and professional audiences. The series showcases studies that illuminate the successes, as well as the problems, of policy formulation and implementation.

Hazardous Waste Siting and Democratic Choice

EDITED BY
Don Munton

GEORGETOWN UNIVERSITY PRESS / WASHINGTON, D.C.

Georgetown University Press, Washington, D.C. 20007

10 9 8 7 6 5 4 3 2 1 1996
THIS VOLUME IS PRINTED ON ACID-FREE ∞ OFFSET BOOK PAPER

Library of Congress Cataloging-in-Publication Data

Hazardous waste siting and democratic choice / edited by Don Munton.
p. cm.—(American governance and public policy)
Based on papers presented to a workshop at the University of British Columbia in September 1993.
1. Hazardous waste sites—Location—Political aspects—Congresses. 2. Hazardous waste sites—Location—Citizen participation—Congresses. 3. NIMBY syndrome—Congresses. I. Munton, Don. II. Series.
TD1050.P64H38 1996
363.72′875—dc20 96-11858
ISBN 0-87840-625-5 (cloth)

Contents

Acknowledgements ix

DON MUNTON

Introduction: The NIMBY Phenomenon and Approaches to Facility Siting 1

NATURE OF NIMBY 2, WASTE FACILITIES VERSUS WASTE REDUCTION: THE DEBATE 4, NEED FOR FACILITY SITING 6, OBJECTIVES 9, SITING STRATEGIES 10, THE VOLUNTARY-CHOICE PROCESS 19, NATURE OF THIS VOLUME 23, NOTES 28, REFERENCES 38

Part 1

Getting Beyond NIMBY: Cases 55

GEOFFREY CASTLE, DON MUNTON

Voluntary Siting of Hazardous Waste Facilities in Western Canada 57

BACKGROUND: THE WESTERN CANADIAN CASES 58, THE ALBERTA CASE 59, THE BRITISH COLUMBIA CASE 62, THE MANITOBA CASE 66, WHY VOLUNTARY PROCESSES SUCCEED 74, APPLICABILITY OF THE VOLUNTARY MODEL 76, EQUITY CONSIDERATIONS 77, NOTES 79, REFERENCES 82

BARRY G. RABE, WILLIAM C. GUNDERSON, PETER T. HARBAGE

Alternatives to NIMBY Gridlock: Voluntary Approaches to Radioactive Waste Facility Siting in Canada and the United States 84

THE CASE OF LOW-LEVEL RADIOACTIVE WASTE 86, EARLY STAGES OF LLRW MANAGEMENT IN CANADA AND THE UNITED STATES 86, THE EVOLUTION OF CANADIAN LLRW MANAGEMENT 89, FEATURES COMMON WITH RELATED SITING AGREEMENTS 96, THE EVOLUTION OF AMERICAN LLRW MANAGEMENT 98, AN ATTEMPT AT VOLUNTARISM: THE CASE OF NEBRASKA 101, CONCLUSIONS 105, NOTES 106, REFERENCES 106

Michael E. Kraft

Democratic Dialogue and Acceptable Risks: The Politics of High-Level Nuclear Waste Disposal in the United States 108

THE POLITICAL CONTEXT OF PUBLIC OPPOSITION 109, U.S. NUCLEAR WASTE POLICY: FROM POOR DESIGN TO FLAWED IMPLEMENTATION 111, PUBLIC REACTIONS TO NUCLEAR WASTE REVISITED 128, CONCLUSIONS: THE PROMISE OF DEMOCRATIC DIALOGUE 133, NOTES 136, REFERENCES 137

David Morell

Hazardous Waste Management and Facility Siting in California 142

INTRODUCTION 142, THE CALIFORNIA HAZARDOUS WASTE MANAGEMENT ACT 143, IMPLEMENTATION ACTIONS AFTER PASSAGE OF AB 2948 149, DOW CHEMICAL COMPANY CASE STUDY (CONTRA COSTA COUNTY) 157, RHONE-POULENC BASIC CHEMICAL COMPANY CASE STUDY (CONTRA COSTA COUNTRY) 162, CHEMICAL WASTE MANAGEMENT CASE STUDY (KINGS COUNTY) 170, CONCLUSION 174, NOTE 178, REFERENCES 178

Don Munton

Siting Hazardous Waste Facilities, Japanese Style 181

INDUSTRIAL AND HAZARDOUS WASTE IN JAPAN 182, TREATMENT AND DISPOSAL OF HAZARDOUS WASTES 184, CONSTRAINTS AGAINST PUBLIC OPPOSITION 186, NIMBY IN JAPAN 190, EFFECTS OF NIMBY IN JAPAN 195, STRATEGIES FOR SITING HAZARDOUS WASTE FACILITIES 198, CONCLUSION 207, NOTES 212, REFERENCES 222

Robert Seeliger

Siting Hazardous Waste Incinerators in Germany: From Political Imposition to Public Involvement? 230

INTRODUCTION 230, HAZARDOUS WASTE MANAGEMENT IN GERMANY: TECHNICAL CHALLENGES AND POLITICAL CONSTRAINTS 231, THE LEGAL FRAMEWORK OF WASTE AND CONFLICT MANAGEMENT 235, SITING OF HAZARDOUS WASTE INCINERATORS: CASE STUDIES 238, PATTERNS AND PROBLEMS 252, THE HAZARDS OF SITING AND THEIR POLITICAL RECOGNITION 255, CONCLUSION: FROM POLITICAL IMPOSITION TO PUBLIC INVOLVEMENT? 259, NOTES 262, REFERENCES 265

Part 2
Public Perceptions 271

Clyde Hertzman, Aleck Ostry

Community Risk Perception and Waste Management in Three Communities at Different Stages in the Siting Process 273

INTRODUCTION 273, METHODS 275, RESULTS 277, DISCUSSION 286, REFERENCES 288

Susan J. Elliott, S. Martin Taylor

Worrying about Waste: Diagnosis and Prescription 290

INTRODUCTION 290, PSYCHOSOCIAL EFFECTS: A REVIEW OF THE EVIDENCE 294, THE ONTARIO STUDY 300, APPLICATIONS AND IMPLICATIONS 310, REFERENCES 315

Part 3
Options and Strategies 319

Alun Richards

Using Co-management to Build Community Support for Waste Facilities 321

MANITOBA'S SITING PROCESS 322, THE COMMUNITY'S ROLE IN A VOLUNTARY PROCESS 325, THE PATTERN OF ISSUES 326, STAGES, ISSUES, AND CO-MANAGEMENT STRUCTURES 327, THE MONTCALM AGREEMENTS 328, A NEED FOR FLEXIBILITY 330, LESSONS LEARNED 331, CONCLUSION 336, REFERENCES 337

Howard Kunreuther

Voluntary Procedures for Siting Noxious Facilities: Lotteries, Auctions, and Benefit Sharing 338

INTRODUCTION 338, WELL-SPECIFIED STANDARDS FOR DEALING WITH TRANSBOUNDARY RISKS 339, ROLE OF COMPENSATION OR BENEFIT SHARING 344, A PROPOSED VOLUNTARY PROCEDURE 345, STRATEGIES FOR SITING LOW-LEVEL RADIOACTIVE WASTE IN ONTARIO 350, OPEN QUESTIONS 353, NOTES 354, REFERENCES 355

CHRISTOPHER ZEISS

Directions for Engineering Contributions to Successfully Siting Hazardous Waste Facilities 358

INTRODUCTION 358, SETTING THE STAGE—A CASE STUDY OF A MAROONED SITING ATTEMPT 360, ENGINEERING SITING INTERVENTIONS 364, SUMMARY OF SITING INTERVENTION PRINCIPLES 379, CONCLUSIONS—DESIGN AND USE OF ENGINEERING SITING INTERVENTION MEASURES 389, REFERENCES 391

Contributors 395

Index 397

Acknowledgments

All but one of the chapters in this volume are revised versions of papers presented to a workshop held at the University of British Columbia (UBC) in September 1993. Much of the conference organizational work was carried out by Geoffrey Castle, then with the UBC Sustainable Development Research Institute (SDRI). The conference was financially supported by a grant from the Social Sciences and Humanities Research Council of Canada, by the UBC Faculty of Graduate Studies, and by the SDRI, under a grant from the British Columbia Ministry of Environment, Lands and Parks. The editor and all of the contributors express their deep gratitude for this assistance. The final editorial work on this volume was carried out at York University while the editor was a visiting research associate at the Centre for International and Strategic Studies. The support of the Centre's director, David Dewitt, and of its staff, particularly Rose Edgecombe, Wendy Kubasik, and Heather Chestnutt, was magnificent and is gratefully acknowledged.

Introduction: The NIMBY Phenomenon and Approaches to Facility Siting*

DON MUNTON

Opposition to siting of waste facilities has become commonplace throughout the industrial world.[1] Much of it arises from what have come to be called NIMBY movements (Not in My Backyard), organized local groups that oppose the siting of facilities in their neighborhoods or communities.[2]

Indeed, significant opposition to LULUs (locally unwanted land use) in general has become one of the hallmarks of the era.[3] It is by no means restricted to siting of waste facilities, although studies have shown such facilities are among the very most unwanted additions to a community, ranking with nuclear power plants.[4] NIMBY responses continue to develop, opposing not only hazardous and non-hazardous waste facilities but also a range of other developments, from nuclear and non-nuclear power plants to prisons, highways, and airports, to public housing, group residences, and research laboratories. Interestingly, but not surprisingly, local opposition has also confronted amusement theme parks, including a Disneyland-to-be in Virginia and a planned Lego theme park in California.[5] If the masses can be inflamed by the threat of Disney's America and Lego World coming to town, it is perhaps not surprising that Waste Management, Inc., meets with some opposition.

Whether or not "NIMBY" is the correct term for all these activities is debatable; many individuals are opposing the construction of waste facilities in general, not just ones that might be located in close proximity. The term has nonetheless taken on a generic meaning and achieved a strong degree of popular recognition and usage. It has become commonplace to find articles offering advice on the whys and the hows of dealing with NIMBY-style opposition aimed at an ever-widening range of interested parties.[6] Much of this applied literature and some of the academic writings on the NIMBY phenomenon portray it explicitly or implicitly as an unambiguously negative force. In extreme variants it is seen as the triumph of a sort of irrational local tribalism or at least of a misinformed self-interest over the wider public good. This volume does not share that bias.

NATURE OF NIMBY

The term "NIMBY" likely had a pejorative connotation when it was first coined, before coming into common parlance. Like many other terms, it lost much of that negativism some time ago as a result of widespread usage.[7] Its use now, a few critics aside, is more neutral and analytical than negative and value-laden. It is used here to refer generically to opposition to the siting of unwanted facilities, largely by persons in the local area.

What are the roots of the NIMBY phenomenon? Critics have tended to see it arising from emotionalism, narrow self-interest, misinformation, and manipulation by unscrupulous groups. It is surely both more and less than that. Most generally, local opposition to some planned or possible land use such as a waste facility is a manifestation of the classic, and well-known, public goods (or collective goods) problem. Goods that benefit all, and from which no one can reasonably be excluded, are difficult to provide because there is an incentive for beneficiaries to be free riders. Even if everyone agrees that the goods should be provided, therefore, the costs tend to be born by a few. And hence trying to provide public goods is difficult and conflict-prone. Some of the solutions developed to deal with public goods problems are discussed in this volume. But the NIMBY phenomenon also has a special character because of the nature of the goods in question. Its particular roots are complex but have at least three major components.[8]

A vital part of the opposition to planned facilities results from what is perceived to be an attack on basic democratic principles. NIMBY is in this sense an assertion of the rights of individuals and communities to control matters that affect them. A decision to site in a particular location, often not made in any sense democratically, is opposed then in part because of the process by which it was made. As one NIMBY group representative put the matter: "We were told literally that everything was chipped in stone. There were no alternatives. . . . [We] weren't fighting a waste disposal site. . . . We were fighting for our democratic rights."[9] "The roots of NIMBY," Matheny and Williams argue, "are in the failure of meaningful public participation" (1988, 43). Or, as Audrey Armour notes similarly, it results from failure to acknowledge the legitimacy of concerns and opposition (1988).[10]

Another part of the opposition that emerges clearly stems from concerns about potential health and environmental problems arising especially from hazardous waste facilities. They represent what might be called the "dread" factor (Gerrard 1994, 100). These concerns and fears are now widespread among the public and are more than evident in much of the nonacademic literature on hazardous waste problems, and in some academic studies (e.g., Bord and O'Connor 1992; Vittes, Pollock, and Lillie, 1993). They need no elaboration here. It should suffice to note, in the words of one firsthand observer of a siting controversy, that the fear he found "was deep and it was complex and it was certainly far more entrenched than simple hysteria."[11] While some experts argue that the perceived health risks are often greater than the actual risks,[12] the fact is that the actual risks remain largely unknown, and perhaps unknowable. Toxicological studies do not, and cannot, provide definitive answers to questions about long-term risks from contaminants. It is nonetheless arguably the case that the perceived risks of hazardous waste treatment facilities are often exaggerated relative to those of such widely accepted industrial neighbors as chemical plants and oil refineries (Salcedo, Cross, and Chrismon 1989, 50).

Yet another part of the NIMBY phenomenon is concerns about fairness and equity. These, many analysts agree, are less potent than those about the risks (Morell 1984; Portney 1988, 1991) but by no means unimportant. The questions asked are common enough. Why should a facility be located *here,* rather than somewhere else? Why should *we* have to bear the costs of it? Why not, well, someone *else*? The costs implied in such questions are both economic and social in nature, such as loss of property value and a negative community image.

Our assumption in this book is that all three of these types of concerns are important sources and components of the NIMBY phenomenon. All of them need to be addressed if agreement is to be reached on siting facilities. In particular, opposition to facility siting has to be recognized as a political and social response, and a legitimate one, not seen as blind ignorance about technical matters.

However neutral the term has become, there is still a prevailing sense in much of the literature on siting that NIMBY is basically a bad thing. Analysts search for and advocate ways to "overcome" it, implying that it is a hostile enemy to be defeated. At least one waste facility proponent labeled opposition to a chosen site as a case of "environmental immorality."[13] As with most phenomena, NIMBY movements cannot be simply characterized as bad or good. Some of their supporters argue that "public concern and fear of waste management facilities is both well-founded and healthy" (quoted in Armour 1984, 77). "NIMBY is a natural, rational reaction to a possible threat to one's community" (Jackson 1990, 304). NIMBY movements, they argue, have forced a more serious examination of inadequate facility plans, have prevented some environmentally unsafe projects, and have forced a more thoughtful search for alternatives such as waste reduction measures.[14] Economists have also debated whether local opposition to siting may lead to economically efficient outcomes, by ensuring that facilities will be located where externalities are low (Becker 1983; Hamilton 1993).

NIMBY movements have also, arguably, in some cases, saved taxpayers millions of dollars that might have been spent on ill-conceived facilities in the long run, as well as cost them more modest sums in ultimately failed siting processes in the short run. One such case involves the elaborate and expensive plan developed in the early 1980s to provide a large integrated waste treatment and disposal facility in Ontario, Canada. After much opposition and long delays, the provincial government decided in early 1995 to scrap the plans developed by the Ontario Waste Management Corporation (OWMC). The government concluded that "much had changed since the OWMC had started its hunt for a plant site," and there was now great doubt whether such a stand-alone facility was the best option, as opposed to waste reduction and treatment on site, and whether public funds should be spent doing what industry ought to do itself.[15] While the province was left, in the short-term, without a ready solution to its hazardous waste problems, the opposition and delays arguably had saved the taxpayers from paying for an expensive, and likely underutilized, facility. As one environmental advocate has observed, "The problem is not NIMBY." Whether such groups are, on balance, forces for good or evil, "the problem is finding safe and effective means of waste management."[16]

WASTE FACILITIES VERSUS WASTE REDUCTION: THE DEBATE

Two major arguments have long dominated the ongoing debates about hazardous waste management facilities. One is that more treatment and disposal facilities are urgently needed to cope with ever-increasing volumes of hazardous wastes. The other is that waste minimization and pollution prevention should take priority over the provision of facilities for waste treatment.[17] The two arguments have long divided those who are concerned about waste management issues. The latter tends to be an argument of environmental groups, some of which oppose all new waste treatment and disposal operations on the grounds that all such facilities provide an excuse for continued pollution and undermine waste minimization and pollution prevention efforts. The former argument is often made by some policy officials and technically oriented experts in the field and is accepted by some environmental groups.[18]

These two arguments have been seen as expressing contradictory conceptions of how to deal with the problem of managing hazardous wastes. The premise of this volume, in contrast, is that both arguments are valid. To pose these as alternatives is to frame the debate wrongly, in a way that may have been valid once but is no longer. The choices in waste management are not either-or. Both facility siting and waste reduction are needed. To some extent, both are happening, though recently there has been more of the latter than the former.

Over the 1980s, the amounts of hazardous wastes generated in most industrial countries grew steadily. In the early 1990s, in contrast, generation levels began to decline. The volumes of wastes accordingly being sent to commercial hazardous waste disposal facilities and landfills began to drop.[19] There have also been numerous closings of commercial incinerators, and plans for the construction of various new facilities have been put on hold. These cancellations and closings are not primarily due to local citizen opposition but, rather, to the reduced demand for the facilities themselves.[20]

Some analysts and waste disposal operators tend to attribute this situation to recession-based industrial-production declines and to the lack of effective regulation of waste generators—that is, to short-term overcapacity due to the slumping economy and the failure of authorities to prevent illegal dumping. The lack of progress in cleaning up contaminated sites in most countries also means that little waste is being moved from these locations to approved disposal facilities.[21]

At least as important as these factors, though, is the fact that business thinking and practices (as well as those of government) have clearly begun to change. Wastes are increasingly being handled on site rather than off site, being treated rather than simply disposed of, and being minimized rather than unthinkingly generated. The concept of waste minimization or waste reduction was largely a vision a decade ago.[22] It has now taken hold and is becoming a reality. Waste minimization had become "an article of faith for corporate environmental managers" by the late 1980s (Winton et al. 1987). It was common practice by the early and mid-1990s. Aggressive efforts to reduce waste generation are under way across the corporate globe. The success of these efforts largely explains the declining demand for waste treatment and final-disposal sites.[23]

The increasing popularity and attractiveness of waste minimization measures is partly, but only partly, a function of increased waste disposal costs, which are, in turn, partly a function of higher costs for the treatment now routinely required by governments.[24] Increased waste disposal costs and politically imposed costs aside, more and more waste-generating companies are recognizing that cutting waste cuts expenses, and not just waste disposal expenses. Perhaps to the surprise of some in corporate boardrooms, it is becoming clear that waste reduction, in short, is cost-effective.[25] One of the pioneers,

Minnesota-based 3M Company, had saved itself hundreds of millions through its waste reduction PPP (Pollution Prevention Pays) projects by the early 1980s (Collins and Saukin 1981; Hirschhorn and Oldenburg 1987; Hirschhorn 1988; Kelly 1990). The corresponding Dow Chemical minimization program, though less successful so far, is called WRAP (Waste Reduction Always Pays). The word has gotten out: "There is money to be made in waste reduction" (Nelson-Horchler 1988, 58). Process changes to reduce wastes can mean greater efficiency and render companies more competitive. "It's not just a matter of costs," an industry official has noted. "It is a matter of a plant's viability" (Goldbaum and Rotman 1989, 48). Waste reductions also improve worker safety and limit the extent to which companies might be legally liable for their wastes in the future.[26] Corporate executives who used to "think of waste as a noun" are now realizing, as one observed, that "waste is more of a verb."[27]

NEED FOR FACILITY SITING

The clear and continuing success of hazardous waste reduction efforts has significantly undercut one of the key arguments for the siting of treatment and disposal facilities, namely, that some means had to be found to deal with the apparently ever-increasing amounts of hazardous waste. The assumption that these amounts will be ever-increasing now seems questionable, to say the least. But that is not to say such new facilities are no longer needed. Waste reduction efforts, however successful, will not entirely eliminate the necessity for adequate treatment and safe disposal. "Even if the push for source reduction and waste minimization were realized," Mazmanian and Morell (1992, 179) argue, "a substantial amount of hazardous waste . . . would still remain to be treated and disposed of."[28] A concrete example (no pun intended) of this fact may be found in the chapter herein by Morell; despite waste reductions of about 80%, Dow Chemical in California still needs access to a treatment facility for wastes that must otherwise be shipped out of state, and across much of the country, for disposal.

Why do more waste facilities need to be sited if those currently in operation have surplus capacity and if waste generation is declining? There are a number of reasons. For one thing, the number of available facilities in most industrial countries has been dropping rather than holding constant, let alone increasing. And many of these available sites are environmentally unsound, even dangerous. They simply need to be replaced. The long-standing trend of closing sites has continued in the early 1990s. From around fifty such sites in the United States in the early 1980s, the total had shrunk to around twenty in the 1990s. Other countries have experienced a similar trend.[29] This decline has been due to tightening regulations governing waste disposal sites and the closure by governments of many inadequate or unsafe sites. It has also been due to the difficulties facing the opening of new facilities.[30] One of these difficulties, of course, though not the only one, has been the opposition of local residents through NIMBY movements.[31]

Another factor creating a new demand for waste facilities is the large number of contaminated sites that need to be cleaned up. These include not only former waste storage, disposal, and treatment operations and former military-weapons manufacturing and storage sites but also, potentially, a large number of supposedly nonhazardous waste disposal sites, such as municipal dumps.[32] These contaminated sites, we now recognize, can have serious environmental and public-health consequences and also impose significant economic costs on neighborhoods.[33] If the wastes from these sites are not to be moved merely from one inadequate disposal location to another, then proper facilities are required. The lack of such facilities and the high costs of those still operating also encourage illegal dumping and thus the creation of more and new contaminated sites.[34] Increased demand coupled with insufficient disposal capacity "threatens to increase illegal disposal, a result which would exact a terrible price in terms of environmental degradation and danger to human health" (Mazmanian and Morell 1992, 184).

Much opposition to waste facilities seems premised on the assumption that the status quo is more or less satisfactory. There is a prevailing sense that such facilities create new environmental problems where none existed, that they represent a step back from an acceptable status quo. For evidence that the status quo is far from satisfactory, we need look no further than the growing problem of abandoned contaminated sites, the ongoing use of municipal sewers for dumping hazardous wastes, and the dangers presented by waste storage. The environmental problems that have arisen recently at seemingly "safe" waste storage sites need little elaboration. Explosions, fires, and chemical reactions are by no means uncommon. Fires have broken out in hazardous waste storage facilities in Jacksonville, Florida (in 1984), and in St. Basile-le-Grande, Quebec (in 1988), in both cases sending toxic PCB fumes across populated areas (Gerrard 1994; Paquin 1989). However dangerous to human health waste incinerators may prove to be, such uncontrolled open fires are not a preferable disposal technique. A facility fire that got out of control almost destroyed Elizabeth, New Jersey. Another fire in a New Jersey landfill burned for over thirty years (Nelson-Horchler 1988). Arsonists set ablaze a dump filled with used tires in Hagersville, Ontario. It burned for days, defying attempts to put it out and requiring evacuations (Corr 1990). In a disaster that did not happen but might have, the European chemical giant Bayer was forced to create a football-field–

sized waste storage area when one of its incinerators was closed for extended repairs and no new facilities could be opened (Gilges 1991). This list of examples could be extended. The point is that temporary storage of dangerous wastes is often far from secure and safe. Adequate modern treatment and disposal facilities are needed.

Regulations in many countries governing waste management are becoming more stringent. One result is that much of the hazardous waste stream now comes from pollution control processes, not from normal industrial production, a fact often ignored. Disposing safely of these wastes is becoming more difficult. The ocean dumping of sewage sludge off the U.S. East Coast was prohibited in the early 1990s, as was incineration of wastes on ships in the North Sea. Concerns about emissions of dioxins and other toxins from combustion processes, including waste treatment processes, is also leading to stiffer regulatory action. The U.S. Environmental Protection Agency (EPA) began, in the early to mid-1990s, to tighten the emissions regulations for incinerators and industrial furnaces and to delay approvals for new units. Even in Japan, where incineration has been the municipal waste treatment option of choice for decades, a great deal of research is being conducted into toxin levels in incinerator emissions. This method of treating at least hazardous wastes may be in the process of being phased out, or at least scaled back drastically, across the industrial world.[35] If so, other types of treatment and disposal will have to be developed and put in place. And again, that means new sites for new facilities or more on-site capacity must be found.

Another factor arguing for the siting of some selected new waste facilities, albeit a less important one, is that the present situation is fostering growing levels of interjurisdictional conflict. Those jurisdictions that have operating facilities are under pressure not to accept wastes from those that do not, or at least not to accept the risks of transporting those wastes. Conflicts have emerged over both these issues and the problems of transboundary effects from waste facilities in the United States, Germany, Japan, Canada, and elsewhere (Kemezis 1990, 1991; Walters 1991; Steinzor and Smith 1988; Munton, this volume). For all these reasons, then, we would argue that the need to site new treatment and disposal facilities is real and has not disappeared.

Our usage of the term "facilities" does not necessarily mean permanent, large-scale, integrated plants. Nor, for that matter, should the term "siting" be taken to mean permanent placement.[36] There seems to be a trend away from such large-scale plants, of the kind that were the model in the 1980s, and this trend will likely continue. One fairly common alternative to such permanent facilities is mobile units, such as relatively small transportable incinerators. New technologies are also emerging as alternatives to incineration, such as chemical decomposition.[37] Even here, however, there are often siting problems. Some communities in the United States and Canada have opposed even the temporary location of such units in their areas to deal with, for example, existing stocks of PCBs. The siting of facilities thus should be seen as a matter of both long-term and short-term waste management through a wide variety of options.

To sum up, while there is now no doubt that waste minimization measures are significantly reducing the demand for treatment and disposal of newly produced hazardous wastes, there remains a need for additional modern, safe facilities, and thus a need to ensure either temporary or permanent siting of these. Industrial societies need both reduction and treatment/disposal. It is still the case that "the combined failure to move forward more rapidly on cleanup, treatment, and source reduction and on siting of new hazardous waste treatment facilities has become a significant stumbling block to realizing the overall goal of managing . . . hazardous waste more safely" (Mazmanian and Morell 1992, 179–80). The question of *how many* waste facilities need to be sited in the coming decade or so is not a question this volume can address. That is ultimately a matter for political jurisdictions in all industrial societies to resolve for themselves. Our view, though, is that the number needed is not zero and thus the question of siting is germane.

OBJECTIVES

One of the key objectives of this volume is to explore various new approaches to siting facilities and to bring these to the attention of all those concerned with waste management. We do not assume that there is a "magic bullet" to the "siting problem." But we do suggest that there are some promising alternatives to existing thinking, and approaches that have not yet become widely understood and practiced. Although we recognize the primacy of waste reduction, that is not the focus here. Nor do we focus much on the broader questions or overall framework of the waste management problem of industrial societies. Our attention to matters of siting does not presuppose that this problem is going to be solved simply by siting more hazardous waste facilities. It does assume, as already noted, that siting is part of the solution.

We also aim to help correct certain misconceptions. One of these is that opposition to new facilities is not a common phenomenon in industrial countries other than the United States. In Germany, Mazmanian and Morell note, "siting negotiations have lacked the NIMBY characteristics so prevalent in the United States, where confrontation is accepted as the norm." They also suggest that local opposition is not much of a factor in Japan.[38] In fact, NIMBYism is prevalent in both. Opposition, if not in-the-street protest, is common. It is a prominent feature of waste management politics not only in Germany and Japan but also in most other industrial countries (Yakowitz 1988; Seeliger and Munton, this volume). Protestors in the thousands marching in opposition to proposed waste incinerators are found in the streets of Sindelfingen and Herrenburg as well as in Ohio and outside the White House. NIMBY has long been a phenomenon in Canada (Armour 1984). It is also emerging in the newly industrializing countries of the Asia Pacific region. Extensive plans to build a substantial number of waste incinerators in Taiwan, for example, have been met with strong opposition (*ENR* 1994a).

To place the contents of this volume in appropriate perspective, we need to review the many different strategies that have been promoted and applied to site facilities. The contributions herein will then be outlined.

SITING STRATEGIES

There is arguably no shortage of strategies that have been advanced for siting facilities. Both adopted strategies and thinking have changed, often in rather short order. Some of the changes have come about as pragmatic reactions to NIMBY-type opposition. Some ideas have been developed to optimize certain criteria, such as economic efficiency. Others have been in response to challenges of environmental equity as critics pointed to a pattern of "environmental racism" by which facilities tend to be sited in areas of lower income or with more persons of color and in areas that are less able to organize against these facilities.[39] Often facility proponents or governments, concerned that some approach was not working, shifted to another without noting that similar negative conclusions had been drawn about these "new" approaches by those applying them elsewhere (Andrews 1988, 120). While there is a temptation to characterize the shifts in terms of phases or generations, it would be a mistake to imply there has been any single temporal sequence of strategies. These approaches, moreover, should not be considered mutually exclusive. Many specific siting attempts have involved two or more elements. A brief review of some of the more prominent strategies follows.[40]

Decide, Announce, Defend Approach

What can be fairly termed the first generation of siting efforts is characterized by a "top-down" approach. They featured decisions, often unilateral, by facility proponents or governments to construct a facility in a chosen location. Most siting decisions until the 1980s or so were of this kind and met with little or no opposition. One variant was what could be called the "low-profile" approach. Essentially, and not to put too fine a gloss on it, the strategy was to slip quietly into a community and build a facility before potential critics, even neighbors, noticed (Morell and Magorian 1982, 126). The largest hazardous waste landfill in the United States was thus sited in Emelle, Alabama (Gerrard 1994). And this process was recommended to the EPA by consultants at the end of the 1970s (cited in Salcedo, Cross, and Chrismon 1989, 64 n. 21).

The most common top-down process is known as the DAD approach (Decide, Announce, Defend).[41] It can be distinguished by a nonpublic process of choosing a particular site, announcing this choice publicly, and then mounting whatever effort seems necessary to defend and secure approval of, or at least acquiescence to, this choice. Increasingly often in the 1980s, DAD led to neither approval nor acquiescence but, instead, to confrontation and stalemate. The siting decision process itself was usually dominated by technical determinations of what made for a "safe" site. It seldom gave adequate thought to social and political factors, much less to public acceptance.[42]

A sophisticated version of the DAD approach might be called the "New Jersey model." This is a "centralized planning process," in which a blue-ribbon panel, applying largely engineering criteria, identifies a short list of "best sites" and holds hearings on these, and then prospective firms bid on specific sites.[43] New Jersey developed and then utilized this process in the early to mid-1980s to choose a site for a major hazardous waste incinerator (Ely 1987). The Canadian province of Ontario opted for a modified New Jersey model when it created in 1981 the Ontario Waste Management Corporation (OWMC), a public agency (or what is called in Canada a crown corporation) charged with siting and constructing what would have been the heavily industrialized province's first major integrated waste treatment and disposal facility. The OWMC spent the remainder of the decade and into the mid-1990s looking for potential sites for its facility, drawing up a short list, narrowing down the short list to one favored site, and then embarking on a lengthy, combative, and expensive hearing process to defend its choice and its plans. As noted above, the quest eventually came to naught about a decade and a half and $140 million later. Local NIMBY-type opposition and broader criticism from environmental groups had a strong influence.

The DAD strategy for siting assumes, first, that the establishment of facilities is largely a technical issue, to be planned and decided by experts and on the basis of technical knowledge and objective standards. It also assumes that people living in the locale of these planned facilities either will not have any serious objections to them or can and will be convinced, after the fact, that the decision is a good or reasonable one. The bias of the former assumption is given little credence today. And clearly, given the regular emergence of determined local opposition, the second assumption is equally invalid. The history of NIMBY movements in most countries is that a sustained campaign by an aroused citizenry which gains political allies will likely emerge victorious. Waste-siting conflicts may pit the little people against big government and big corporations, but these days the little people almost always win. That the Davids have again slain the Goliaths is hardly even news anymore.

Preemption

Trying to stem a tide of NIMBY movements, some governments moved to preempt local authority in siting matters. Eliminating local authority's power to make land use decisions regarding the siting of facilities or to adopt local environmental laws that might restrict such siting, it was thought, would make it possible to force "needed" facilities on reluctant localities.[44] One assumption here, as in the DAD approach, was that senior governments had the technical and policy expertise to decide what was best for all concerned. Legislation allowing such preemption, often favored by bureaucrats (Davis 1984–85), was a feature in some American states. It was also rejected in others. In any case, the strategy of preemption had a fatal flaw. "While a state can strip local officials of their permitting power," noted Morell and Magorian, "it cannot strip them of their political power" (1982, 94). That political power was often sufficient to obstruct an unwanted facility. Local governments proved quite resourceful in thinking up novel and semi-legal ways to halt projects they opposed. One American town, for example, dug a substantial trench across the only access road into an unwanted facility, as part of so-called flood control measures (p. 96).

Local Veto

The alternative response to preemption was to leave authority to local governments and even to ensure through legislation the rights of local communities to approve or reject siting proposals. It was argued that local communities ought to be able to control their own destinies, including turning down developments they did not want, and especially developments such as hazardous waste facilities. If local communities could mount a de facto veto, perhaps it made sense to give them a de jure veto. The local powers formally provided by this sort of legislation, in effect, confirmed the basic principle of local authority for land use decisions, a well-established one in the United States and many other countries. The principle was hard to dispute. The practical problem with the local veto, on the other hand, was that few communities were inclined to look favorably on hazardous waste facilities proposed for their areas. Thus, little facility siting could be anticipated. The record of siting attempts under local-veto conditions, however, was no worse than where preemption was allowed (Andrews 1988).

Public Education

Another response to growing local resistance was to try to engage in public education. One example was a 1983 act adopted in Florida that was designed both to protect the state's water resources and to solve the siting stalemate. A clear assumption of the law was that "if citizens could be shown the extent of the local hazardous waste problem" and if, simultaneously, "state government could demonstrate its capacity to deal with such wastes safely, then citizens would be less resistant to facility siting in their localities" (Matheny and Williams 1988, 46). Some jurisdictions organized household hazardous waste cleanup drives. Others tried advertising campaigns. Yet another example was the extensive series of workshop-style meetings on the hazardous waste problems held in communities around the Canadian province of Alberta following the defeat of an initial effort at siting a facility there (and prior to an eventually successful siting). Efforts to increase public trust were an important part of this sort of public education (Portnoy and Rich 1987). Given the difficulties facing, and the limited success of, public-education campaigns in general, it is questionable whether these efforts had much impact. (They may, in fact, have been counterproductive; some research [Wright 1993] suggests knowledge and opposition are positively related.) A related tendency was to see siting and risk management as communications problems and to think that the secret to success lay in the way siting proponents communicate with the public (e.g., Sandenburgh and Cox 1987; Coleman 1989–90).

Public Participation

Yet another response to NIMBY movements was to recognize the important and legitimate role of local citizens in these decisions and to allow for that, if not encourage it. This was also a response to the conflict engendered by DAD and preemptive processes and to the acceptance of the right of local veto. The aim of the public-participation approach, of course, was to facilitate cooperation rather than engender opposition. The "inclusion of public officials and local citizens early on in the site selection process," some advocates argue, "has greatly enhanced the chances for successful siting" (Cahill and Holman 1981).[45] Others are less sure but recognize public participation as necessary nonetheless. While "public involvement programs may become a gamble against public prejudices or an act of faith in the possibilities of public education," Kraft and Kraut observed, "a strategy . . . that allows active participation by citizen groups may be the only feasible alternative" (1988, 78). The fear of siting proponents here, of course, is that more active public participation may increase the likelihood of rejection. Portney (1991) concludes that it does so, at least in cases where the participation is only short-term. One of the oldest reports on the NIMBY phenomenon similarly argued that "if given too large a substantive role, opponents will probably block all siting attempts."[46] Also among the skeptics are many government officials, who are customarily reluctant to encourage greater public participation because of the perceived threat to their own bureaucratic objectives.

Override

A less accommodating strategy advanced and adopted to overcome local opposition and counter the local veto problem is that of the legislated override.[47] If, after becoming involved, local residents insisted on refusing a facility under the offered conditions, governments, it was argued, should then override that opposition. The underlying assumption was that the greater and broader public good, represented, of course, by senior governments, should prevail in the end over narrow local self-interest and emotionalism, represented, of course, by local councils. When communities ultimately reject compromise, Morell and Magorian suggested, "state-enforced arbitration and use of state override authority may be desirable to ensure that a balanced siting process can yield a final site determination." Indeed, "some amount of state coercion probably cannot be avoided" (1982, 176, 135). By the 1990s, at least twenty-four American states had assumed some override powers in the siting of hazardous waste facilities (Gerrard 1994, 142).

Override has faired little better than earlier approaches, which is to say, not well at all (Andrews 1988, 123). When applied, it probably seemed to those affected no different from the preemption approach. The difference was merely one of timing, of *when* the local community lost control. And like preemption, it failed to address the essential political problem of local opposition. There were also political pressures in many jurisdictions against adopting override laws and against utilizing those that existed. The idea was thus given limited opportunity in practice.[48]

Negotiation

Negotiation is, of course, a general approach to resolving conflict, but one no less applicable to siting conflict than others. It is not so much a specific alternative strategy, like the others discussed above, as it is a generic process. It became particularly popular in American states that had tried and failed with preemptive strategies (Andrews 1988, 125). Negotiation processes, with government bodies as formal or informal mediators, were adopted by at least six American states by the end of the 1980s. This was part of a broader movement in the United States during this period toward the use of environmental-dispute resolution and away from the more common confrontations and litigation. As a prescription for the siting stalemate, negotiation has been widely endorsed, explicitly or implicitly.[49] And it is even more widely applied.

In contrast to the preemption and override strategies, negotiation assumes that interests are not permanently fixed and necessarily irreconcilable. While often held to be an alternative to more coercive techniques such as preemption or override, it can, at the same time, also be a complement to them. In some jurisdictions the override option was seen not as an alternative to negotiation but as a measure of last resort after negotiation had failed. Although a statute such as the 1981 Massachusetts Siting Law is often pictured as emphasizing negotiation, and did to a considerable extent, it also had an override provision. If siting talks with a community reached an impasse, a state board could require binding arbitration, which could, of course, decide against the community. Negotiation is also at the core of comprehensive siting strategies such as the "Keystone Siting Process" (Andrews and Lynn 1989) and the "Facility Siting Credo" (Kiser 1992; Kunreuther and Susskind 1993; Kunreuther, Fitzgerald, and Aarts 1993).

The strengths of negotiation and the related process of mediation lie in the fact, among others, that they are probably the best way to build up between two parties the kind of trust necessary for a resolution of a conflict (Kiser 1992). Despite the time and energy required, they can also prove more efficient. "Negotiation, consensus building, and joint problem solving can lead to quicker, less costly, and more successful resolution of conflicts in cases of facility siting than might be obtainable through more conventional forms of decision making" (Kraft and Kraut 1988, 74).

Negotiation has few critics[50] but some fairly obvious weaknesses. Wherever it is prescribed and at whatever issues it may be aimed, negotiation assumes that the (usually two) opposing sides each have at least some common interests to discuss and a degree of shared interest in working toward some sort of an agreement, albeit perhaps a minimal degree. For negotiation to succeed, there must be a mutual willingness to reach a compromise solution. These assumptions are often not met. They do not apply to many international disputes, for example, and seem even more questionable in the case of facility siting. Siting proponents clearly want a particular solution and are usually willing to compromise on compensation packages, benefits, and the like. The opponents' ultimate objective, on the other hand, is usually to prevent the siting. Often that objective is one on which compromise is unthinkable. The "solution" preferred is for the proponents to abandon their siting effort. Negotiation thus provides a process but no guarantees that the process will provide a mutually agreeable outcome. This weakness is more than evident in the experience of Massachusetts, which pioneered negotiation processes and where siting efforts have still run into continual failure (Portney 1991; Wheeler 1994).

Compensation and Incentives

For many who advocate negotiation over siting, a key issue to be discussed and agreed upon is compensation of those affected by a facility. The idea arises out of the observation that facility siting creates an asymmetry of costs, on the one hand, most of which are paid by the community affected, and benefits, on the other, only a small part of which accrue to the community. Compensation is thus intended to narrow that gap between costs and benefits. It is also asserted, not unreasonably, that offering incentives is more conducive to reaching agreement than is threatening coercion. Compensation can take a variety of forms (Gregory and Kunreuther 1990; Portney 1991). Governments or potential waste facility developers might offer guarantees that individuals' property values will not decline, or offer direct monetary payments to affected residents on an individual basis.[51] Either operators or governments might agree to provide affected communities with desired infrastructure (e.g., new highway access), social services (e.g., a new hospital), or amenities (e.g., new parks). This type of compensation has been reasonably common in North America and Europe and even more common in Japan (see Munton, this volume). Facility operators might agree to pay special taxes or a share of the facility's profits to the affected municipality. Arrangements of this sort, for example, were negotiated for the huge Emelle Landfill in Alabama and for the hazardous waste treatment and disposal facility built in the prophetically named town of Last Chance, Colorado, the only such siting in the United States during the 1980s.

The compensation approach has some advocates and some devout adherents.[52] It also has fierce critics. There are two underlying assumptions in strategies that depend on compensation packages. The first is that individuals in a potential host community are essentially—and mainly—rational economic actors. The second is that the disparity between the diffuse social benefits from a facility and the disproportionate individual costs borne by those living close by is the basic problem in a siting controversy, and that compensation can close this gap. As one negotiation/compensation proponent put the matter, "The availability of compensatory payments addresses *the underlying causes* of public opposition" (Holznagel 378; italics added). Both assumptions are highly questionable. The lack of compensation for economic costs imposed by a nearby facility is seldom *the* underlying cause of local opposition, and virtually never the *only* cause. Clearly these costs are relevant to some extent, and at some point. But they may well not be primary or even central. There are some experts who argue, sometimes flatly, that compensation cannot and does not work, that is, it is not at all an effective measure in siting facilities.[53] Indeed, offering it may be counterproductive if people perceive the compensation offers by a facility's proponents as attempts to buy them off or bribe them. This perception can be heightened if compensation is being offered by a facility proponent but other concerns, for example, about health risks, are not being addressed.

To say economic costs are neither the most important nor the only underlying cause of local opposition is not to say that individuals' or a local community's perceptions about an unfair distribution of costs is of no importance to whether they accept or oppose a facility. Some evidence suggests that compensation may in fact have a role to play in reaching an agreement on siting. It has been a part, and perhaps an important part, of some successful siting cases.[54] These packages almost always deal with more than the financial and economic concerns of those affected, however.

Impact Management

Yet another set of strategies subject to negotiation between facility proponents and the local community focuses on managing impacts, both real and perceived.[55] The assumption here is that opponents are concerned not only about the economic costs being imposed on them but also, perhaps first and foremost, about the risks posed by the facility to public health and environment. One option here involves reducing the risks, perceived and actual, from operation of the facilities in question through engineering measures and regulation. Risk reduction can be achieved, for example, by requiring strict emission control standards on waste incinerators or by collecting and treating leachate from a landfill. This sort of measure is what Davis and Lester refer to as "risk mitigation" (1988) and what Kasperson (1986a) calls "risk reduction."

Another option is to have fully agreed-upon plans for what to do if some operational problem emerges. For example, if plant air emissions were to go higher than specified levels, the standard operating procedure (SOP) could be to shut down the plant. These are what Armour (1988) refers to as contingency measures. Yet another is to provide continuous monitoring of the site through an open, "transparent" process in which the local community is fully involved. Gerrard has argued that "the ability of a municipality to monitor a facility, to participate in its management, and to shut it down if necessary, has a far greater impact on local acceptance than any other measure" (1994, 109). These elements, taken together, amount to what some siting analysts and participants call "co-management."[56]

A rather different risk management strategy is to develop facilities on sites or in locations where there already is, or was, a noxious or even dangerous facility of some kind. One of the large German waste facilities, for example, is located on the former site of a munitions factory. Portney (1991) calls this "risk substitution." The assumption here is that there can be some sort of acceptable trade-off between an existing serious risk and a replacement, presumably less serious, risk. Another alternative would be to develop commercial hazardous waste treatment facilities on the site of an existing waste-producing plant and perhaps open the surplus capacity of these facilities to waste from other operations. This idea assumes the local population will make a connection between the manufacturing, and its attendant waste generation, and the need for waste treatment and will see a trade-off between the benefits of the former and the costs of the latter. There can be other advantages of such arrangements. They may facilitate waste reduction, at least by the host plant; reduce the risks of waste transportation and, presumably, of inadequate or illegal disposal; and reduce potential corporate-liability problems. To some extent, on-site treatment (but not necessarily importing of wastes for commercial purposes) seems to be the option adopted by many large chemical companies in connection with waste minimization strategies.[57]

Cost and Risk Sharing

Yet another strategy involves sharing widely the social costs of waste treatment, rather than concentrating treatment facilities, and thus these costs, in one location. This has been labeled the YIMBY (Yes, in Many Backyards) approach. The YIMBY philosophy, as employed in California, "asked each county to take responsibility for its fair share of the waste problem, though not necessarily to manage all of its own wastes. . . . [Thus] waste management problems are addressed at a comprehensive regional level" (Mazmanian and Morell 1992, 192). This approach might involve various treatment options, including dispersing the facilities or the transportable hazardous waste treatment equipment. More grandly, YIMBY might underlie a national strategy. Each state or province might be allocated in some sense an equitable share of the overall treatment and disposal responsibilities. Gerrard envisages what he terms a "comprehensive national program that considers all the different kinds of wastes and allocates the burdens equitably among the states." This plan features, in short, "local control, state responsibility, and national allocation" (1994, 5).[58]

Various assumptions underly cost and risk sharing. One is that if the costs and risks are divided somewhat equitably, and seen to be divided equitably, there will be greater public support for accepting some of this overall burden. Another assumption seems to be that the approach will encourage a sense that everyone is involved in the problem and everyone needs to be involved in the solution. While the experience with this strategy is too limited to judge its results, it is clear from the California experience that it can easily be undermined by governmental missteps (see Morell, this volume).

THE VOLUNTARY-CHOICE PROCESS

The recent history of siting waste facilities in industrial countries suggests that the above strategies and approaches have not worked satisfactorily. While many of them have some merits, these approaches, taken one by one or even together, have not provided adequate solutions to the siting element of the waste management problem. Nor have they reconciled social needs and the principle of local decision making. "In the face of growing evidence that the current wisdom is failing," Roger Kasperson has argued, "new approaches are needed" (1986a, 142).

One relatively new approach that may be able to overcome some of these problems in earlier strategies is what might be called "voluntary-choice siting." It appears to hold at least some promise. (This approach is the focus of many, but not all, of the contributions to this volume.) In contrast to the top-down nature of the DAD and preemption strategies, it is quintessentially a bottom-up approach. It also avoids some of the more unrealistic assumptions underlying other approaches. And it deals both with democratic rights and with perceptions.

Where most of the other strategies take as a given a prior decision to site in a particular locale, the voluntary approach does not.[59] It assumes that this decision will be made only after invitations have been issued to potential host communities, after the expressions of interest by volunteer communities have been considered, after the communities have had the opportunity to study the proposal in depth and secure expert advice, and after key provisions of the arrangements for the siting and operation of the facility have been negotiated between its proponents and one or more communities. Rather than aim to secure local approval for a particular site chosen by a private developer or government agency or panel, it seeks first to identify potentially interested communities. Rather than keeping the local community at arm's length, it seeks to involve them directly and intimately, not only in the siting decision itself but also in the management and operation of the facility. The voluntary approach not only assumes that no siting decision is made *a priori* but also posits that, when made, the decision will be a genuinely mutual one. At the outset, potential communities must choose to opt in. At every stage, from exploration of possible physical locations through community acceptance (for example, by means of a referendum), communities can drop out of the siting process. And prior to acceptance, they can negotiate with the facility proponent the terms of both the siting and the operation of the facility, and do so from a position of some strength.

Chief among the advantages of the voluntary-choice approach is that it is nonthreatening. As it does not involve an *a priori* choice of any particular site and local communities must volunteer to be considered as potential sites, it avoids the anxieties often created by traditional DAD-type processes, those arising from the shock and dismay of discovering that one's town or neighborhood has already been chosen and that the decision was final and cannot be appealed. The distinction between being presented with what appears to be a fait accompli and voluntarily accepting certain risks is a crucial one. There is convincing evidence that people perceive risks as lower, or are often more willing to accept risks, if the risks are assumed voluntarily.[60] Acceptance of the risks of lung cancer from smoking is a case in point.

Quite apart from the potential for facilitating agreement of host communities through perceived-risk reduction, a voluntary process also seems ethically preferable. Among Roger Kasperson's ethical principles for siting, along with the need to reduce and share risks, is the stipulation that "the imposition of risk should be as voluntary as possible" (1986a, 139). The approach outlined here aims to achieve that goal. Assuming that risks are minimized and that there is no manipulation or coercion behind the choice, it is surely preferable to an imposed solution, no matter how well considered that solution. The voluntary approach acknowledges that there is an important moral difference "between jumping and being pushed" (Mark Sagoff, quoted in Gerrard 1994, 200).

Another advantage is that a voluntary-choice process is usually seen to be legitimate. It affords local control and respects local decisions. While the importance of legitimacy in siting has long been recognized and often been sought (Morell and Magorian 1982, 97), few of the alternative strategies or approaches ensure it in the degree needed.

A third advantage of the voluntary-choice approach is the efficiency of its self-selection process. Those communities more inclined to assume what others regard as the substantial risks of facilities select themselves. Proponents do not have to go searching for them in the usual, confrontation-producing manner. And despite the considerable consultative effort required in dealing with prospective communities, voluntary approaches may save time. They may identify interested communities more easily and more quickly than would be possible in a more top-down process.

What we call the voluntary-choice approach has been tried in both the United States and Canada but has not yet become commonplace in the practical world. It might seem at first glance to be the approach certain to be voted least likely to succeed, in the sense that one might expect few or no volunteers to come forward. Indeed, it might seem so even on reflection.[61] That, however, has not been the experience when it has been tried. Calls for volunteers to act as host communities have been answered, at least in the cases considered here. The approach has thus been successful in identifying communities and in siting facilities in a number of cases. Improbably enough, there have even been competitions between communities vying for a facility. (Even with this approach, however, there is still only a slight danger that "NIMBY" might come to stand for "No! [put it] in My Backyard!")

No doubt one of the primary reasons communities volunteer for, and agree to, hosting waste facilities is the expectation of economic gains from the facility itself and related business. While the numbers of direct jobs are seldom very high, they are often of considerable importance to smaller communities. Moreover, the nature of the employment is also attractive. Facility jobs are steady and often at a professional level. Thus, quite apart from incentives and compensation packages, the anticipated economic benefits of a facility lead to community interest. In the case of one host community, in Alberta, Canada, there was also an unanticipated benefit—a significant tourist boom. In the first year of its waste plant's operation, over 1,000 visitors came to have a look (McQuaid-Cook and Simons 1989, 227).

The importance of providing economic benefits is also often overrated. The assumption many draw from the history of community after community rejecting waste facilities is that no one is willing to tolerate living anywhere near these plants under any circumstances. That assumption would seem to be incorrect. As Groothuis and Miller

(1994) demonstrate, there is not only an avoidance factor in individuals' attitudes about these facilities but also a tolerance factor. Whereas top-down processes seem to bring out the former, voluntary-siting processes seem to facilitate the mobilization of the inherent potential of the latter, the willingness of some individuals in some communities to accept what others would reject.

The voluntary notion was introduced into the siting literature in the late 1980s with firsthand accounts by Jennifer McQuaid-Cook and Eugene Kupchanko of the siting of an integrated hazardous waste treatment plant in the Canadian province of Alberta by a then novel voluntary process (McQuaid-Cook 1987; McQuaid-Cook and Simpson 1986; McQuaid-Cook and Simons 1989; Kupchanko 1989). Various facilities in the United States have also been sited in, or at least considered for, communities that have volunteered as hosts (Gerrard 1994, 132–38). The processes of voluntary siting are further elaborated, and some relevant case studies analyzed, in this volume.

The ideas underlying the voluntary approach described and analyzed herein are not entirely new or unique. Many of its elements can be found in earlier siting strategies.[62] While in key respects novel, the voluntary approach also incorporates various elements of the ideas already discussed and, as practiced, represents a combination of the more complementary of these approaches. It embodies the local-veto idea, of course, although in a process somewhat differently structured than that traditionally assumed in facility siting. It requires public participation and full communication with potential host communities. It features negotiation between facility proponents and those communities. It accepts the need for risk mitigation through community involvement and a transparency in facility operation. It has, at its core, community meetings and a formal community decision, usually a referendum. While both are frequently part of siting cases, they have more often in the past been used as a tool to defeat proposals than to approve them.[63] Voluntary siting processes also typically involve compensation at the community or individual level. What the voluntary approach does not encompass is the top-down aspect of the strategies of DAD, override, and preemption. In philosophical terms, a voluntary approach moves us away from a utilitarian (good-for-most) criterion to a more populist one, in which people decide for themselves. Legitimacy lies in the process adopted more than in the outcome.

There are some close links between the voluntary-choice approach and some of the auction-based ideas that have been developed by advocates of compensation strategies. Howard Kunreuther and colleagues proposed in 1987 the notion of a "low-bid auction," in which communities would submit a sealed bid indicating the minimum amount they would need to host a facility (Kunreuther et al. 1987). S. Hayden Lesbirel (1990) has advanced a scheme for siting power plants in Japan in which locations would be chosen by auction. Herbert Inhaber (1991a, 1992a, 1992b) has outlined a very similar proposal to find waste sites using what he terms a "reverse Dutch auction." Inherent in all these proposals is the notion that communities would volunteer to host a waste site, for a price, of course.[64] The emphasis here is more on costs and compensation and less on a negotiation process than in the case of the voluntary-choice approach, but there are some important parallels in the underlying structure.

Voluntary-choice processes have been described or advocated by a few analysts since the Swan Hills, Alberta, siting. Indeed, that case is now frequently cited in the hazardous waste management literature, and was covered in the trade and business press (MacKenzie 1984; Tomsho 1991). Academic researchers have investigated the Alberta process in some depth (Rabe 1990, 1991, 1992; Castle 1992). Adoption and elaboration of the Alberta model in the siting of a facility in Manitoba has also been examined (Connell 1990; Rabe 1991; Castle 1993 and this volume). Audrey Armour (1991) has attributed the success achieved in one case with which she was directly involved to the fact that it adopted a similar voluntary process. Michael Gerrard's compelling proposal for hazardous waste management in the United States, based on his extensive experience as an attorney in siting controversies, features a voluntary approach to finding potential host communities (1994). His plan, which couples local control with a national system of allocating sites, is heavily based on the Alberta and other voluntary examples.

Voluntary processes could also be tried with other issues. What has worked in the case of hazardous and toxic chemicals seems applicable to the other siting controversies noted earlier (see note 3).

NATURE OF THIS VOLUME

The contributors to this volume are as interdisciplinary a group as could be imagined. They bring a wide variety of perspectives and orientations to their tasks. The disciplines represented include engineering (Zeiss), public policy (Harbage), public health (Hertzman, Ostry, and Rabe), business economics (Kunreuther), planning (Richards), political science (Gunderson, Munton, Seeliger, Kraft, and Rabe), geography (Elliott and Taylor), environmental studies (Castle), and environmental consulting (Richards and Morell). Despite this diversity of backgrounds, however, their studies have some strong common themes. Using case studies, they examine problems with well-known approaches and explore recent experiences with siting waste facilities, particularly those utilizing what are called here voluntary-choice approaches. They assess how and why the approaches adopted lead to decisions to site facilities of various kinds. They all tend to agree that, in the words of one contributor, facilities need to find supportive communities, not just sites. All would agree with Audrey Armour, who observed that "the name of the game" in facility siting has become "gaining public acceptance" (1988, 383). While some of their contributions are analyses more of failure than of success in this regard, all agree on the centrality of that goal. We can learn from failures as well as from successes, and maybe learn more than from easy successes.

The chapter by Geoffrey Castle and Don Munton describes and analyzes three attempts to site, through voluntary processes, provincial hazardous waste treatment facilities in western Canada. There were differences in the way in which these processes were implemented in these cases, but overall they bear some strong similarities. The authors offer an explanation for the successful sitings that occurred in two of these cases and the failed siting in the other case. While there are certain advantages to voluntary procedures simply in terms of the number of potential communities considered, the argument is made that voluntary siting is ultimately successful because it relies on a noncoercive, local-level democratic process. As a consequence, an important degree of legitimacy is established. Local opinion leaders who back the proposed facility are able to secure support for it under the voluntary approach. It is argued in conclusion that equity-based criticisms seem ill founded in these cases: the facilities were not sited in low-income areas. Employment benefits were, however, an important incentive to the communities. Voluntary processes may thus be both effective and nondiscriminatory, if not equitable.

The chapter by Barry Rabe, William Gunderson, and Peter Harbage compares the current and historical approaches to siting low-level radioactive waste management facilities in Canada and the United States. Neither country has sited a new radioactive waste management facility in more than two decades, despite the continuous generation of new waste and the paucity of reliable disposal capacity. Both nations have experienced considerable political controversy in attempting to site such facilities, with aggressive local collective action consistently blocking proposals. Rabe, Gunderson, and Harbage argue that efforts in Ontario and Nebraska to deal with low-level radioactive waste, through variations of the voluntary, comprehensive processes used successfully in Alberta and Manitoba for hazardous waste facility siting, both show signs of deviating from the classic NIMBY response. Their investigation supports powerfully the contention that a voluntary process for site selection is more workable and effective than a top-down process.

Disposal of high-level radioactive waste, as Michael Kraft notes, is the ultimate siting controversy. The problems faced by the U.S. Department of Energy (DOE) in finding a repository for high-level radioactive waste reinforce some of the consensus lessons about NIMBY movements: public opposition is strong and deep, facility proponents have little credibility and are widely mistrusted, and strictly top-down processes tend to generate bottom-up opposition. Kraft also describes recent efforts by the DOE to make its siting process more open and democratic, and some modest headway made to date under the new approaches. He argues that greater participation may also improve the technical quality of the decisions being made.

David Morell looks at the origins, implementation, and breakdown of California's innovative county-based hazardous waste management planning system under an innovative 1986 state law that received wide support. The core of this law was an equity–"fair share" principle for the siting of needed facilities—a variant of the voluntary-siting principle. Morell shows how the way in which the state agency decided to implement this law substantially undercut that principle and, hence, public and local government support for the envisaged process. Three case studies of siting failure show the extent to which NIMBY responses, as a result, still prevail, and suggest certain shortcomings of the California plan and especially its implementation. (All the cases, it might be noted, concerned incinerators. One of the cases prominently featured charges of "environmental racism.") Morell concludes that the law encouraged public participation (though not, he says, at the level of the voluntary-siting processes discussed here by others). It also prompted serious study of waste management problems and led to significant cooperative efforts. The fair-share principle thus seems important in encouraging cooperation and progress. Ultimately, however, the California law did not facilitate siting solutions. Because of the implementation failures, it is difficult to judge whether the law might have led to a more cooperative facility-siting process. The jury is still out, therefore, on whether YIMBY can be realized. A more voluntary element perhaps still needs to be considered.

It is sometimes assumed that NIMBY is a peculiarly North American phenomenon. It is also, perhaps even more often, argued or implied that facility-siting problems are significantly less in evidence in Europe or Japan. Neither is the case. NIMBY opposition, siting issues, and efforts and strategies to deal with this opposition are international, if not universal, phenomena, as the chapters by Don Munton and Robert Seeliger suggest.

Munton provides an analysis of hazardous waste siting in Japan, perhaps the first such English-language treatment. He argues that despite certain constraining elements in the Japanese political culture, there is indeed a widespread NIMBY phenomenon in Japan and that the Japanese hazardous waste problem is likely much more serious than officially recognized. The chapter then describes and analyzes the approach inherent in a new Japanese policy response to providing waste management facilities. The package of government regulation and siting processes now being implemented in Japan is very much patterned to that situation, but at the same time, the various individual strategies mirror those being proposed for, or adopted in, North American and European jurisdictions. The emerging Japanese approach comprises efforts to increase public trust, to mitigate and substitute risks, to provide compensation, and to allow greater public participation, including local approval of siting.

Robert Seeliger examines three key questions about the often misunderstood case of Germany. First, how were a number of hazardous waste treatment and disposal facilities successfully sited (e.g., in Bavaria and Hesse) in the 1970s and early 1980s? Second, have there been more recent successful siting attempts based on the government directed, top-down approach? And third, what have been the dynamics characteristic of recent siting attempts that added negotiation and/or mediation strategies to the legalistic procedures set out by German law, and how has this affected popular attitudes toward the siting of hazardous waste incinerators? Seeliger explores four recent case studies of hazardous waste incinerator sitings in different areas of the country. The case studies are placed in the larger context of hazardous waste management in Germany. In his conclusion, he offers some observations on the prospects for innovative siting strategies in Germany, such as voluntary processes, and on the discourse over obstacles to facility siting.

Two chapters in the book focus specifically and systematically on public perceptions of waste facilities. Both help to illuminate why voluntary processes seem to be successful. In search of determinants of facility acceptability, Clyde Hertzman and Aleck Ostry offer an innovative research design based on surveys in three communities in British Columbia that were slated to host or did host a waste management facility. The communities (two urban and one rural) were at different stages of the siting process, and different facilities were involved; a medical waste incinerator was proposed for one, another rejected a hazardous waste facility but accepted a municipal landfill, while the third hosts a municipal solid waste incinerator. Hertzman and Ostry contest some common assumptions about NIMBY, concluding that awareness of proposed waste management facilities in communities does not come easily and that respondents unaware of a facility are even more negative than those who know one exists and are concerned about it. They suggest that if majority support for facility siting is to arise, it will be built among unconcerned groups rather than among concerned groups.

Susan Elliott and S. Martin Taylor examine perceptions of people in Ontario living near three municipal waste facilities (one municipal incinerator and two landfills, one old and one new). Although none of these facilities was dedicated to managing hazardous waste, many of the concerns aroused nevertheless related to threats to health and environment from potential hazardous emissions. Elliott and Taylor focus on the psychosocial effects of environmental contamination, utilizing a framework adopted from the literature on environmental stress and coping, public-facility location theory, and attitude theory. Interestingly, they find that the major factors explaining the levels of both individual concerns and response actions are not personal characteristics but, rather, site and social characteristics. That is, the factors relate to the contaminant source and to social networks and the wider community. Analyzing longitudinal changes in perception of respondents near one landfill site, they find little change in the types of concern reported but a modest decrease in general concerns and health-related concerns over time and in the frequency of individual response actions. Differences in concern due to site proximity largely disappeared over time; that is, those nearest sites showed the greatest drop in expressed concerns. The lowest levels of concern were expressed by those near a facility that was responsive to complaints and had a very open operation; much higher levels of concern were expressed about one that refused public access.

Three chapters offer somewhat different, but still complementary, perspectives on voluntary siting processes. Alun Richards, a planner, offers a practical philosophy for siting, based on his involvement in the siting of the Manitoba Hazardous Waste Management Corporation facility. He begins from two premises: facility proponents need communities, not sites; and communities need control over their destinies, particularly when dealing with potential risks or unknowns. Proponents should therefore be prepared to share some of their decision-making power. His approach is based on a number of principles concerning the community's role: the process must be public; the community can negotiate terms and conditions (with the understanding that the proponent may also be negotiating with other communities); a negative decision by a community at any point will be honored; comanagement will characterize not only siting but facility operation. Formal agreements between host communities and facility operators should set out stringent operating conditions and joint-management understandings.

Howard Kunreuther comes to siting questions from the perspective of an economist and business decision-making analyst. His approach thus differs from that of Richards, but in key respects is quite parallel. Kunreuther explores and advocates a voluntary-siting process that includes well-specified environmental standards and explicit operational arrangements to reduce risks and transboundary effects, compensation to ensure benefits are shared, voting processes to determine willingness, and a lottery-auction mechanism to specify the ultimate terms of an agreement. He applies these ideas to the case of siting a low-level radioactive-waste facility in Ontario. The lottery-auction mechanism for choosing a host community, as outlined by Kunreuther, is very similar in principle to the voluntary siting process described and analyzed in other chapters.

Christopher Zeiss, an engineer who has studied facility siting extensively, develops and explores a set of four broad "siting intervention" principles: justify the need; manage both physical and nonphysical impacts; provide benefits, compensation, and incentives; and manage the process. He relates these to key theories about public responses to facility siting, including risk assessment and prospect theory. Zeiss looks at a number of past siting cases to determine the extent to which these intervention principles were applied. He finds that the successes involved the application of a combination of these principles. He also examines in some detail the voluntary process adopted in Alberta to site waste transfer stations. His general argument is that engineering design needs to address environmental impacts and host community concerns.

NOTES

1. I am grateful to Michael Kraft, Howard Kunreuther, David Morell, Barry Rabe, and two anonymous reviewers of this volume for comments on this introductory chapter. The staff and resources at the Bladen Library, Scarborough College, University of Toronto, were also especially helpful.

2. Some of the basic references on the NIMBY phenomenon are U.S. General Accounting Office 1978; US EPA 1979; Seley 1983; Morell and Magorian 1982; Armour 1984; Marshall 1989; Mazmanian and Morell 1992; Cook 1988; Guerra 1991. On local opposition to siting facilities in Europe, not always successful, see Piasecki and Davis 1987, chap. 8; Farago 1989; Farago, Vari, and Vecsenyi 1989; Visocki 1989; Lidskog and Elander 1992; Petts 1992. The Citizens' Clearinghouse on Hazardous Waste estimates there are about 6,300 active NIMBY groups in the United States (cited in Jessup 1992). For various estimates of the number of citizen groups in Japan, see Munton, in this volume.

3. The term "LULU" was coined by Frank Popper in 1981 (Popper 1981, 1991).

4. See, for example, the results summarized in Kasperson 1986a. Kauffmann, Serie, and Gesalman (1990) note that few people in one U.S. survey wanted even a park next to their houses and that only libraries and elementary schools were welcomed in the neighborhood. A national U.S. survey in the early 1990s showed 62% of respondents opposed the siting of a landfill in

their communities (Kiser 1992). The percentage of Americans supporting the siting of a nuclear power plant within five miles of their homes halved in ten years, from 42% in 1976 to 23% in 1986, while the proportion opposing it rose from 43% to 73% over the same period (Rose and Dunlap 1994, 311). Opposition to waste disposal seems to have increased at least as dramatically. A 1973 report to the U.S. Congress found the American public generally positive about the idea of disposal sites, perceiving them to be "relatively nondangerous" (cited in Dawson and Mercer 1986). Interestingly, a surveyed group of 2,000 experts, mostly with a chemical background, in 1985 collectively ranked both a hazardous waste incinerator and a hazardous waste physical-chemical treatment plant ahead of (i.e., more desirable than) an oil refinery as facilities to be located near their homes (Salcedo, Cross, and Chrismon 1989, 50).

5. The literature on NIMBY movements continues to grow. Recent studies explore opposition not only to the siting of hazardous waste facilities but also to chemical-weapons incineration (Silton 1993; Carr 1988; Gerrard 1994), nuclear power plants (Navarro 1988; Christensen 1979; Pligt 1992), steam and electric power plants (Kuehn 1993; Endries 1989; Maize and McCaughey 1992; Lesbirel 1990, 1987–8; Hamilton 1979; Foley and Prepdall 1990), hydropower plants (Russo 1995), transmission lines (Risk 1981; Moran 1980), wind turbine farms (in the Netherlands, Wolsink 1994), oil refineries (Popper 1991), research laboratories (Krimsky 1986), airports (Allison 1986; Popper 1991; Shrader-Frachette 1991), highways (Seley 1983; Allison 1986), public housing (Cowden 1991; Opalski 1992; Susskind and Ozawa 1985), community residences for special groups, such as developmentally challenged persons (O'Looney 1993; Caves 1990; Smith and Jaffe 1986); prisons (Seley 1983; Rasmussen 1992; Shichor 1992); and amusement theme parks (O'Brien 1994) and to siting where endangered species were affected (Rhodes and Wilson 1995).

6. Articles offering advice on the necessity and methods of dealing with NIMBY-style opposition include those for planners (Dear 1992), engineers (Connor 1988), developers (Yarzebinski 1992), public-relations specialists (Young 1990), and—in what might be termed the hard-line approach—utility executives (Maize and McCaughey 1992).

7. See, for example, the manner in which the term was used by almost all participants, including NIMBY groups, at the 1984 multistakeholder conference organized by Armour (1984).

8. For a rather different typology of concerns, see Connor 1990.

9. Pat Hayes, quoted in Armour 1984, 16. The validity of this point notwithstanding, the fact that the planned facility which Hayes's group opposed was a waste disposal operation, however, clearly did make a difference. Later in these same proceedings she disputed a comparison (made by someone else) between siting waste treatment centers and siting day-care centers. What we were fighting, she noted pointedly, was the former, not the latter, thereby acknowledging that more than democratic rights were seen to be at stake.

10. As Matheny and Williams (1988, 39) perceptively note, waste facility siting questions are "at the nexus" of what Alford and Friedland (1985) term the democratic, managerial, and capitalist views of the state: that is, they evoke debates over the operation, weaknesses, and viability of the capitalist system (i.e., market failure to deal with hazardous waste externalities), over the management of issues by technical experts in bureaucracies, and over democratic decision making and participation. Gregory McAvoy (1993) argues that contrary to much conventional wisdom, siting controversies show citizens are not too emotional and uninformed to influence waste management policy and, in fact, can play an important role in the democratic control of the bureaucracy, one that is essential to rational and democratic decision making.

11. Rev. Ted Creen, quoted in Armour 1984, 54. Creen notes that Love Canal and other incidents offered the "precedent for linking the fears of cancer, miscarriage and genetic damage to toxic waste and to specific sites where such waste has been dumped," an assessment of the Love Canal disaster with which many would agree.

12. For the argument that public perceptions of health risks exceed actual risks, see, for example, Ewel and Lewis 1991. On perceptions of health-related risks in general, see Bord and O'Connor 1992. For a Freudian view of public concerns about wastes, see Walker 1989.

13. A Canadian official, quoted in Jackson 1990, 303.

14. For the case that NIMBY movements are in the front ranks of opponents of market capitalism, see Lake 1993.

15. The final decision in the OWMC case is covered by Rusk 1995. The once widely touted OWMC siting process was, in happier days, offered as an instructive model case for developing countries (Batstone, Smith, and Wilson 1989, 122).

16. Phil Weller, a Canadian environmentalist, quoted in Armour 1984, 78. John Jackson makes the same point, that "the real problem is the wastes, not the NIMBYs" (Jackson 1990, 308). Various supporters of NIMBY movements also argue that while these groups may emerge out of concerns with purely local siting issues, they usually expand these to broader interests, such as opposing waste facilities in general or pushing for minimization and recycling (see Schwab 1991; Gibbs 1989; Jackson 1990). At the same time, local opposition has mounted to oppose facilities, such as medical waste incinerators, where there is an obvious and clear health risk from lack of treatment (Sellers 1993) and where a new incinerator is to replace an older, less safe one (Hertzman and Ostry, this volume).

17. This was, for example, one of the arguments made against the "blue ribbon" New Jersey siting commission (Ely 1987) and against the planned OWMC facility (Jackson 1990). See also, for example, Heiman (1990, 359).

18. Among the groups that have at one time, at least, accepted the argument that more sites were needed are the former Conservation Foundation and the U.S. National Wildlife Federation.

19. On the decline in waste volumes due to decreased waste generation, see *Chemical Week* (1994a, 1994b); Heller (1992); Hirschhorn (1988). U.S. chemical companies, for example, have been effecting significant cuts in waste generation since the late 1980s (Goldbaum and Rotman 1989). Declining waste generation, especially of chemical-industry wastes, in various European countries has significantly reduced demand there for hazardous waste treatment facilities. Many are operating below capacity (Chynoweth 1994, 26). In Sweden, for example, shipments of organic chlorines have dropped by 50%. (See also Williams 1987.) Waste reduction and recycling initiatives are also taking hold elsewhere, for example, in hospitals (Weber 1991a).

20. Various states have abandoned plans for hazardous waste facilities for which there seemed insufficient demand (Gerrard 1994). U.S.-based Chemical Waste Management has reportedly canceled plans for two commercial incinerators, one in California and another in New York. Partly in response to the dropping demand for the company's facilities, it was placed on a "credit watch" list by one investment agency (Gerrard 1994, 11). Planned commercial incinerators have also been canceled by Dow Chemical and Westinghouse (*ENR* 1994b). The EPA estimates that only about 55% of the available capacity of commercial liquid-waste incinerators was being used as of the early 1990s, after a period of expansion in the industry. A higher proportion, about 90%, of combustion capacity for solid wastes is being utilized (Cavanaugh 1993; Rotman 1993; Rotman and Begley 1994). A trend evident in the United States in the 1980s toward large diversified companies entering the hazardous waste treatment market has been reversed; most have withdrawn (Harris 1994).

21. The cyclicity argument is made by Plishner 1991 and Heller 1990. It is also possible that the reason for the reduced volumes being shipped to disposal facilities is that more waste is being disposed of illegally. While this may be partly true, there are enough corporate waste reduction success stories to suggest that fewer wastes are indeed being generated. Moreover, the phenomenon of declining waste amounts has also been noted in the much more tightly structured and controlled system in Germany.

22. For arguments that pollution prevention can achieve significant reductions and ought to be pursued more vigorously, see, among others, Campbell and Glenn 1982; National Research Council 1985; U.S. OTA 1986: U.S. EPA 1986; Hirschhorn and Oldenburg 1987; Hirschhorn 1988; Commoner 1989; Jackson 1990; Nelson-Horchler 1988; Bryant 1989; Freeman 1990. Harris, Want, and Ward (1987) suggested that reductions of about 20% were obtainable. The 1986 U.S. Office of Technology Assessment estimate was 50%. Warren (1990) listed various estimates of the effects of potential waste reduction measures, ranging from 11% to 75% of total wastes. Davis, Huisingh, and Piasecki (1987) examined practices in Europe. Alberti (1992) was much more skeptical about the success of European waste reduction programs. Progress in this area may well slow, as the easy measures are taken first (Lucas 1994b). Hirschhorn and Oldenburg noted that over the 1980s fully 99% of U.S. government spending on the environment was for pollution control, or after-the-fact measures (1987, 11). Programs in other jurisdictions were generally similar in emphasis; Jackson (1990, 307) calculated that the amount spent on pollution prevention was only about 3% of the total amount spent on hazardous waste facility siting efforts in the province of Ontario (at that point, $100 million or so).

23. On the impact of the trend toward developing and implementing corporate waste minimization (and elimination) strategies, see Winton et al. (1987); Goldbaum and Rotman (1989); Underwood and Dorfman (1990); Skinner (1991); Ember (1992); Holman (1993); Cavanaugh (1993); Grund and Kemezis (1993); Kemezis (1993); Harris (1994); Rotman and Begley (1994); Fouhy and Ondrey (1994); Shanoff (1994); Fairley and Chynoweth (1994). A growing list of international chemical companies, having already begun waste reduction programs in the 1980s, have set themselves the target of reducing waste generation by 50% through the 1990s including Dow, DuPont, ICI, and Sterling (Kemezis, Morris, and Rotman 1991; Gilges 1991; Lucas 1994a; LaBar 1990; Shireman 1993). General Dynamics and PPG Industries have been pioneers in striving for zero discharge (Beroiz 1990; Kemezis 1991). The U.S. federal policy approach to waste reduction, formalized by the 1990 Pollution Prevention Act, has been to encourage and promote, but not generously support, voluntary reductions by industry (Kelly 1990; Hazen 1992; *Environmental Manager* 1994a; *Small Business Reports* 1993; Darcey 1994; Hoffman 1993). Some regulatory requirements are nevertheless beginning to creep in (Begley 1993; Rubin et al. 1993). State governments began to set and pursue seriously waste reduction targets in the late 1980s (Trewhitt and Rich 1986; Hansen 1989; Schwab 1988; Barron 1993; Walka and Harris 1991–92). Schecter (1990) shows an exponential growth in American state waste reduction programs over the later 1980s and provides a summary, now somewhat dated, of all state programs. Eight states had adopted waste reduction laws by 1992 (Glenn 1992). These laws generally encouraged or required generators of hazardous wastes to adopt practices that reduced their waste production.

24. European incineration costs doubled in the late 1980s (Gilges 1991). On costs, see also Rotman 1990; Sternberg 1990; Winton et al. 1987.

25. The list of corporate waste reduction success stories, where substantial costs have been reduced, is now lengthy (see Nelson-Horchler 1988; Hunt 1990; Willenbrink 1990; Water Pollution Control Federation, Task Force 1990; *Environment Today* 1994a). Northrop had reduced 1989 waste levels by almost 80% by the mid 1990s (*Environment Today* 1994a). A joint study by chemical giant DuPont and the EPA concluded that $6 million in capital expenditure would not only reduce company hazardous wastes by 4 million lbs/year but also save $15 million/year in costs at a single DuPont site (Kirschner 1993, 16). A pioneering program in Ventura County, California, had by the late 1980s reduced by more than 80% the volume of wastes shipped off site from the seventy-five companies that together generated 95% of the wastes previously destined for county landfills. For every dollar of public money invested in the program, the companies saved $67 (Sherry 1990; Koepp 1990). The success stories notwithstanding, some evidence suggests that waste minimization programs have "not fully penetrated . . . industry" (Duke 1994).

26. This is a particularly strong incentive in the United States under the liabilities imposed by the Superfund legislation. Another factor is the increasing use of environmental auditing and of environmental assessment of company property as part of any sale of assets. It may come as a surprise to many environmentalists, but the insurance companies and accountants are becoming quite effective allies.

27. Bill Beck of DuPont, quoted in LaBar 1990, 52.

28. In their pioneering 1982 book on America's hazardous waste "crisis," Morell and Magorian suggest that "in the immediate future, . . . measures to reduce waste generation are likely to remain so modest that they will not constitute a solution to our hazardous waste problem" (p. 9). Current measures have become much more than modest, but the point stands: reductions do not constitute a complete solution to the problem of hazardous wastes.

29. One recent survey, by McCoy and Associates, puts the 1992 total of commercial hazardous waste landfills at 24. Another survey, by Environmental Information Ltd., found 20 (Holman 1993). The number of general waste landfills operating in the United States, which was around 20,000 in 1978, dropped by half in the following five years, and by half again in the next five (Cook, 1988). Repa and Sheets 1992 report that in the early 1970s, 300–400 municipal landfills were built in the United States each year. By the 1980s, the number was 50–200. During the 1980s, 70% of the municipal landfills in the United States closed. Closures outnumber new openings ten to one. The declining numbers are expected by most experts to continue. In 1992 Canada had roughly one-third the landfills it had in 1983. Most other industrial countries report similar patterns.

30. In the United States during the early and mid-1980s, only eleven new hazardous waste treatment facilities were given permits and only seven were built (Mazmanian and Morell 1992, 180).

31. Pressure against a proposed new facility, no matter how well engineered, seems easier to mount than pressure to close an existing, demonstrably unsafe site. As prospect theory predicts, people are more likely to mobilize to prevent perceived losses than to garner potential gains (Kahneman and Tversky 1979; Levy 1992). For some examples of locally forced closures, however, see Schwab 1991.

32. On the contamination problem at defense facilities, see Skryness, Swenson, and Pananos 1994 and Gerrard 1994. Of 9,300 U.S. municipal landfills tested by the EPA, 2,300 were found to be releasing substances to air, land, or water (Machlin 1987). Hazardous wastes make up about 3% of the municipal waste stream, not 1% as often assumed and previously estimated (Zeiss 1991). Among wastes in this category are such common items as discarded fluorescent lamps (Jankowski 1995). On household hazardous waste, see Robertson et al. 1987 and Kelleher 1991. Contaminated sites, of course, are not just a problem in the high-income industrial countries; see, for example, Amador 1992 on buried pesticide waste in Poland.

33. One study has found that the annual damages from Superfund sites for just a sample of 151 U.S. counties was over $14 billion (Walker 1992).

34. One analyst, Michael Gerrard, argues, to the contrary, that illegal dumping is not related to the lack of proper treatment facilities and disposal sites and to the costs charged by those that do exist (1994, 81). While it is certainly true that some illegal dumping will (and does) go on regardless of prices, it does not follow that price increases will not encourage more illegal operations and unsafe dumping. His case thus seems unconvincing, and was perhaps even to Gerrard himself. Later in the same book he asserts not only that "a shortage of disposal supply will drive up the price, and thus reduce demand" but also that higher prices can have "other undesirable effects," such as "an increase in illegal dumping" (p. 207). The argument that illegal dumping is not related to the supply and cost of disposal also runs contrary to evidence, for example, in Italy (Yakowitz 1988) and in California, where the number of reported instances of illegal dumping increased as the number of treatment and disposal facilities decreased (personal communication to author from David Morell). On illegal treatment and dumping, see *Chemical Marketing Reporter* 1994a.

35. The issue of incinerator emissions is controversial (see, for example, *Chemical Marketing Reporter* 1994b). On the tightening of U.S. policy on incineration, see *Environmental Manager* 1995a; *Chemical Week* 1994b. While incinerators as a means of waste treatment are opposed in principle by some experts and groups, such as Greenpeace, certain studies suggest that modern, well-controlled hazardous waste incinerators emit *less* dioxins and furans than regular municipal waste incinerators and emit less of other emissions than a conventional coal-fired power plant (Oppelt 1990; Dean 1987). Incineration still has its defenders. "We can minimize, we can reduce, we can reuse," Marcus Gleis, an official in the Bundesumweltamt, the Germany environment ministry, has said. "But every industry will eventually produce something that must be burned" (Gilges 1991).

36. Some analysts seem to assume any facility will become permanent. Once people "have the new facility," Gerrard suggests, "they will never be rid of it" (1994, 139).

37. The alternatives include, besides mobile incinerators (Hunt 1990; Kupchanko 1988), high-tech incineration (Fouhy and Ondrey 1994; Fairley and Chynoweth 1994), supercritical water oxidation, pyrolysis, plasma torch and "artificial lightening," molten salt (Fairley and Chynoweth 1994), chemical and thermal treatment, and bioremediation (Weber 1991c; Fairley and Chynoweth 1994; Chynoweth 1990; Krukowski and Simpson 1995). The complexities of dealing with specific toxic substances through marketlike mechanisms would be considerable. Different policies would have to be adopted depending on the nature of the substance and its production and use (Macauley, Bowes, and Palmer 1992). On overhauling the corporate tax system and/or instituting a system of deposit fees to encourage recycling and on-site treatment by firms, see Mazmanian and Morell 1992, 218–19.

38. Mazmanian and Morell note that "Japan, too, has successfully sited numerous hazardous waste treatment facilities tailored to specific industries" (1992, 137) and also imply that public sector–private sector cooperation in Japan means there is not much of a waste disposal problem there (pp. 137–39). Others have suggested that Germany has found some technical and organizational keys to solving the hazardous waste problem, but they do not deal with its emerging NIMBY movements (Piasecki and Davis 1987). See also Moskowitz 1984; Paparian 1983.

39. There is now a considerable literature on this issue. See, for example, Bryant and Mohai 1992; Ember 1994; Gerrard 1994; Kusek 1994; Boerner and Lambert 1995.

40. The typology of strategies adopted here is based in part on the work of Andrews (1988), Kasperson (1986a, 1986b), Mazmanian and Morell (1992), and others. A significantly different approach to classification is taken by Hisschemoller and Midden (1989). Their four categories (technical, participation, market, and distributive-justice approaches) are admirably parsimonious but tend to lump together rather different ideas and assumptions and to create some artificial distinctions. One strategy not considered here is the siting of an international waste facility on a remote island somewhere that would be purchased specially for this purpose (see Suess 1992).

41. The term "DAD" has been attributed to Dennis Ducsik.

42. The dominance of technical criteria in this approach, together with a lack of serious attention to social and political factors, is clear in such volumes as Wentz (1989) and Dawson and Mercer (1986). The explicit siting criteria of Clark-McGlennon Associates (1987) include, for example, potential impact on aquifers, contaminant flow period to water table, possibility of site flooding, safety and proximity of transportation routes, proximity to residential areas, potential for earthquake activity, etc., but not public acceptance.

43. See Matheny and Williams (1988, 39–40). Ely (1987) suggests that public acceptance of the commission was enhanced by its perceived objectivity, coupled with its autonomy from the legislature.

44. On preemption, see Bowman and Lester 1985; Andrews 1988. Among the advocates of the preemption approach, see Duffy 1984, who leans heavily to this strategy, albeit with a nod toward greater public participation. Morell and Magorian 1982 and Morell 1984 are responses to, rather than advocates of, preemption and its application.

45. Among others, Morell and Magorian (1982) discuss participation in New Jersey, Heckley (1985) in New York, Bacow and Milkey (1984) in Massachusetts, Kraft and Kraut (1988) in Ohio, Matheny and Williams (1988) in Florida, Morell (1984) in California, Davis (1984–85) in the United States generally, Connor (1990) and Armour (1984, 1988) in Canada, and Munro et al. (1990) in Europe and the United States. Hadden, Veillette, and Brandt (1983) explore various forms and degrees of public participation across U.S. states in the early 1980s. The Reagan administration systematically cut program funding for public participation, but it once again became an accepted part of environmental policy making, including permitting, in the Clinton administration (*Environmental Manager* 1994b).

46. From the "Centaur" report to the EPA, quoted by Morell and Magorian (1982, 136). Mazmanian and Morell (1992) also note evidence that U.S. jurisdictions high in participation tended to have few, if any, facilities sited during the 1980s.

47. On override, see Morell and Magorian 1982, Morell 1984, Andreen 1985. Richard Andrews discusses an approach, which he terms "procedural restraint," that involves an override authority in certain situations or under certain constraints (1988, 124–25). See Wolsink 1994 on the limitations on local governments and publics in the Netherlands.

48. One of the problems is that override procedures themselves violate the principle, noted by Morell and Magorian, to "ensure that the decision-making process is perceived as legitimate by the residents of the local area" (1982, 97). In some states, override-based siting statutes with negotiated-compensation provisions nevertheless appear to have been more successful than others. Wisconsin was one such jurisdiction (Nieves et al. 1992).

49. Advocates of negotiation processes in siting conflicts include O'Hare 1977, 1993; O'Hare, Bacow, and Sanderson 1983; Bacow and Wheeler 1984; Holznagel 1986; Kraft and Kraut 1988; Wheeler 1994. Those arguing for mediation (as opposed to arbitration) include Bingham 1986; Rovet 1987; Carr 1991; Mank 1991; Susskind and Ozawa 1985.

50. Among the critics is Heiman, who argues that "many of the public education, participation, and negotiation programs now being implemented

. . . in an effort to overcome entrenched local opposition are a poorly executed attempt to undermine a populist environmental movement at precisely the point where it is poised to challenge exclusive private control of the production and pollution process" (1990, 361).

51. On the impact waste facilities have on property values, see Smolen, Moore, and Conway 1992; Greenberg and Hughes 1993; McClelland, Schulze, and Hurd 1990; *World Wastes* 1995. Zeiss and Atwater 1988 find no consistent evidence across studies on property value effects. Almost $7 million in damages was awarded in 1995 by an Ohio court to about 1,700 property owners on the grounds of their proximity to a hazardous waste site even though none of the properties was directly contaminated (*Environmental Manager* 1995b).

52. Among the devotees, see, for example, O'Hare (1977); O'Hare, Bacow, and Sanderson (1983); Morell and Magorian (1982); Kunreuther (1986); Kunreuther and Easterling (1990); Sullivan (1990); Inhaber (1991a, 1991b, 1992a, 1992b); Schlossberg (1993). Among others, Armour (1988), Crim (1990), and Connor (1990) include compensation as one important part of a comprehensive approach to siting but do not focus on it as a singular measure.

53. See, for example, Portney (1985, 1988, 1991). Gerrard, while advocating compensation as a way of reducing the costs imposed on host communities and individuals living therein, asserts that "compensation has never been used successfully in siting a [hazardous waste or radioactive waste] disposal facility in a *reluctant* community in the United States" (1994, 126; italics added). That may be so, but the test of what is or is not a "reluctant" community is unclear.

54. For general arguments that compensation is an important, and presumably effective, part of siting agreements, see Armour 1988 and Connor 1990. Portney's argument in his 1991 book that compensation is ineffective seems based more on survey data than on case studies. On cases where compensation has been part of a siting agreement, see, for example, Swartzman, Croke, and Swibel 1985; Portnoy and Rich 1987; Shuff 1988; Paige and Owens 1984; Visocki 1988; U.S. OTA 1989; Lynch 1995. Zeiss and Atwater 1988 find that property value guarantees do not deal adequately with individual concerns about compensation.

55. Arguments for risk mitigation include Armour 1988; Kurland 1992; Swartzman, Croke, and Swibel 1985; Crim 1990; Petts 1992.

56. See for example, Richards, this volume. Petts (1992) argues similarly the case for good information on facility operations, ongoing liaison, and close, effective monitoring in U.K. siting processes.

57. The siting of facilities at existing industrial locations, which are often already contaminated, is a key part of the overall hazardous waste management plan of Gerrard (1994). A strategy of developing on-site waste treatment has, like other options, some potential dangers as well. Morell and Magorian (1982) discuss a case where an already operating chemical firm had a demonstrated need for waste treatment capacity and the company threatened to relocate if permission for an on-site treatment facility was not granted. The local community approved the facility. Perhaps the same case is noted by Davis and Lester (1988, 7–8). Public attitudes about on-site waste facilities are discussed in Zimmerman 1989.

58. Gerrard's (1994) approach envisages a United States–wide allocation of facilities. A similar approach, although less explicitly designed and implemented, seems to be emerging in western Canada. The provinces are moving toward a form of regional cooperation whereby each takes responsibility for establishing certain types of facilities and treating parts of the waste stream (see Castle and Munton, this volume).

59. In most of the siting literature, a corporate or government decision to site a facility in a specific location is taken as a given. A prominent example is the well-known work of O'Hare, Bacow, and Sanderson, who note, "The distinguishing feature of our work . . . is its focus on the problem *after* the point at which one participant or another has identified what appears to be a technically acceptable or at least preferred site" (1983, 40; italics added). They thus essentially assume away the critical site decision process, regarding it as a fait accompli and relegating everything that happens up to that point to the realm of technical determinations about site suitability and the like. This is also a bias underlying much of the rest of the literature, including Morell and Magorian (1982, 100). Swallow, Opaluch, and Weaver (1992) propose a siting process, based on a case in Rhode Island, that is a variant of the "decide, negotiate, and defend" process. It involves negotiations with a limited number of communities on a (top-down determined) shortlist and the choice of a final site after these negotiations. The bias to be found in the old-fashioned planning- and engineering-oriented siting literature was, of course, rather different. There what is emphasized is the investigatory effort that precedes the decision about location itself. Once that decision was made on technically good grounds, its implementation was then most often assumed away. The commonality between this technical literature and most of the process-oriented siting literature is that, at some point, the siting decision is made unilaterally by the facility proponent.

60. On this important point, the perception of voluntary risks, see Davis and Lester 1988, 6; Matheny and Williams 1988, 49; Fischhoff et al. 1981; Gerrard 1994.

61. The belief that communities will always oppose waste facilities is common. Even those who advocate both participation and voluntarism to the extent "reasonably achievable" usually assume it will not suffice; "Purely voluntary means are not possible," Kasperson suggests (1986a, 140). Mayor Bradley of Los Angeles spoke for many local politicians when he concluded, "No city, no local community is going to gladly or willingly accept disposal" (Morell and Magorian 1982, 102). Analysts drew the same conclusion: "It must be recognized that the public may never accept hazardous waste disposal sites." There may be absolutely no way to convince the average citizen that a well-designed, -operated, and -regulated site poses minimal risk" (Robert Pojasek, quoted in Morell and Magorian 1982, 183).

62. Andrews (1988, 121) notes in passing that a siting attempt in Minnesota in the early 1980s adopted this approach, at least in the sense that it called for volunteer communities.

63. The use of referenda is discussed, for example, by Morell and Magorian (1982, 129–30), who tend to emphasize the problems inherent in this mechanism. There have been positive referenda, even in top-down processes; see, for example, the 1985 case of a municipal waste incinerator siting in

Lisbon, Connecticut, where the referendum passed but the town government rejected the idea regardless (Portney 1991).

64. A different "chance" process was proposed by Roger Kasperson (1986a, 138). It envisaged a lottery mechanism that would choose or exempt potential sites along the lines of the U.S. military draft process, which was based on birth date.

REFERENCES

Alberti, Marina. 1992. Minimization of Hazardous Waste in Western Europe: Policy Implementation and Harmonization. Ph.D. diss., Massachusetts Institute of Technology.

Alford, Robert R., and Roger Friedland. 1985. *Powers of Theory: Capitalism, the State, and Democracy*. Cambridge: Cambridge University Press.

Allison, Lincoln. 1986. On Dirty Public Things. *Political Geography Quarterly* 5 (July): 241–51.

Amador, Alphons A. C. U. 1992. Buried Pesticide Waste Hazard to Poland. *Waste Management and Research* 10:387–98.

Andreen, William L. 1985. Defusing the "Not In My Back Yard" Syndrome: An Approach to Federal Preemption of State and Local Impediments to the Siting of PCB Disposal Facilities. *North Carolina Law Review* 63 (June): 811–47.

Andrews, Richard N. L. 1988. Hazardous Waste Facility Siting: State Approaches. In *Dimensions of Hazardous Waste Politics and Policy*, edited by Charles E. Davis and James P. Lester, 117–28. New York: Greenwood Press.

Andrews, Richard N. L., and Frances M. Lynn. 1989. Siting of Hazardous Waste Facilities. In *Standard Handbook of Hazardous Waste Treatment and Disposal*, edited by Harry Freeman, 3.3–3.17. New York: McGraw-Hill.

Armour, Audrey, ed. 1984. The Not-In-My-Backyard Syndrome. Symposium, February, Faculty of Environmental Studies, York University.

———, ed. 1984. The Not-In-My-Backyard Syndrome. Symposium, February, Faculty of Environmental Studies, York University.

———. 1988. The Social Aspects of Devising an Impact Management Program. In *10th Canadian Waste Management Conference*, 379–85. Ottawa: Environment Canada.

———. 1991. Opting for Co-operation: A Case Study on Siting a Low-level Radioactive Waste Management Facility. *Progress in Planning* 35:49–67.

Bacow, Lawrence, and James R. Milkey. 1982. Overcoming Local Opposition to Hazardous Waste Facilities: The Massachusetts Approach. *Harvard Environmental Law Review* 6:265–305.

Bacow, Lawrence, and Michael Wheeler. 1984. *Environmental Dispute Resolution*. New York: Plenum Press.

Barron, Tom. 1993. A Guide to Pollution Prevention Assistance Programs: What They Can–and Cannot–Do for You. *Environment Today* 4, no. 3 (March): 33, 35+.

Batstone, Roger, James E. Smith, and David Wilson, eds. 1989. *The Safe Disposal of Hazardous Wastes*. World Bank Technical Paper, no. 93. Washington, D.C.: World Bank.

Becker, G. S. 1983. A Theory of Competition Among Pressure Groups for Political Influence. *Quarterly Journal of Economics* 98: 371–400.

Begley, Ronald. 1993. Surprise New Incinerator Policy Has "Industry's Blood Boiling." *Chemical Week* 152, no. 20 (May 26): 13.

———. 1993. Incinerator Firms Turn Up the Heat As Markets Shrink. *Chemical Week* 153, no. 24 (December 22): 9.

———. 1994. Incineration Cuts, Permitting Goals Laid Out by EPA in New Strategy. *Chemical Week* 155, no. 21 (November 30): 8.

Beroiz, Denny. 1990. The General Dynamics Zero Discharge Program. In *Hazardous Waste Minimization*, edited by Harry Freeman, 263–71. New York: McGraw-Hill.

Bingham, Gail. 1986. *Resolving Environmental Disputes: A Decade of Experience*. Washington, D.C.: Conservation Foundation.

Boerner, Christopher, and Thomas Lambert. 1995. Environmental Injustice. *Public Interest*, no. 118 (winter): 61–82.

Bord, Richard J., and Robert O'Connor. 1992. Determinants of Risk Perceptions of a Hazardous Waste Site. *Risk Analysis* 12, no. 3:411–16.

Bowman, Ann O., and James P. Lester. 1985. Hazardous Waste Management: State Government Activity or Passivity? *State and Local Government Review* 17 (winter): 155–61.

Boyle, Susan B. 1983. *Siting New Hazardous Waste Management Facilities through a Compensation and Incentives Approach: A Bibliography*. Council of Planning Librarians, CPL Bibl., no. 122.

Bryant, Bunyan, and Paul Mohai, eds. 1992. *Race and the Incidence of Environmental Hazards*. Boulder, Colo.: Westview Press.

Bryant, Jeffrey S. 1989. Strategies for Source Abatement of Industrial Hazardous Wastes. In *11th Canadian Waste Management Conference*, 43–58. Ottawa: Environment Canada.

Cahill, Lawrence B., and Walter R. Holman. 1981. Siting Waste Management Facilities. In *Hazardous Waste Management*, edited by J. Jeffrey Pierce and P. Aarne Veslind, 87–97. Ann Arbor: Ann Arbor Science.

Campbell, Monica E., and William M. Glenn. 1982. *Profit from Pollution Prevention: A Guide to Industrial Waste Reduction and Recycling*. Toronto: Pollution Probe Foundation.

Carr, F. Housley. 1988. Chemical Weapons Are for Burning. *Chemical Week* 142, no. 11 (March 16): 84, 88.

———. 1991. Heading Off Environmental Disputes. *ENR* 226, no. 6 (February 11): 32.

Castle, Geoffrey. 1992. *Siting the Alberta Special Waste Treatment Centre*. Sustainable Development Research Institute Discussion Paper Series. Vancouver: SDRI.

———. 1993. Siting the Manitoba Hazardous Waste Treatment Facility: Case Study of a Success. *Journal of the Air and Waste Management Association* 43 (July): 1–7.

Cavanaugh, Tim. 1993. Waste Management '93: Incineration–Slim Pickin's. *Chemical Marketing Reporter* 244, no. 22 (November 29): SR14–SR15.

Caves, Roger C. 1990. "Community and land use restraints in siting group homes for the mentally retarded." *Glendale Law Review* 9 nos. 1–2: 1–19.

Chemical Marketing Reporter. 1994a. Borden Chemicals Lashes Back at EPA. 246, no. 19 (November 7): 5, 19.

———. 1994b. Kiln Incineration under Fire. 246, no. 26 (December 26): 7, 32.

Chemical Week. 1994a. Forecast 94: No Relief in Hazwaste. January 5–12, 43–44.

———. 1994b. Hazwaste Firms Struggle to Survive Market Doldrums. August 17, 26–27.

Christensen, Steven A. 1979. Nuclear Power Plant Siting: A Comparative Analysis of Public Interaction in the Siting Process in France and the United States. *Denver Journal of International Law and Policy* 8 (Winter): 343–66.

Chynoweth, Emma. 1990. New Options Take on Incineration. *Chemical Week* 147, no. 7 (August 22): 49–50.

———. 1994. Waste Reduction Takes Toll on European Treatment Business. *Chemical Week* 155, no. 6 (August 17): 26.

Clark-McGlennon Associates. 1987. Criteria for Evaluating Sites. In *Hazardous Materials, Hazardous Waste*, edited by Raymond D. Scanlon, 133–45. Washington, D.C.: International City Managers Association.

Coleman, Cynthia-Lou. 1989–90. What Policy Makers Can Learn from Public Relations Practitioners: The Siting of a Low-Level Radioactive Waste Facility in Cortland County, New York. *Public Relations Quarterly* 34, no. 4 (Winter): 26–31.

Collins, John, and Walter P. Saukin, eds. 1981. *The Hazardous Waste Dilemma: Issues and Solutions*. New York: American Society of Civil Engineers.

Commoner, Barry. 1989. *Making Peace with the Planet*. New York: Pantheon Books.

Connell, Barbara. 1990. The Consultative Process for Siting a Hazardous Waste Management Facility in Manitoba: Experience from Manitoba Communities. In *12th Canadian Waste Management Conference*, 59–71. Ottawa: Environment Canada.

Connor, Desmond. 1988. Breaking through the "Nimby" Syndrome. *Civil Engineering* 58, no. 12 (December): 69–71.

———. 1990. Managing the "NIMBY Syndrome" in the 1990s: Principles and Cases for Waste Managers. In *12th Canadian Waste Management Conference*, 1–8. Ottawa: Environment Canada.

Cook, James. 1988. Not in Anybody's Backyard. *Forbes* 142, no. 12 (November 28): 172–82.

Corr, Denis. 1990. The Tire Fire. In *37th Ontario Waste Management Conference*, 73–75. Toronto: Ontario Ministry of the Environment.

Cowden, Dick. 1991. The Search for Affordable Housing. *Regulation* 14, no. 4 (Fall): 32–35.

Crim, Sarah. 1990. The NIMBY Syndrome in the 1990s: Where Do You Go after Getting to No? *Environment Reporter* 21, no. 1, part 2:132–38. Washington, D.C.: Bureau of Public Affairs.

Darcey, Sue. 1994. Hazwaste Industry Responds to New EPA Policy. *World Wastes*. 37, no. 1 (January): 22.

Davis, Charles E. 1984–85. Substance and Procedure in Hazardous Waste Facility Siting. *Journal of Environmental Systems* 14, no. 1:51–62.

Davis, Charles E., and James P. Lester. 1988. Hazardous Waste Politics and the Policy Process. In *Dimensions of Hazardous Waste Politics and Policy*, 1–34. New York: Greenwood Press.

Davis, Gary, Donald Huisingh, and Bruce Piasecki. 1987. Waste Reduction Strategies: European Practice and American Prospects. In *America's Future in Toxic Waste Management: Lessons from Europe*, edited by Bruce Piasecki and Gary Davis, 17–42. Westport, Conn.: Quorum Books.

Dawson, Gaynor W., and Basil W. Mercer. 1986. *Hazardous Waste Management*. New York: Wiley.

Dean, Robert B. 1987. Control of Emissions from Municipal Incinerators. In *9th Canadian Waste Management Conference*, 153–59. Ottawa: Environment Canada.

Dear, Michael. 1992. Understanding and Overcoming the NIMBY Syndrome. *Journal of the American Planning Association* 58, no. 3 (Summer): 288–300.

Duffy, Celeste P. 1984. State Hazardous Waste Facility Siting: Easing the Process through Local Cooperation and Preemption. *Boston College Environmental Affairs Law Review* 11:755–804.

Duke, L. Donald. 1994. Hazardous Waste Minimization: Is It Taking Root in U.S. Industry? *Waste Management* 14, no. 1:49–59.

Elliott, Michael L. Poirier. 1984. Improving Community Acceptance of Hazardous Waste Facilities through Alternative Systems for Mitigating and Managing Risk. *Hazardous Waste* 1:398–410.

Ely, E. S. 1987. New Jersey's Incinerator Will Be No. 1. *Chemical Week* 141, no. 12 (September 16): 47, 51.

Ember, Lois R. 1992. Pollution Prevention at Chemical Plants Benefits Environment at Low Cost. *Chemical and Engineering News* 70, no. 25 (June 22): 18–19.

———. 1994. Minorities Still More Likely to Live Near Toxic Sites. *Chemical and Engineering News* 72, no. 36 (September 5): 19.

Endries, John M. 1989. Not in My Backyard: Curing a Syndrome. *Public Utilities Fortnightly* 124, no. 8 (October 12): 13–14.

ENR. 1994a. Florida Burner Plans Sent to Scrap Heap. 233, no. 25 (December 19): 20.

———. 1994b. NIMBY Problems Slow Taiwan Building Plan. 233, no. 16 (October 17): 21.

Environment Today. 1994a. Northtrop, Jamro Win NAEM's 1994 Enviroawards. 5, no. 10 (October): 13.

———. 1994b. Small Firm Praised for Big E-Policies. 5, no. 11 (November): 11.

Environmental Manager. 1991. Pollution Prevention Permits Debut in New Jersey. 3, no. 1 (August): 1–2, 14.

———. 1994a. More Public Involvement in Permitting. 6, no. 5 (December): 1–3.

———. 1994b. Where EPA Is Headed with Hazwaste. 6, no. 2 (September): 8–9.

———. 1995a. EPA's Final Waste Min/Combustion Strategy. 6, no. 7 (February): 5–6.

———. 1995b. In the States. 6, no. 8 (March): 15.

Ewel, Dexter, and Tonya Lewis. 1991. Public Perception of Health Risks. In *Wastetech '91: Canadian Waste Management Conference*, 381–86. Ottawa: Environment Canada.

Fairley, Peter, and Emma Chynoweth. 1994. Treatment Technologies Promise an Alternative to Incineration. *Chemical Week* 155, no. 6 (August 17): 28–32.

Farago, Klara, Anna Vari, and Janos Vecsenyi. 1989. Not in My Town: Conflicting Views on the Siting of a Hazardous Waste Incinerator. *Risk Analysis* 9, no. 4:463–71.

Fischhoff, Baruch, S. Lichtenstein, P. Slovic, S. L. Derby, and R. L. Keeney. 1981. *Acceptable Risk*. New York: Cambridge University Press.

Foley, Michael, and Dan Prepdall. 1990. Siting New Powerplants: Challenges for the 1990s. *Electrical World* 204, no. 10 (October): 58, 60.

Fouhy, Ken, and Gerald Ondrey. 1994. Incineration: Turning Up the Heat on Hazardous Waste. *Chemical Engineering* 101, no. 5 (May): 39–43 .

Freeman, Harry, ed. 1989. *Standard Handbook of Hazardous Waste Treatment and Disposal*. New York: McGraw-Hill.

———. 1990. *Hazardous Waste Minimization*. New York: McGraw-Hill.

Gerrard, Michael. 1994. Whose Backyard, Whose Risk. Cambridge: MIT Press.

Gibbs, Lois. 1989. The Movement on the Move. In *Everyone's Backyard*. Arlington, Va.: Citizen's Clearinghouse for Hazardous Waste.

Gilges, Kent. 1991. Europe's Hazardous Waste Dilemma. *Chemical Engineering* (August): 30–39.

Glenn, Jim. 1992. 1991 Waste Reduction Laws. *BioCycle* 33, no. 5 (May): 35–36.

Goldbaum, Ellen, and David Rotman. 1989. Hazardous Waste: Faced with Dwindling Choices, Companies Must Seek New Ways to Manage It. *Chemical Week* 145, no. 8 (August 23): 18–48.

Greenberg, Michael, and R. Anderson. 1984. *Hazardous Waste Sites: The Credibility Gap*. New Brunswick, N.J.: Center for Urban Policy Research.

Greenberg, Michael, and James Hughes. 1993. Impact of Hazardous Waste Sites on Property Value and Land Use: Tax Assessors' Appraisal. *Appraisal Journal* 61, no. 1 (January): 42–51.

Gregory, Robin. 1991. Incentives Policies to Site Hazardous Waste Facilities. *Risk Analysis* 11:667–75.

Gregory, Robin, and Howard Kunreuther. 1990. Successful Siting Incentives. *Civil Engineering* (April): 73–75.

Groothuis, Peter A., and Gail Miller. 1994. Locating Hazardous Waste Facilities: The Influence of NIMBY Beliefs. *American Journal of Economics & Sociology* 53, no. 3 (July): 335–46.

Grund, Howard P., and Paul Kemezis. 1993. Despite New EPA Rules, ISP Upbeat on Hazardous Waste Incinerator. *Chemical Week* 153, no. 13 (October 13): 12.

Guerra, Sarith. 1991. NIMBY, NIMTOF, and Solid Waste Facility Siting. *Public Management* 73:11–12+.

Hadden, Susan G., Joan Veillette, and Thomas Brandt. 1983. State Roles in Siting Hazardous Waste Disposal Facilities: From State Preemption to Local Veto. In *The Politics of Hazardous Waste Management*, edited by James

P. Lester and Ann O'M. Bowman, 196–211. Durham, N.C.: Duke University Press.

Hamilton, James T. 1993. Politics and Social Costs: Estimating the Impact of Collective Action on Hazardous Waste Facilities. *RAND Journal of Economics* 24, no. 1:101–25.

Hamilton, Michael S. 1979. Power Plant Siting: A Literature Review. *Natural Resources Journal* 19 (January): 75–95.

Hanke, Jon. 1994. Hazwaste Landfill Operators Face Difficult Times. *World Wastes* 37, no. 9 (September): 10–12 .

Hansen, Fred. 1989. Pollution Prevention Planning: A New Mandate for Oregon's Environment. *Environmental Forum* 6:30–34.

Harris, Christopher, William Want, and Morris Ward. 1987. *Hazardous Waste: Confronting the Challenge.* New York: Quorum Books.

Harris, Paul. 1994. "Survival of the fittest" Is TSD Field's New Credo. *Environment Today* 5, no. 10 (October): 25–28.

Hazen, Susan. 1992. EPA's 33/50 Program Second Progess Report: Reducing Risks through Voluntary Action. In *Proceedings of the 39th Ontario Conference on the Environment*, 233–73. Toronto: Ontario Ministry of Environment.

Heckley, Jane. 1985. Hazardous Waste Treatment Facility Siting in New York State. Discussion paper, Department of Geography, Syracuse University.

Heiman, Michael. 1990. From "Not in My Backyard!" to "Not in Anybody's Backyard!": Grassroots Challenge to Hazardous Waste Facility Siting. *Journal of the American Planning Association* 56, no. 3 (Summer): 359–62.

Heller, Karen. 1990. Commercial Firms: Will the Boom Continue? *Chemical Week* 147, no. 7 (August 22): 51–52.

———. 1992. Toxics Use Reduction. *Chemical Week* 151, no. 7 (August 19): 52–56.

Hersch, Paul. 1988. Management in Action–Three Options; But How Many Choices? *Water Engineering and Management* 135, no. 12 (December): 22.

Hirschhorn, Joel S. 1988. Cutting Production of Hazardous Waste. *Technology Review* 91, no. 3 (April): 52–61.

Hirschhorn, Joel S., and Kirsten U. Oldenburg. 1987. Preventing Pollution Is No End-of-Pipe Dream. *Across the Board* 24, no. 6 (June): 11–15.

———. 1991. *Prosperity without Pollution: The Prevention Strategy for Industry and Consumers.* New York: Van Nostrand.

Hisschemöller, Matthijs, and Cees J. H. Midden. 1989. Technological Risk, Policy Theories, and Public Participation in Connection with the Siting of Hazardous Facilities. In *Social Decision Methodology for Technological Projects*, edited by Charles Vlek and George Cvetkovich, 173–194. Dordrecht: Kluwer.

Hoffman, John. 1993. Waste Management '93: Minimization–a Goal of Zero. *Chemical Marketing Reporter* 244, no. 22 (November 29): SR12–SR13.

Holman, Curt. 1993. Hazardous Waste Landfills Continue to Lose Ground. *World Wastes* 36, no. 6 (June): 6.

Holznagel, Bernd. 1986. Negotiation and Mediation: The Newest Approach to Hazardous Waste Facility Siting. *Boston College Environmental Affairs Law Review* 13, no. 3:329–78.

Hunt, Gary E. 1990. Waste Reduction Techniques and Technologies. In *Hazardous Waste Minimization*, edited by Harry Freeman, 25–54. New York: McGraw-Hill.

Hunt, Stephanie. 1990. The Federal PCB Destruction Program: An Approach to the Siting of Mobile PCB Incinerators in Canada In *12th Canadian Waste Management Conference*, 49–57. Ottawa: Environment Canada.

Inhaber, Herbert. 1991a. Can We Find a Volunteer Nuclear Waste Community? *Public Utilities Fortnightly* 128, no. 2 (July 15): 19–21.

———. 1991b. NIMBY and LULU. *Regulation* 14, no. 4 (Fall): 22–24.

———. 1992a. How Much for This Dump? *Across the Board* (December): 38.

———. 1992b. Of LULUs, NIMBYs, and NIMTOOs. *Public Interest*, no. 107 (Spring): 52–64.

Jackson, John. 1990. NIMBY: A Positive Influence. In *12th Canadian Waste Management Conference*, 303–9. Ottawa: Environment Canada.

Jankowski, Wanda. 1995. Cut Costs and Hazardous Waste by Recycling Lamps. *Facilities Design and Management* 14, no. 1 (January): 37.

Jessup, Deborah H. 1992. *Waste Management Guide: Laws, Issues, and Solutions*. Washington, D.C.: Bureau of National Affairs.

Kahneman, Daniel, and A. Tversky. 1979. Prospect Theory: An Analysis of Decision under Risk. *Econometrica* 47: 263–91.

Kasperson, Roger E. 1986a. Hazardous Waste Facility Siting: Community, Firm, and Governmental Perspectives. In *Hazards: Technology and Fairness* by National Academy of Engineering, 118–44. Washington, D.C.: National Academy Press.

———. 1986b. Six Propositions on Public Participation and Their Relevance for Risk Communication. *Risk Analysis* 6, no. 3:275–81.

Kauffman, Jennifer, Patricia J. Serie, and Claire M. Gesalman. 1990. US Lessons Learned: Selecting Sites for Controversial Facilities. In *12th Canadian Waste Management Conference*, 13–32. Ottawa: Environment Canada.

Kelleher, Maria. 1991. Household Hazardous Waste—How Much Is Really Out There? In *Wastetech '91: Canadian Waste Management Conference*, 457–69. Ottawa: Environment Canada.

Kelly, Margaret. 1990. New Trends in Clean Technologies. In *The Management of Hazardous Substances in the Environment*, edited by K. L. Zirm and J. Mayer, 115–24. London: Elsevier.

Kemezis, Paul. 1990. Court Nixes Alabama Waste Ban. *Chemical Week* 147, no. 7 (August 22): 48–49.

———. 1991. States Fight for Rights. *Chemical Week* 149, no. 4 (August 21): 56–57.

———. 1992. Among the States: Free Trade. *Chemical Week* 151, no. 7 (August 19): 50–52.

———. 1993. A Cease-Fire in the Hazardous Waste War. *Chemical Week* 153, no. 4 (August 4): 34.

———. 1993. Interstate Transport Is Still a Fair Game. *Chemical Week* 153, no. 6 (August 18): 30–31.

Kemezis, Paul, Gregory Morris and David Rotman. 1991. Implementing the Program: PPG Industries; Lyondell; Quantum. *Chemical Week* 149, no. 20 (December 11): 53–57.

Kirschner, Elisabeth. 1993. DuPont and EPA Find Pollution Prevention Profits. *Chemical Week* 153, no. 15 (October 27): 16.

Kiser, Elizabeth. 1992. Getting to YIMBY: Yes in My Backyard. *World Wastes* 35, no. 12 (December): 38–41.

Koepp, Donald. 1990. The Ventura Country Waste Minimization Program. In *Hazardous Waste Minimization,* edited by Harry Freeman, 239–52. New York: McGraw-Hill.

Kraft, Michael, and Ruth Kraut. 1988. Citizen Participation and Hazardous Waste Policy Implementation. In *Dimensions of Hazardous Waste Politics and Policy,* edited by Charles E. Davis and James P. Lester, 63–80. New York: Greenwood Press.

Krimsky, Sheldon. 1986. Research under Community Standards: Three Case Studies. *Science, Technology, and Human Values* 11 (Summer): 14–33.

Krukowski, John, and Catherine A. Simpson. 1995. Pollution Technologies Get Research Boost. *R&D* 37, no. 1 (January): 46.

Kuehn, Steven E. 1993. Steam Turbine Technology Keeps Pace with Demands. *Power Engineering* 97, no. 3 (March): 17–24.

Kunreuther, Howard. 1986. Hazard Compensation and Incentive Systems: An Economic Perspective. In *Hazards: Technology and Fairness* by National Academy of Engineering, 145–63. Washington, D.C.: National Academy Press.

Kunreuther, Howard, and Douglas Easterling. 1990. Are Risk-Benefit Tradeoffs Possible in Siting Hazardous Facilities? *American Economic Review* 80, no. 2 (May): 252–56.

Kunreuther, Howard, Kevin Fitzgerald, and Thomas D. Aarts. 1993. Siting Noxious Facilities: A Test of the Facility Siting Credo. *Risk Analysis* 13 (June): 301–18.

Kunreuther, Howard, Paul Kleindorfer, Peter J. Knez, and Rudy Yaksick. 1987. A Compensation Mechanism for Siting Noxious Facilities: Theory and Experimental Design. *Journal of Environmental Economics and Management* 14, no. 4 (December): 371–83.

Kunreuther, Howard, and Lawrence Susskind. 1993. Facility Siting Credo. Manuscript. University of Pennsylvania.

Kupchanko, Eugene. 1988. Mobile Destruction Systems. In *10th Canadian Waste Management Conference,* 107–30. Ottawa: Environment Canada.

———. 1989. Case Study of a Successful Facility Siting. In *Standard Handbook of Hazardous Waste Treatment and Disposal,* edited by Harry Freeman, 3.39–3.49. New York: McGraw-Hill.

Kurland, Orin M. 1992. Risk Communication, Mitigation, and Uncertainty. *Risk Management* 39, no. 12 (December): 60.

Kusek, Kristyn. 1994. Pollution and prejudice. *Government Executive* 26, no. 10 (October): 36–41.

LaBar, Gregg. 1990. Du Pont: Watching Its Waste. *Occupational Hazards* 52, no. 7 (July): 51–55.

Lake, Robert W. 1993. Rethinking NIMBY. *Journal of the American Planning Association* 59, no. 1 (Winter): 87–93.

Lesbirel, S. Hayden. 1987–88. Project Implementation in Japan: Origins of Delay in Developing Energy Facilities. *Journal of Business Administration* 17, nos. 1–2:249–86.

———. 1990. Implementing Nuclear Energy Policy in Japan. *Energy Policy* 18, no. 3 (April): 267–82.

Levy, Jack S. 1992. An Introduction to Prospect Theory. *Political Psychology* 13, no. 2:171–86.

Lidskog, Rolf, and Ingemar Elander. 1992. Reinterpreting Locational Conflicts: NIMBY and Nuclear Waste Management in Sweden. *Policy and Politics* 20:249–64.

Lucas, Allison. 1994a. DuPont: Fighting Its Way to a Cleaner Standard. *Chemical Week* 154, no. 25 (June 1): 41–42.

———. 1994b. Spending, Waste Reduction Slow. *Chemical Week* 154, no. 25 (June 29): 10.

Lynch, Rodney C. 1995. Host Agreements for Siting Waste Disposal Facilities. *Government Finance Review* 11, no. 1 (February): 46–47.

Lynn, Frances M. 1987. Citizen Involvement in Hazardous Waste Sites: Two North Carolina Success Stories. *Environmental Impact Assessment Review* 7: 347–61.

Macauley, Molly K., Michael D. Bowes, and Karen L. Palmer. 1992. *Using Economic Incentives to Regulate Toxic Substances.* Washington, D.C.: Resources for the Future.

MacKenzie, R.C. 1984. Canadian City Welcomes Hazwaste Facility. *World Wastes,* July, 45–50.

Machlin, Jennifer. 1987. A System Born of Crisis: Hazardous Waste Management and Cleanup in the United States. In *9th Canadian Waste Management Conference,* 241–300. Ottawa: Environment Canada.

Maize, Kennedy P., and John McCaughey. 1992. NIMBY, NOPE, LULU, and BANANA: A Warning to Independent Power. *Public Utilities Fortnightly* 130, no. 3 (August 1): 19–21.

Major, Michael J. 1992. Exchanging Surplus Materials Rather Than Dumping. *Modern Paint and Coatings* 82, no. 12 (November): 10, 12.

Mangan, William R. 1988. A Comparative Analysis of Hazardous Waste Management Policy in Western Europe. In *Dimensions of Hazardous Waste Politics and Policy,* edited by Charles E. Davis and James P. Lester, 205–21. New York: Greenwood Press.

Mank, Bradford C. 1991. The Two-Headed Dragon of Siting and Cleaning Up Hazardous Waste Dumps: Can Economic Incentives or Mediation Slay the Monster? *Boston College Environmental Affairs Law Review* 19: 239–85.

Marianov, Vladimir, and Charles ReVelle. 1991. The Standard Response Fire Protection Siting Problem *INFOR* 29, no. 2 (May): 116–29.

Marshall, Patrick G. 1989. Not in My Back Yard! *Editorial Research Reports,* June 9, 306–19.

Mason, G. 1989. Closure and Rejection of Waste Facilities: What Effect Has Public Pressure? *Hazardous Material Control,* July–August, 54–58.

Massam, Bryan H. 1993. *The Right Place: Shared Responsibility and the Location of Public Facilities.* Harlow, Essex, England: Longman.

Matheny, Albert R., and Bruce A. Williams. 1988. Rethinking Participation: Assessing Florida's Strategy for Siting Hazardous Waste Disposal Facilities. In *Dimensions of Hazardous Waste Politics and Policy,* edited by Charles E. Davis and James P. Lester, 37–52. New York: Greenwood Press.

Mazmanian, Daniel, and David Morell. 1992. *Beyond Superfailure: America's Toxics Policy for the 1990s.* Boulder: Westview Press.

———. 1994. The "NIMBY" Syndrome: Facility Siting and the Failure of Democratic Discourse. In *Environmental Policy in the 1990s,* edited by Norman J. Vig and Michael E. Kraft, 233–42. 2d ed. Washington, D.C.: CQ Press.

McAvoy, Gregory. 1993. State Autonomy and Democratic Accountability: The Politics of Hazardous Waste Policy. Unpublished Ph.D. Dissertation, University of Minnesota.

McClelland, Gary H., William D. Schulze, and Brian Hurd. 1990. The Effect of Risk Beliefs on Property Values: A Case Study of a Hazardous Waste Site. *Risk Analysis* 10, no. 4: 485–97.

McQuaid-Cook, Jennifer. 1987. Hazardous Waste Management System Development: Alberta, Canada. *Wastes Management,* April, 220–26.

McQuaid-Cook, Jennifer, and C. S. Simons. 1989. Development and Operation of a Waste Management System in Alberta, Canada. *Waste Management and Research* 7: 219–27.

McQuaid-Cook, Jennifer, and Kenneth J. Simpson. 1986. Siting a Fully Integrated Waste Management Facility in Alberta. *Journal of the Air Pollution Control Association* 36 (September): 1031–36.

Moran, Mary A. 1980. Transmission Line Siting: Local Concerns versus State Energy Interests. *Urban Law Annual* 19: 183–203.

Morell, David. 1984. The Siting of Hazardous Waste Facilities in California. *Public Affairs Report* 25: 1–10.

Morell, David, and Christopher Magorian. 1982. *Siting Hazardous Waste Facilities.* Cambridge, Mass.: Ballinger.

Moskowitz, Joel S. 1984. Hazardous Waste Management in West Germany: An Identification of Issues. *Environmental Forum* 3 (June): 20–24.

Mumme, Stephen P. 1988. Complex Interdependence and Hazardous Waste Management along the U.S.-Mexico Border. In *Dimensions of Hazardous Waste Politics and Policy,* edited by Charles E. Davis and James P. Lester, 223–39. New York: Greenwood Press.

Munro, John, Ginger King, Katherine Kornfield, and Viola McAvey. 1990. Public Communication and Participation Activities in Siting Controversial Facilities in the US and Western Europe. Paper prepared for Waste Management '90, April, Tucson.

National Research Council. 1985. *Reducing Hazardous Waste Generation: An Evaluation and a Call for Action.* Environmental Studies Board. Washington, D.C.: National Academy Press.

Navarro, Peter. 1988. Comparative Energy Policy: The Economics of Nuclear Power in Japan and the United States. *Energy Journal* 9, no. 4 (October): 1–15.

Nelson-Horchler, Joani. 1988. Hazardous Waste: Who Wants It? *Industry Week* 236, no. 4 (February 15): 56–58.

Nieves, Leslie, J. J. Himmelberger, S. J. Ratick, and A. L. White. 1992. Negotiated Compensation for Solid-Waste Disposal Facility Siting: An Analysis of the Wisconsin Experience. *Risk Analysis* 12, no. 4:505–11.

O'Brien, Tim. 1994. NIMBY Groups: A Pain in the Amusement Industry's Side. *Amusement Business* 106, no. 10 (March 7–13): 3, 29.

O'Hare, Michael. 1977. Not on My Block You Don't: Facility Siting and the Strategic Importance of Compensation. *Public Policy* 25, no. 4 (Fall): 407–58.

———. 1993. Waste Disposal: Not in My Back Yard. *Vital Speeches* 59, no. 4 (December 1): 121–25.

O'Hare, Michael, Lawrence Bacow, and Debra Sanderson. 1983. *Facility Siting*. New York: Van Nostrand.

O'Hare, Michael, and Debra Sanderson. 1993. Facility Siting and Compensation: Lessons from the Massachusetts Experience. *Journal of Policy Analysis and Management* 12 (spring): 364–76.

O'Looney, John. 1993. Framing a Social Market for Community Responsibility: Governing in an Age of NIMBYs and LULUs. *National Civic Review* 82 (Winter): 44–62.

Opalski, Douglas V. 1992. Jobs, Housing, Transportation: A Balance through Mediated Fair Shares. *Journal of Housing* 49, no. 6 (November–December): 290–95, 298–300.

Oppelt, E. Timothy. 1990. Air Emissions from the Incineration of Hazardous Waste. In *Health and Environmental Hazards from Toxic Chemicals in the Air*, edited by L. Grant et al., 23–52. Princeton: Scientific Publishing.

Paige, H. W., and J. E. Owens. 1984. Assessment of National Systems for Obtaining Local Acceptance of Nuclear Waste Management Siting Activities. In *Radioactive Waste Management: Proceedings of an International Conference*,1:319–31. Vienna: IAEA.

Paparian, Michael. 1983. In Europe, Dump Is a Dirty Word: Hazardous Waste Disposal in Denmark and West Germany. *Planning* 49:16–17.

Paquin, Jean. 1989. Clean-up Following the PCB Fire at St. Basile-Le-Grand in *36th Ontario Waste Management Conference*, 61–64. Toronto: Ontario Ministry of the Environment.

Petts, Judith. 1992. Incineration Risk Perceptions and Public Concern: Experience in the UK Improving Risk Communications. *Waste Management and Research* 10:69–82.

Piasecki, Bruce, and Gary Davis, eds. 1987. *America's Future in Toxic Waste Management: Lessons From Europe*. Westport, Conn.: Quorum Books.

Pligt, Joop van der. 1992. *Nuclear Energy and the Public*. London: Blackwell.

Plishner, Emily S. 1991. Cyclicity Rears Its Ugly Head. *Chemical Week* 149, no. 4 (August 21): 44–48.

Pojasek, Robert. 1980. Developing Solutions to Hazardous Waste Problems. *Environmental Science and Technology* 14 (August): 924–29.

Popper, Frank J. 1981. Siting LULUs. *Planning* (April): 12–15.

———. 1991. The Great LULU Trading Game. *Planning* (May): 15–17.

Portney, Kent E. 1985. The Potential of the Theory of Compensation for Mitigating Public Opposition to Hazardous Waste Treatment Facility Siting: Some Evidence from Five Massachusetts Communities. *Policy Studies Journal* 14: 81–89.

———. 1988. The Role of Economic Factors in Lay Perceptions of Risk. In *Dimensions of Hazardous Waste Politics and Policy*, edited by Charles E. Davis and James P. Lester, 53–62. New York: Greenwood Press.

———. 1991. *Siting Hazardous Waste Treatment Facilities: The Nimby Syndrome*. Westport, Conn.: Auburn House.

Portnoy, Kristine. 1988. Making Waste Sites Seismically Safe. *Chemical Week* 142, no. 6 (February 10): 22–23 .

Portnoy, Kristine, and Laurie A. Rich. 1987. A "NIMBY" Cure: Aid and Education. *Chemical Week* 140, no. 18 (May 13): 41–42.

Provost, Denise. 1982–83. The Massachusetts Hazardous Waste Facility Siting Act: What Impact on Municipal Power to Exclude and Regulate? *Boston College Environmental Affairs Law Review* 10 (December): 715–95.

Rabe, Barry G. 1990. The Hazardous Waste Dilemma and the Hazards of Institutionalizing Negotiation. In *Conflict Resolution and Public Policy*, edited by Miriam K Mills, 3–20. New York: Greenwood Press.

———. 1991. Beyond the NIMBY Syndrome in Hazardous Waste Facility Siting: The Albertan Breakthrough and the Prospects for Cooperation in Canada and the United States. *Governance* 4, no. 2: 184–206.

———. 1992. When Siting Works, Canada-Style. *Journal of Health Politics, Policy, and Law* 17, no. 1 (Spring): 119–42.

Rasmussen, Thomas H. 1992. Not in My Backyard: The Politics of Siting Prisons, Landfills, and Incinerators. *State and Local Government Review* 24 (fall): 128–34.

Repa, Edward W., and Susan Sheets. 1992. 1991 Landfill Capacity Report. In *14th Canadian Waste Management Conference*, 37–50. Ottawa: Environment Canada.

Rhodes, Thomas C., and Paul N. Wilson. 1995. Sky Islands, Squirrels, and Scopes: The Political Economy of an Environmental Conflict. *Land Economics* 71, no. 1 (February): 106–21.

Risk, R. C. B. 1981. A Long, Sad Story: Siting Transmission Lines in Ontario. *University of Toronto Law Journal* 31 (Winter): 27–71.

Ristoratore, Mario. 1985. Siting Toxic Waste Disposal Facilities in Canada and the United States: Problems and Prospects. *Policy Studies Journal* 14, no. 1:140–48.

Robertson, David K., et al. 1987. Liquid Household Waste in the United States. *Environmental Management* 11, no. 6:735–42.

Rose, Eugene A., and Riley E. Dunlap. 1994. Nuclear Power: Three Decades of Public Opinion. *Public Opinion Quarterly* 58:295–325.

Rotman, David. 1990a. Chemical Firms Move to Turn Waste to Profits *Chemical Week* 147, no. 7 (August 22): 46–47.

———. 1990b. Hazardous Waste: Tightening Rules—and Options—Up the Ante. *Chemical Week* 147, no. 7 (August 22): 34–44.

———. 1993. Du Pont Incinerator Cut Signals Industry Slump. *ENR* 231, no. 11 (September 13): 20–21.

———. 1993. Responsible Care: The quest for reduced emissions, greener processes. *Chemical Week*, 153, no. 1 (July 7/14): 117–118.

Rotman, David, and Ronald Begley. 1994. Hazardous Chemical Producers Face Uncertain Prospects. *Chemical Week* 155, no. 6 (August 17): 24–25.

Rotman, David, and Ian Young. 1991. Hazardous Waste. *Chemical Week* 149, no. 4 (August 21): 40–44.

Rovet, Ernest. 1987. Legitimized Structured Public Involvement in the Environmental Assessment Process. In *9th Canadian Waste Management Conference*, 301–12. Ottawa: Environment Canada.

Rubin, Debra K., Hazel Bradford, Mary B. Powers, and Paul Kemezis. 1993. EPA Gets Tougher on Facilities. *ENR* 230, no. 21 (May 24): 8.

Rusk, James. 1995. Toxic-Waste Treatment Plant for Niagara down the Drain. *Globe and Mail*, February 8.

Russo, Thomas N. 1995. Making hydro sustainable. *Public Utilities Fortnightly*, 133, no. 1 (January 1): 14–19.

Rzadzki, John. 1993. Waste Management '93: In the Dumps. *Chemical Marketing Reporter* 244, no. 22 (November 29): SR3–SR6.

Salcedo, Rodolfo N., Frank Cross, and Randolph Chrismon. 1989. *Environmental Impacts of Hazardous Waste Treatment, Storage, and Disposal Facilities*. Lancaster, Penn.: Technomic.

Sandenburgh, Robin, and Marion Cox. 1987. Communicating Risk Assessment Findings to the Public. In *Hazardous Materials, Hazardous Waste*, edited by Raymond D. Scanlon, 163–70. Washington, D.C.: International City Managers Association.

Schecter, Roger N. 1990. Profile of State Waste Reduction Programs. In *Hazardous Waste Minimization*, edited by Harry Freeman, 223–38. New York: McGraw-Hill.

Schlossberg, Howard. 1993. HCBs Designed to Convert NIMBYs to YIMBY-FAPs. *Marketing News* 27, no. 13 (June 21): 2.

Schwab, Jim. 1988. Hazardous Waste: What Goes Around Comes Around; Federal Legislation Prompted State Involvement in Hazardous Waste Management. *Planning* 54 (February): 18–2.

———. 1991. Blue-Collar Groups Are Saying, "Not in Our Backyard." *Planning* 57, no. 10 (October): 8–11.

Seley, John E. 1983. *The Politics of Public Facility Planning*. New York: Macmillan.

Sellers, Martin P. 1993. Nimby: A Case Study in Conflict Politics. *Public Administration Quarterly* 16, no. 4 (Winter): 460–77.

Shanoff, Barry. 1994. States Seek Long-Lerm Security for Hazwaste Sites. *World Wastes* 37, no. 10 (October): 14–19.

Sherry, Susan. 1990. Reducing Hazardous Wastes: Strategies and Tools for Local Governments. In *Hazardous Waste Minimization*, edited by Harry Freeman, 203–21. New York: McGraw-Hill.

Shichor, David. 1992. Myths and Realities in Prison Siting. *Crime and Delinquency* 38: 70–87.

Shireman, William K. 1993. Beyond 50% Recycling—the Next Recycling Agenda: To Grow an Economy That Source-Reduces Waste. *Executive Speeches* 7, no. 4 (February–March): 45–48.

Shrader-Frechette, Kristin. 1991. *Risk and Rationality*. Berkeley: University of California Press.

Shuff, Richard G. 1988. "Bribes" Work in Wisconsin. *Waste Age*, March, 51–55.

Silton, Triana. 1993. Out of the Frying Pan: Chemical Weapons Incineration in the United States. *Ecologist* 23:18–24.

Skinner, John. 1991. Hazardous Waste Treatment Trends in the US. *Waste Management and Research* 9:55–63.

Skryness, Richard S., Karl Swenson, and William Pananos. 1994. Spoils of War. *Civil Engineering* 64, no. 11 (November): 51–53.

Small Business Reports. 1993. Preventing Pollution at the Source. 18, no. 11 (November): 6.

Smith, Thomas P., and Martin Jaffe. 1986. *Siting Group Homes for Developmentally Disabled Persons*. American Planning Association.

Smolen, Gerald E., Gary Moore, and Lawrence V. Conway. 1992. Hazardous Waste Landfill Impacts on Local Property Values. *Real Estate Appraiser* 58, no. 1 (April): 4–11.

Steinzor, Rena, and Velma Smith. 1988. The Toxic Combat Zone: Is a Ceasefire Possible? *Environmental Forum* 5 (July–August): 5–9.

Sternberg, Ken. 1990. Environmental Services. *Chemical Week* 146, no. 17 (May 2): 20–23.

Suess, Michael J. 1992. Safe Management of Hazardous Waste. *Waste Management and Research* 10:93–97.

Sullivan, Arthur M. 1990. Victim Compensation Revisited: Efficiency versus Equity in the Siting of Noxious Facilities. *Journal of Public Economics* 41, no. 2 (March): 211–25.

Susskind, Lawrence, and Connie Ozawa. 1985. Mediating Public Disputes: Obstacles and Possibilities. *Journal of Social Issues* 41 (Summer): 145–59.

Swallow, Stephen K., James J. Opaluch, and Thomas F. Weaver. 1992. Siting Noxious Facilities: An Approach That Integrates Technical, Economic, and Political Considerations. *Land Economics* 68, no. 3:283–301.

Swartzman, Daniel, Kevin Croke, and Sheri Swibel. 1985. Reducing Aversion to Living near Hazardous Waste Facilities through Compensation and Risk Reduction. *Journal of Environmental Management* 20:43–50.

Thayer, Ann M. 1992. Pollution Reduction. *Chemical and Engineering News* 70, no. 46 (November 16): 22–52.

Tomsho, Robert. 1991. Small Town in Alberta Embraces What Most Reject: Toxic Waste. *Wall Street Journal*, December 27, 1.

Trewhitt, Jeffrey, and Laurie A. Rich. 1986. Putting Up More Money for Less Waste. *Chemical Week* 139, no. 26 (December 24): 26–27.

Underwood, Joanna D., and Mark Dorfman. 1990. Stemming the Tide of Chemical Waste *Business and Society Review*, no. 75 (Fall): 63–66.

Unger, Donald G., Abraham Wandersman, William Hallman. 1992. Living near a Hazardous Waste Facility: Coping with Individual and Family Distress. *American Journal of Orthopsychiatry* 62, no.1:55–70.

U.S. Congress. 1986. Office of Technology Assessment (U.S. OTA). *Serious Reduction of Hazardous Waste: For Pollution Prevention and Industrial Efficiency*. LC 86–600571. Washington, D.C.: Superintendent of Documents.

———. 1989. *Facing America's Trash: What Next for Municipal Solid Waste*. Washington, D.C.: U.S. Government Printing Office. October.

U.S. Environmental Protection Agency (U.S. EPA). 1979. *Siting of Hazardous Waste Facilities and Public Opposition*. SW–809, November.

———. 1986. *Minimization of Hazardous Waste*. EPA 530-SW–86–0330. Washington, D.C.: U.S. Government Printing Office.

———. 1987. Office of Solid Waste and Emergency Response. *The Hazardous Waste System*. Washington, D.C.

U.S. General Accounting Office. 1978. *How to Dispose of Hazardous Waste—a Serious Question That Needs to Be Answered*. CED–79–13, December.

Visocki, Kathryn. 1988. *Incentives, Compensation, and Other Magic Tricks: Will They Help in Establishing New Waste Disposal Sites?* Raleigh, N.C.: Southeast Compact Commission for Low-Level Radioactive Waste Management.

———. 1989. Siting Controversial Facilities. *Popular Government* 55 (Summer): 34–38.

Vittes, M. E., P. H. Pollock, and S. A. Lillie. 1993. Factors Contributing to NIMBY Attitudes. *Waste Management* 13, no. 2: 125–29.

Walka, Richard M., and Andriani Harris. 1991–92. New Hazardous Waste Reduction Strategies for Business. *Journal of Environmental Regulation* 1, no. 2 (Winter): 123–35.

Walker, David Ray A. 1992. Method for Estimating the Local Area Economic Damages of Superfund Waste Sites. Ph.D. diss., Michigan State University.

Walker, Richard. 1989. Return of the Repressed: Freudian Theory, Hazardous Waste Siting, and Public Resistance. In *Psychosocial Effects of Hazardous Toxic Waste Disposal on Communities*, edited by Dennis L. Peck, 239–78. Springfield, Ill.: Charles C. Thomas.

Walters, Jonathan. 1991. The Poisonous War over Hazardous Waste. *Governing* 5:32–35.

Warren, John L. 1990. Potential for Waste Reduction. In *Hazardous Waste Minimization*, edited by Harry Freeman, 15–23. New York: McGraw-Hill.

Water Pollution Control Federation, Task Force on Hazardous Waste Treatment. 1990. *Hazardous Waste Treatment Processes*. Alexandria, Va.: Water Pollution Control Federation.

Weber, David O. 1991a. Friends of the Earth. *Healthcare Forum Journal* (March–April): 21–26.

———. 1991b. Reclaiming Our Polluted Sites. *Insurance Review* 52 (April): 26–27.

———. 1991c. Supermess. *Insurance Review* 52, no. 10 (October): 24–34.

Wentz, Charles A. 1989. *Hazardous Waste Managment*. New York: McGraw-Hill.

Wheeler, Michael. 1994. Negotiating NIMBYs: Learning from the Failure of the Massachusetts Siting Law. *Yale Journal on Regulation* 11, no. 2 (Summer): 241–91.

Willenbrink, R. V. 1990. Waste Minimization in the Petroleum Refining Industry. In *Hazardous Waste Minimization*, edited by Harry Freeman, 303–12 New York: McGraw-Hill.

Williams, Alan C. 1987. A Study of Hazardous Waste Minimization in Europe: Public and Private Strategies to Reduce Production of Hazardous Waste. *Boston College Environmental Affairs Law Review* 14 (Winter): 165–255.

Winton, John M., Laurie A. Rich, Elizabeth S. Kiesche, Conrad B. MacKerron, Kristine Portnoy, Kay Pinckney, Ann Slakter, Lisa Lazorko, Jacqueline Campbell, and Laurie Henderer. 1987. Hazardous Waste: Minimization Eases the Cleanup. *Chemical Week* 141, no. 8 (August 19): 22–53.

Wolsink, Maarten. 1994. Entanglement of Interests and Motives: Assumptions behind the NIMBY-Theory on Facility Siting. *Urban Studies* 31:851–66.

World Wastes. 1995. Facility Siting Affects Housing Prices. 38, no. 1 (January): 74.

Wright, Stuart A. 1993. Citizen Information Levels and Grassroots Opposition to New Hazardous Waste Sites: Are NIMBYists Informed? *Waste Management* 13:253–59.

Wynne, Brian. 1987. *Risk Management and Hazardous Waste*. London: Springer-Verlag.

Yakowitz, Harvey. 1988. Possible Future Trends for Management of Hazardous Wastes in OECD Member Countries. In *10th Canadian Waste Management Conference*, 25–32. Ottawa: Environment Canada.

Yarzebinski, Joseph A. 1992. Handling the "Not in My Backyard" Syndrome: A Role for the Economic Developer. *Economic Development Review* 10, no. 3 (Summer): 35–39.

Young, Stewart. 1990. Combatting NIMBY with Risk Communication. *Public Relations Quarterly* 35, no. 2 (Summer): 22–26.

Zeiss, Chris. 1991. Estimates and Sources of Hazardous Waste Loading in a Municipal Landfill in a Resource-Based Community. In *Wastetech '91: Canadian Waste Management Conference*, 441–54. Ottawa: Environment Canada.

Zeiss, Chris, and James Atwater. 1988. The Significance of Property Value Guarantees as a Tool for Siting Waste Disposal Facilities. In *10th Canadian Waste Management Conference*, 355–71. Ottawa: Environment Canada.

Zimmerman, Rae. 1989. Public Acceptability of Alternative Hazardous Waste Management Services. In *Psychosocial Effects Of Hazardous Toxic Waste Disposal On Communities*, edited by Dennis L. Peck, 197–237. Springfield, Ill.: Charles C. Thomas.

Part 1

Getting beyond NIMBY: Cases

Voluntary Siting of Hazardous Waste Facilities in Western Canada

GEOFFREY CASTLE, DON MUNTON

Reconciling the overall societal need for hazardous waste facilities, on the one hand, and the rights of local citizens who oppose the siting of such facilities in their areas, on the other, is one of the more difficult problems of modern democratic government. Very few attempts to site hazardous waste facilities in North America have been successful since the 1980s, and most have failed because of opposition from the would-be host community. Proposals to develop what have come to be called generically "locally unwanted land uses" (LULUs) almost invariably foster the emergence of grassroots movements to oppose them. This type of opposition has become known, accurately or not, as NIMBY, for "Not in My Backyard." Hosts of prescriptions have been advanced to deal with this type of opposition, many offered as solutions to "overcome the NIMBY problem." But most siting attempts continue to fail. The reconciliation of societal needs and local interests remains elusive.

This chapter explores three cases of the use of a "voluntary" process for waste facility siting. As it has been developed and applied, this process has two main features. First, consideration as potential sites is given only to those communities that volunteer to be possible hosts. Second, decisions about siting and subsequent facility operation are subject to negotiation and to local community approval. This basic process was employed in the Canadian province of Alberta, and then in Manitoba and British Columbia, during the 1980s. Two of these provinces managed to establish facilities, and all three succeeded in securing favorable referenda. These results suggest that this type of process is a potential solution both to siting gridlock in other jurisdictions and to the underlying democratic dilemmas. Voluntary processes may also be a partial response to legitimate concerns about the equity of siting decisions.

Although there are certain differences among the voluntary-siting processes followed by agencies and task forces in Alberta, Manitoba, and British Columbia, they nevertheless contained a number of common elements. These are summarized here. (See Fig. 1.) The specific

1. Local Interest Required: No siting investigations are made in localities that do not initially express interest.
(No expression of interest = Local veto)
2. Open Houses: Instead of public debating forums, citizens are invited to open houses. At these events, stations are set up where citizens can raise questions about various aspects of the facility or about potential risks.
(Poor response at open houses = Local veto)
3. Citizen Advisory Committee: Local committees, funded by the proponent, are formed and given the means to investigate the proposal from each community's perspective. The local committees have considerable power, as their recommendation is critical in proceeding to a local referendum.
(Negative committee report = Local veto)
4. Referendum: Formal public support for the project is required.
(Negative vote = Local veto)
5. Negotiation: The facility proponent and local community representatives negotiate the terms of the siting and the nature of community input into facility management.
(No agreement = Local veto)

Figure 1. Stages of the Voluntary Siting Process

applications of this general pattern will be dealt with in the description and analysis of each case, a task forming the bulk of this chapter. The accounts of the siting attempts here are based on field research, personal interviews with participants, and reviews of available news clippings and public documents, all carried out by the first author.

BACKGROUND: THE WESTERN CANADIAN CASES

As of 1980, none of the provinces concerned here had publicly accessible hazardous waste treatment facilities large enough to handle significant volumes. In that year, the four westernmost provinces in Canada (Manitoba, Saskatchewan, Alberta, and British Columbia), in association with the federal government, jointly commissioned a study that explored and outlined a policy agenda for managing hazardous wastes. The three-volume report, prepared by consulting engineers, proposed that the provincial governments act cooperatively to establish a regional treatment and disposal system for hazardous wastes (Reid-Crowther 1980). A number of options were outlined, but the benefits of regional cooperation and the need to avoid duplication in establishing facilities were a strong theme of the report. Regional cooperation, however, is always easier to plan or to advocate than it is to implement.

In the succeeding years, the provinces seemed to be going their separate ways.[1] Most began to develop plans to establish their own facilities. In each case, the provincial government was involved in the siting process. In Manitoba and Alberta, an agency of the provincial government was the facility proponent. In British Columbia, the province selected a private proponent and provided support in its search for a site. The processes followed in British Columbia and Manitoba owed much to that adopted in Alberta. Thus it is there the stories should begin.

THE ALBERTA CASE

Alberta, which produces most of Canada's gas and oil, was the first of the three provinces to begin a government-sponsored program to establish a hazardous waste treatment facility.[2] Two private siting efforts in the towns of Two Hills and Fort Saskatchewan had failed in 1979. Alberta's environment minister then commissioned an internal task force to study the problem of hazardous waste treatment and disposal and to recommend a course of action. An external board, the Environment Council of Alberta, was subsequently asked to conduct public hearings on the recommendations of the environment ministry's task force, which it did. Following the study and hearing process, the minister decided to establish an integrated hazardous waste treatment facility somewhere in the province. The decision, however, may well have been a foregone conclusion.

Responsibility for finding an acceptable site was given to an expanded ministry task force led by Assistant Deputy Minister for Pollution Control Eugene Kupchanko. It approached the problem of siting from what was then a novel perspective, one that acknowledged the overarching importance of gaining community support for a waste facility proposal. "The task force developed a step by step process, known as the invitational process," Kupchanko later wrote. "Entrance into the community was on invitation only" (1989, 3.42). Moreover, a town or area could withdraw from the process at any time. Local interest, or "social acceptability," was coupled with environmental acceptability. The latter took the form of a system of mapping ecologically unsuitable areas to determine which were unsuitable for any of a variety of reasons and which had not been ruled out on some grounds.[3] This system of "constraint mapping" was thus a process more of site elimination than of site selection (3.42–3.45).

Siting activities commenced in January 1981. The task force stipulated from the outset that potential host communities had to volunteer explicitly before environmental studies, let alone siting investigations, would commence in their areas. In order to volunteer, city and town councils were asked to pass resolutions of interest in the facility. Ultimately, a local plebiscite would be required to demonstrate local support.

As a first step, a "regional assessment" document was drawn up for each interested volunteer community. Ministry experts, including geographers and cartographers, constructed a detailed assessment of the soils, hydrology, and terrain in the area. Since there was no obligation to proceed further and the service was free, the assessment was seen as a useful document for local needs (e.g., in locating appropriate sites for landfills or for zoning purposes). Many municipalities participated in this phase solely out of interest in the assessment itself. Citizen committees monitored the local technical studies, including drilling operations. Ultimately, invitations to do detailed site studies were received from fifty-two out of seventy potential communities.

Communities that maintained an interest in hosting the waste facility continued into the second stage, that of public involvement. This phase comprised creating a local advisory committee to study the proposal and organizing a series of informational meetings. In all, sixty-five were held across the province. These meetings were general in nature. Under the guidance of staff of the Alberta Rural Education and Development Association, they looked not just at hazardous waste treatment and disposal but at broader questions of waste management. A key objective, at least of provincial government officials, was "to inform the public that the risks associated with a well-designed waste-management system were considerably less than the risks associated with waste mismanagement" (Kupchanko 1989, 3.46).

This public-involvement phase did not lead to the final stage of a community referendum in the majority of cases. As might be expected, the process aroused strong opposition in many communities. Citizen groups formed and lobbied. In several, councils rescinded their initial expression of interest. Numerous others either dropped out of the process or were eliminated by the siting task force for a variety of reasons. But some remained.

Most of the communities that stayed in the process did so in part because of the perceived economic benefits of hosting the facility. Other than the promise of new local jobs and a significant tax assessment that would result directly from a facility development, however, the task force made no explicit offers of community compensation. In general, the strongest expressions of interest came from smaller communities, where the benefits per capita were magnified.

Six communities agreed to proceed to the formal public approval stage, although one subsequently withdrew. The first of the remaining five communities to go to plebiscite was Beaver County, 100 kilometers southeast of Edmonton. A vote was held there in early 1982, following a series of stormy public meetings. Although polls from the town of Ryley supported the facility, other county centers voted strongly against the proposal, and the plebiscite was defeated by a four-fifths majority.

After this defeat, the siting process underwent a number of changes. The nongovernmental agency that had been responsible for organizing the community involvement processes was replaced, and a new strategy for dealing with community groups was adopted. The new strategy included reducing the number of large town-hall-style meetings and replacing these with information seminars and small-group meetings. This approach almost certainly lessened the potential for confrontation.[4] As an early report on the NIMBY phenomenon noted, large-audience hearings provide "for only the most limited substantive discussion and primarily [serve] as forums for expounding positions for or against proposed or operating facilities" (quoted by Morell and Magorian 1982, 129).

The following six months saw a renewed effort by the task force to build support for possible facilities in the remaining volunteer communities. In September 1982, the northern Alberta town of Swan Hills voted 79% in favor of accepting the waste treatment center. Shortly thereafter, two more towns voted to accept the facility, Veteran (70%) and Ryley (77%). Approvals did not come from every potential location, however. A March 1983 plebiscite in southeastern Alberta's Special Areas (which are provincially administered) was defeated handily (only 20% in favor).

Swan Hills, the eventual choice of the task force, is located northwest of Edmonton, the provincial capital. The community (pop. 2,000) evolved from a petroleum exploration camp established by Imperial Oil Limited in the early 1950s. In 1969, the town was moved, literally, to a location somewhat distant from oil drilling operations. The new town site is about 10 kilometers from the original site. It is remarkable to the visitor for its high proportion of mobile homes. Like many northern resource industry towns, Swan Hills is home to a young and transient population with higher than average income.[5] In 1989, various oil companies were the direct employers of 400 full-time workers. When related service employment is included, the oil industry provides about half of the total jobs in town and almost all of the high-paying ones. The dependence on the oil industry, with its cyclical nature, results in a boom-and-bust economy in the town. This, in turn, helps explain why community leaders volunteered as eagerly as they did to be considered as a host community for a hazardous waste facility.

A source of potential opposition was also missing from the Swan Hills situation. Whereas conflict between farmers and nonfarmers was a feature of the facility siting debate in other Albertan would-be host communities, there was little such division in Swan Hills. The poor soils of the region and its harsh northern climate both inhibit agriculture. Consequently, there is no farming around Swan Hills; indeed, there is no farm within 50 kilometers of the town. A small group of local opponents organized under the name FRESH (FRiends of the Environment of Swan Hills) during the site investigations, but it was unable to generate substantial opposition to a facility favored by the mayor, council, and local member of the legislative assembly (MLA).

The Alberta cabinet chose Swan Hills as the host community at a March 1984 meeting. Three years later, the Alberta Special Waste Treatment Centre was opened at a site 10 kilometers from Swan Hills. Expanded in 1992, the facility now consists of two rotary kilns, a transformer decontamination unit, physical-chemical treatment plant, deep well injection unit, and landfill.[6] Public support for the facility has remained high in the town, though it has been criticized by some in neighboring areas.

THE BRITISH COLUMBIA CASE

Starting in 1987, the government of British Columbia tried to reproduce Alberta's siting success by using a similarly designed volunteer process. An attempt to establish a private sector hazardous waste treatment facility in the province with a nonvoluntary process had collapsed in 1984. A significant factor in the demise of the 1984 proposal was a provincial "no net import" policy that reduced the amount of wastes potentially available to the facility. The private sector proponents, Genstar and IT Corporation, pulled out shortly after the provincial waste import policy had been announced.

A few private hazardous waste facilities were proposed between 1984 and 1986. None made much progress.[7] In January 1987, the provincial environment minister, Stephen Rogers, appointed a Special Waste Advisory Committee (SWAC), charged with overseeing proposals for major waste treatment facilities. SWAC was chaired by a former president of the Canadian Cancer Society and included industry and ratepayer (taxpayer) representatives (British Columbia, Special Waste Advisory Committee 1988). The committee reviewed the previous reports and recommendations on the state of hazardous waste management in British Columbia. Its members toured the province in early 1987, seeking public input into the hazardous waste problem, and again in the late winter and spring of 1988. The committee concluded that the province did indeed need additional hazardous waste treatment capacity, thus agreeing with the conclusions of earlier studies by the Ministry of Environment.[8]

SWAC proceeded to advertise for a developer to build a hazardous waste management facility in the province. Its advertisement offered the right to build a privately owned and operated hazardous waste treatment facility on provincial land. This attempt to find a private sector partner was similar to the 1984 siting attempt. It was criticized by one environmental organization as "a repeat of [Rogers's] 1982 call for tenders . . . similarly doomed to failure."[9]

In December 1987, the committee chose a proponent from the several applicants, basing its decision on outside advice.[10] A consortium of four companies, headed by Vancouver-based Envirochem Limited, was selected to construct an integrated treatment facility. This facility was to include an incinerator for organic compounds, a physical-chemical treatment plant for inorganic wastes, and a lined landfill (British Columbia, Special Waste Advisory Committee 1988). The Envirochem-led consortium was incorporated as BC Special Waste. SWAC had originally asked that the companies have a site selected prior to submitting their final proposal. It was subsequently agreed that the Ministry of Environment and SWAC would assist the successful bidder in locating a site. The pledge of siting assistance required SWAC to take on more of a proponent's role, expanding the committee's function from merely providing advice to selling the proposal to potential host communities.

The proposed siting process, hastily put together by SWAC's chair and staff, borrowed heavily from the successful siting precedent in Alberta. The committee would 1) seek expressions of interest from communities around the province, 2) hold informational open houses in interested communities, 3) create local advisory committees to review the proposal, and 4) hold plebiscites in interested areas. The committee planned to carry out these tasks without any expansion in staff. (At the time, there were only the three commissioners, one secretary, and an assistant.)

Following the Alberta siting precedent, the committee requested in January 1988 that interested communities apply to be considered as the host community for a hazardous waste facility. No compensation package was offered other than the promise of the new economic activity that would follow from the construction and operation of the waste facility. Nine firm expressions of interest were received by SWAC by May 1988. Interested communities included the city of Kamloops (pop. 50,000) and a number of smaller towns and regional districts, including Cache Creek–Ashcroft, Princeton, Fort Nelson, Cranbrook, Merritt, East Kootenay Regional District, and Tumbler Ridge.

These expressions of interest were followed up by informational meetings with SWAC committee members. Following its selection as the facility manager, BC Special Waste company representatives participated in SWAC-organized open houses in candidate communities. This second series of meetings in would-be host communities were much more conflictual than the initial fact-finding meetings. Although a community had not yet been chosen nor a facility design proposed, substantial interest in the facility—both pro and con—emerged in the nine candidate areas. Local opposition groups developed in most.[11] Proponents who favored economic development in their community also came forward. Letters to the editor in community newspapers included a number of responses to critics by local leaders or members of advisory committees.[12]

A major difference between the Albertan and the B.C. experiences was the overall time frame. While the Albertan siting task force spent three years pursuing volunteer communities and attempting to develop support for their facility, the B.C. process was much quicker. There were only eighteen months allowed between the creation of SWAC and a provincial deadline for selection of the host community. Of that time, only the final six months were available for candidate communities to discuss the facility. Perhaps as a result of the compressed time frame, the pool of volunteer communities in British Columbia was much smaller. Indeed, there was only one serious contender, another major difference from the Alberta case.

The first and only community referendum held was in Cache Creek–Ashcroft. The two towns, close by each other and possessing a combined population of 2,500, are located approximately 300 kilometers northeast of Vancouver, on British Columbia's dry interior plateau. The area is a center for cattle ranching, forestry, and mining activities. Its dry climate, favorable soil structure, and relative proximity to Vancouver made the location a prime candidate area in the 1980 Reid-Crowther report. It was for this reason that the Cache Creek–Ashcroft area had been chosen as the preferred site for the failed provincial siting attempt in 1984.

Situated on the Trans-Canada Highway, the towns had historically benefited economically from the road traffic traveling between Vancouver and the northern and eastern parts of British Columbia as well as to and from Alberta. In 1986, a new highway opened well to the east of Cache Creek–Ashcroft, drawing a significant portion of the traffic that previously passed through the community and leading to a significant economic slump. Another major impact was the relocation of the Greyhound bus depot, and its attendant sixty jobs, from Cache Creek to Merritt, B.C., nearer the new route. In the eyes of some local supporters of the facility, the hazardous waste treatment plant would offset this and other losses caused by the traffic diversion.

Following meetings and open houses involving SWAC, BC Special Waste, and local citizens, a referendum was organized by the Cache Creek–Ashcroft Joint Special Waste Committee. The vote took place only three weeks before the provincial cabinet's June deadline. In the days leading up to the vote, members of SWAC and representatives from BC Special Waste were involved in a campaign-style process that included radio talk show appearances and debates with facility opponents. The eligible electorate included all municipal voters within a 20-kilometer radius of the communities, but not local native band members.

The result of the Cache Creek–Ashcroft referendum was favorable to the facility proponents. Of 960 voters (representing 58% of those eligible), 60% voted in favor of hosting the facility. Local politicians from Ashcroft and Cache Creek responded with enthusiastic statements in the wake of the referendum. "if a person is interested in cleaning up the environment," declared Ashcroft mayor Mervin Pears, "this is the way to go."[13] An opposition group, the Interior Waste Coalition, however, argued that the voting list should have been expanded to include the 200 to 300 persons on nearby First Nations reserves.[14]

Bound by the May 31, 1988, deadline, SWAC recommended in its final report that the Cache Creek–Ashcroft region be selected as the host area for the BC Special Waste facility. The remaining siting work, which included the selection of the particular site within the region, was to be carried out by BC Special Waste and the Ministry of Environment's Waste Management Branch. In the report the SWAC chairman, Dr. David Boyes, also recommended that the committee's mandate be permanently extended in order to oversee the final stages of the siting process. This recommendation was not acted upon by the province. SWAC was dissolved following the submission of the report.

The following months saw the slow unraveling of the Cache Creek–Ashcroft choice. On June 6, the *Vancouver Sun* reported that ENSCO, the company whose incinerator was to be used by BC Special Waste, had a poor record. On June 20, the *Kamloops Daily News* revealed that BC Special Waste was considering siting its facility on land owned by a former mayor of Cache Creek, a fact that had not previously been made public. In July, an environmental newsletter, the *New Catalyst*, reported that the incidence of Lou Gehrig's disease near ENSCO's El Dorado, Arkansas, facility was "way above normal." In September, a large anti-facility rally was held, addressed by a U.S. environmentalist who accused the facility proponents of "brainwashing."

Although the charges about the facility and related news stories were answered by company officials, support for the project declined on both the Cache Creek and Ashcroft town councils. The communities asked for money to conduct a new study of the proposed hazardous waste treatment facility. In late October 1988, B.C. environment minister Bruce Strachan issued a press release that referred to an "impasse" in the Cache Creek–Ashcroft proposal. Accordingly, the minister proposed a full-scale public review, to be followed by a new referendum that would incorporate groups outside the initial voting boundaries. By November 3, 1988, according to minutes of a BC Special Waste meeting with officials from the provincial Waste Management Branch, the Ministry of Environment was considering reviewing whether a facility was needed at all. The same minutes note that the private consortium was refusing to spend any more money on the project.[15]

Observers point to two additional blows to the BC Special Waste project. The first was a televised forum in November 1988 in which the provincial minister displayed a distinct lack of specific knowledge on the issue.[16] The second was the emergence of a strong lobby representing a group of ranchers from the Cache Creek–Ashcroft area who vehemently opposed the facility. The ranchers reportedly had close political connections to provincial government politicians.[17] As in a number of Alberta communities, the perception of a conflict between waste management and agriculture had become a factor.

On November 29, 1988, the minister's office announced a suspension of the hazardous waste facility project, citing "a lack of the consensus needed to proceed further." No waste facility has been sited in British Columbia in the ensuing years.

THE MANITOBA CASE

The Manitoba Hazardous Waste Management Corporation (MHWMC) was created in August 1987 and began actively planning a hazardous waste management system for that province in January 1988.[18] The corporation's plan for the initial phases of system development included gathering market information and completing design and site selection for a treatment facility (MHWMC 1992, 42–43). The facility planned and eventually sited was a less comprehensive plant than that in Swan Hills, including a combined physical-chemical treatment plant, storage and transfer operations, and bioremediation capability. The facility was not to have an incinerator and was not intended to treat such hazardous wastes as PCBs.

The siting approach followed by MHWMC was consciously modeled, in large part, on the successful siting process used in Alberta in the early 1980s. The key feature of this approach, once again, was that individual cities and towns had to volunteer for consideration as a potential host of the hazardous waste treatment plant. Only after such a declaration of community interest, and only if the community was in an area deemed geologically suitable by MHWMC, would the corporation begin to examine potential sites inside the municipal boundaries. While the Albertan experience was not the only model for the siting process, MHWMC officials acknowledged "it helped a great deal that a facility of this nature had been sited voluntarily and that we could look at how things had been done in Alberta."[19]

Beginning in 1988, MHWMC sponsored a series of informational open houses at selected areas throughout the province, in cooperation with the Manitoba Environment Council.[20] The overall response to the open houses was relatively positive. Approximately half of all visitors declared some interest in further investigations by MHWMC in their community.[21] Eventually, the process resulted in forty separate expressions of interest from Manitoba cities, towns, and rural municipalities. Of these forty expressions of interest, only ten were considered in detail by MHWMC.

Most of the other communities that had expressed interest were either in unsuitable geological areas or showed signs of early opposition sufficiently strong to cause MHWMC to withdraw. For example, Pinawa, home to a large Atomic Energy of Canada research facility, expressed an early interest in the process, and a majority of citizens seemed in support. Even the best Pinawa sites, however, did not have the deep clay soils required for providing a suitable level of environmental protection (MHWMC 1992, 177–78).

In two other towns, Rossburn and Hamiota, referenda were called by local councils prior to the completion of work by the citizen advisory committees. In Rossburn, a new council withdrew its predecessor's support for the project, canceled its sponsorship of the advisory committee, and called a referendum. In close votes, Rossburn and Hamiota both rejected hosting the MHWMC facility. The referendum defeats may have been a result of insufficient time spent in the community by MHWMC. One official later noted: "There seems to be an eighteen-month rule having to do with building a level of support in a community. It just seems to be the amount of time that it takes for a community to become involved and make decisions. In both Rossburn and Hamiota, we had only had about twelve months before we were facing a vote."[22]

Winnipeg, the provincial capital, remained until the end in the running to host the facility. As the source of over 80% of the hazardous waste stream in Manitoba, the city was an obvious candidate. Not only were the Winnipeg mayor and most councilors supportive of considering the waste treatment plant but most Manitoba environmental organizations also favored the provincial capital. Of particular importance to these groups was the issue of transportation risk and the idea that Winnipeg should be responsible for treating its own waste.

Hoping to pre-empt a decision to site the facility elsewhere, the Winnipeg Community Advisory Committee released a draft report of its findings on February 21, 1992. The report was based on investigations by issue-based subcommittees that had toured the Swan Hills plant of the Alberta Special Waste Management Corporation and a physical-chemical waste treatment plant near Minneapolis. In a summary statement, the citizen committee declared its support for a site on the city's Four Mile Road. "The Winnipeg Community Advisory Group has examined the Manitoba Hazardous Waste Management Corporation's proposal in considerable depth," it said. "We have found no technical factors which would make the Winnipeg site unacceptable" (Winnipeg Community Advisory Group on Hazardous Waste 1992).

Neither the support for Winnipeg nor this report ultimately swayed the decision of the corporation. One week after the release of the statement, the MHWMC board of directors elected to proceed with work on an alternative site, in the Rural Municipality (R.M.) of Montcalm. Montcalm, located approximately 50 miles due south of Winnipeg, near the border with the United States, contains three unincorporated villages: St. Jean Baptiste, St. Joseph, and Letellier. The population is predominantly French-speaking and Roman Catholic. Its main industry is agriculture, the chief crops being grain and specialized crops such as soup peas and lentils. A sign on the way into the town of St. Jean proudly tells visitors they are entering the "Soup Pea Capital of Canada."

Aside from its French-speaking majority, Montcalm is not very different from thousands of other agricultural communities on the North American prairies. Built beside railway lines, each of Montcalm's three hamlets are centered around a main street and a grain elevator. In St. Jean Baptiste, a food processing plant lies at the end of the main street. St. Jean also is home to the local French language high school and an impressive stone church. The municipal offices and council chambers are located in Letellier, a ten minute drive south from St. Jean. Five minutes west of Letellier is the hamlet of St. Joseph. Smaller than St. Jean and Letellier, St. Joseph clings most strongly to its religion, French culture, and history. "It's like going back two generations when you go to St. Joe," one St. Jean resident confided.

St. Joseph is said to set itself apart from Letellier and St. Jean. A long-standing feud between St. Joseph and St. Jean revolves around the closure of the St. Joseph school in the early 1970s. "The fact of the matter," some residents say, "was that there weren't enough students to justify the school." Others describe family feuds reminiscent of the Hatfields and McCoys. In addition to personal differences, there seems to be a modest degree of economic rivalry between St. Jean and St. Joseph. Both are home to small-scale vegetable-processing operations.[23]

Like many rural communities in North America, Montcalm is currently in the midst of a long economic slump. With crop prices dropping because of international competition, incomes have fallen and many farmers have gone out of business. Consequently, there are fewer employment opportunities for young people, and the rural municipality has seen a steady decrease in population, from nearly 2,500 in the 1970s to a current total of 1,600. The decline of the farm economy in the region has led to a proliferation of local task forces promoting economic diversification on the prairies. St. Jean organized such a task force, and it was in this organization that interest in the hazardous waste treatment facility developed.

Yves Sabourin, the secretary treasurer of the R. M. of Montcalm, brought the idea of hosting the facility to a meeting of the St. Jean Development Corporation in September 1989. As Sabourin recalled, the idea was not immediately seized upon: "A lot of people at the meeting were unsure about the idea of hazardous wastes. But we talked it over and decided that it was at least worth investigating. . . . From what we understood, they were looking at possibly thirty to thirty-five jobs. If it could be done safely, I thought, why not do it here?"[24]

Following supportive results from the open houses, MHWMC invited the Montcalm council to form a community committee. A resolution was duly passed on May 9, 1990, creating the Montcalm Advisory Committee (MAC). Unlike the Winnipeg committee, MAC had a broad mandate. It was asked to make recommendations in many areas, "including whether to proceed with a community decision such as a referendum."[25] The committee was an all-volunteer committee, comprising Montcalm citizens who came out to an advertised meeting. "No one was asked," said the mayor, Flo Beaudette, "and no one was turned down." The fourteen-member advisory committee was intended to be viewed as a neutral agent in the eyes of the community. As a consequence, MHWMC suggested that Montcalm councilors not be appointed members. Also absent from the committee, though not by design, were any known opponents of the facility. Opponents were later specifically invited to join the committee, but they refused. The absence of known opponents on the committee may have worked against its neutral image.

The MAC chairman at first had some reservations about the facility. His concerns, and those of others, about health and safety were addressed to their satisfaction, however, and subsequently he supported the facility. His support rested, in large part, on the opportunities that the hazardous-waste treatment facility would afford his high-school students: "When opponents would criticize the facility, they would say that no one in Montcalm would be qualified to work there. And I would say to them, 'Not now maybe, but wait. I've got some students in my classes who could go on to college or university and, by the time the plant is built, be ready to work in the lab. It really gives my students something to shoot for.' "[26]

The first formal opposition to the hazardous waste treatment facility emerged in October 1990. The fact that it had taken over a year for the opposition to organize came as a surprise to MHWMC staff. When a drilling team was sent by MHWMC to sample the subsoil at a site in the northern portion of the municipality, local opposition began to mount. "As soon as the drill showed up, we began to hear that there was a petition going round," the mayor later said. "Of course, they would not have come to me, but people tell you what's happening."[27] The petition was organized by a group calling themselves Concerned Citizens for Ecology and Environment. An information sheet handed out by petitioners raised several legitimate concerns about the project, including potential water contamination during floods and the risk from transportation of the wastes. Concerned Citizens also invoked the memory of Chernobyl and Love Canal in an effort to raise concerns about the waste treatment plant.[28] Over two months, the group did manage to collect 450 signatures on a petition calling for an end to the site investigations. This total represented over one-third of the registered voters in Montcalm. While it appeared initially that the petition might be sufficient to end MHWMC's siting activities, the community leadership, who knew personally those who had signed the petition, were skeptical of its value. As the mayor later recalled: "Before the petition was presented to council, I had people coming to me and saying, 'Look, my name is on that list, but I signed it because I couldn't say no because they were a friend or relative.' Or they told me, 'I signed it just to get rid of the person.' . . . The opposition appeared to have started with one St. Joseph area family, which then influenced other families, spreading by the sons and daughters into the parents of the in-laws and into various committees, such as the choir and the PTA. . . . A lady said that 'the choir director was against it and I'm one of the choir.' "[29] There was, in fact, broad support for the facility. This was evident in a small local poll conducted in late March and early April 1991. Of 202 respondents, almost three-quarters (74%) were strongly or somewhat in support. Only 20% opposed it.

The final report of MAC was completed in July 1991. The committee members recommended a number of amendments to the facility plan but were generally supportive of the plant. They recommended that the Montcalm council hold a referendum on whether to accept the hazardous waste facility. On September 25, 1991, the following question was put to the registered voters in the R. M. of Montcalm: "Do you support the location of a hazardous waste management facility by the Manitoba Hazardous Waste Management Corporation in the Rural Municipality of Montcalm?" Most did. The referendum passed, 618 (67.1%) voting yes and 303 (32.9%) no.

This victory, however, did not ensure that the hazardous waste treatment plant would be awarded to Montcalm. In the wake of the vote, the municipality and the local member of the Manitoba Legislative Assembly lobbied hard for the facility. Their effort was assisted by those campaigning against the Winnipeg site. On the other side, a number of environmental groups and Winnipeg residents fought to keep the capital city in consideration. Finally, on February 28, 1992, the MHWMC board of directors announced that they would seek licensing approval for the Montcalm site. A few days later, the provincial government announced it had chosen Montcalm and was going ahead with the plans.

The reactions from Montcalm were generally positive, even jubilant. Jack Penner, the area's provincial legislator, called the decision "a recognition of the tremendous effort put out" by the people and the council of Montcalm. Farmer Gilbert Fontaine, who owned the land chosen by MHWMC, was satisfied. "I'm not opposed that the treatment center will be next door to me," he said. "It will only help [Montcalm] tax- and job-wise."[30]

Some others were less pleased, however. The mayor of Winnipeg objected to the decision. "I'm extremely disappointed," he said; "I think [Montcalm] is the wrong location. It's unfortunate the process we [the city] were in the middle of was not allowed to be completed." This view was echoed by others on the Winnipeg advisory committee who felt that locating the facility in Winnipeg would minimize transportation risks. Naturally enough, the opponents of Winnipeg's Four Mile Road site were pleased by the decision to go ahead with the Montcalm site. "We are relieved we don't have to go through environmental hearings and begin filing injunctions against the site, which is what we ultimately would have to do," offered a spokesperson. Marcel Laurendeau, the St. Norbert politician who had campaigned against the Four Mile Road site, was also happy: "It's one big headache behind me."[31]

Following the referendum victory, Montcalm disbanded its advisory committee and formed a Terms and Conditions Negotiating Committee and an Environmental Impact Assessment Committee. The former group met on several occasions with representatives from MHWMC to negotiate terms and conditions. Several drafts later, a

1. **Project Proposal, EIA, and Licences**
   a) Any future expansion or significant change in technology at the facility will require obtaining community support "in a similar manner as was used with the Corporation's current proposal."

2. **Co-management**
   a) Two regular members of the Manitoba Hazardous Waste Management Corporation's board of directors shall be nominated by Montcalm Council.
   b) A Plant Co-management Committee will be established to review and discuss ongoing activities of the plant. It will consist of senior management at the plant and at least two community representatives nominated by Montcalm Council.
   c) An independent Community Liaison Committee, appointed by Montcalm Council, will be constituted to provide general oversight of the plant and to conduct any investigation with respect to the facility's environmental performance.
   d) A Workplace Safety and Health Committee will be organized and administered by plant management.
   e) $50,000 (indexed to inflation) will be provided annually by the operator to cover the expenses of the community members of the Co-management Committee and Community Liaison Committee.

3. **Community Protection Programs**
   a) The corporation guarantees nearby landowners and farmers financial compensation in the event that the operations of the waste treatment facility have a demonstrably negative impact on property values or crop prices.
   b) The facility will be subject to an environmental monitoring program, the design and scope of which shall be agreed upon by the R.M. of Montcalm and the corporation.

4. **Facility Operation**
   a) The facility shall accept out-of-province wastes, but priority will always be given to Manitoba's wastes.
   b) The facility and the site located in Montcalm will not include any long-term storage of PCB-contaminated material without the express approval of the council.
   c) The operators will not apply for a license amendment to expand its repository capacity beyond the limits established in the EIA (15,000 tonnes/year) without approval from Montcalm Council.
   d) Any future expansion of treatment capacity or change in technology requires the support of Montcalm Council.

5. **Community Benefits**
   a) The facility operator will "maximize opportunities for local residents and businesses to benefit from its hiring, purchasing and contracting requirements."

Figure 2. Highlights of Montcalm-MHWMC Comanagement Agreement

b) The facility operator will operate a household hazardous-waste and used motor oil drop-off at no charge.
c) In the event the facility is sold, the head office of the corporation shall remain in Montcalm.

6. **Arbitration**
a) Any disagreement or controversy between the parties shall be submitted to binding arbitration with the agreement of both parties.

Figure 2. *continued*

"Co-Management Agreement" between Montcalm and MHWMC was reached, in September 1992. The agreement contains a number of significant provisions about the plant. See the summary of this agreement in Fig. 2.[32]

These include the stipulations that Montcalm Council will nominate representatives to the Manitoba Hazardous Waste Management Corporation's board of directors; that a plant comanagement committee will be established to review and discuss ongoing activities of the plant; and that an independent community liaison committee, appointed by Montcalm Council, will be constituted to provide general oversight of the plant and to conduct any investigation with respect to the facility's environmental performance. The facility will also be subject to an environmental monitoring program, the design and scope of which is to be agreed upon by Montcalm and the corporation. Any significant technology change at the facility requires community support, and expanding repository capacity beyond the existing limits will require approval from Montcalm Council. Furthermore, MHWMC guarantees nearby landowners and farmers financial compensation in the event that the operations of the waste treatment facility have a demonstrably negative impact on property values or crop prices. The facility operator will also emphasize local hiring and contracting and will operate a household hazardous waste and used motor oil drop-off at no charge. Finally, any disagreement or controversy between the parties is to be submitted to binding arbitration.

With this agreement, the people of Montcalm established one of the broadest and possibly one of the most effective systems of community input into any corporate structure in North America.

In addition to the Montcalm-MHWMC accord, the municipal council signed an agreement with the Manitoba government concerning the permanent guarantees it was offered during the siting process.

The most significant provision of the Manitoba-Montcalm agreement is that while the province may privatize the waste treatment facility, the land on which the facility is located will always be owned outright by the province of Manitoba. It is thus ensured that the direct responsibility for any liabilities associated with the land lies with the province.

WHY VOLUNTARY PROCESSES SUCCEED

The western Canadian siting experiences are good examples of the use of a voluntary-siting process in establishing "noxious" facilities or LULUs. While the attempts were successful in only two of three provinces, majority local support in a community for a proposed hazardous waste treatment facility was achieved in all three, albeit only fleetingly in the case of British Columbia. One successful siting of a hazardous waste facility is unusual. Two out of three cases is rare indeed. The voluntary nature of these processes is almost certainly a significant reason for the result. The communities themselves made the decisions to participate in the siting processes, instead of these siting decisions being made by a central political agency or a private company, and the communities then could negotiate to defend and promote their interests.

Why is a volunteer bottom-up siting process more likely to gain local support and ultimately to succeed than a top-down process? One reason lies in the strategic advantage obtained by considering as potential sites all communities within a jurisdiction instead of considering only a single site chosen by a proponent. Certain communities are clearly more predisposed to accept LULUs than others. As no standard economic or demographic measure has been developed to predict which communities will accept LULUs, the voluntary process may be the best means available of uncovering these communities. Such a process will at least rule out the socially or politically impossible locations in the same way that geological and hydrological investigations can rule out the technically undesirable locations.

There is thus a sort of mathematical advantage to voluntary siting processes. Whether or not some communities are more likely than others to accept facilities, the voluntary process at least increases the likelihood of a success. Given that facility proposals are affected by uncontrolled and often random or chaos-pattern events at the local political level, the more communities that are considered, the higher the overall likelihood of one siting success. Even the ultimately successful Manitoban and Albertan siting efforts, like those elsewhere, failed to secure approval in the vast majority of communities originally considered. On a community-by-community basis, voluntary processes generally do not work, in the sense that most volunteer communities do not ultimately accept a facility. But from the collective point of view, and that of a facility proponent, the overall process works in the sense that a facility somewhere was eventually approved by a host community. And in these cases, only one favorable result was needed, of course.

The voluntary siting process also yields successful outcomes for reasons other than the law of averages. The most important of these is that it offers a considerable degree of control to the would-be host community, in the form of potential local government vetoes and a public referendum. Although the possibility of manipulation by governments and proponents is not entirely absent—and both have some cards to play even if there is no manipulation—the sense of local disenfranchisement so often evident in traditional LULU processes is substantially reduced by voluntary, local-veto processes. It is reasonably well established that people tend psychologically to be more acceptant of risks that are voluntarily entered into than they are of those over which they have no control.[33] Voluntary siting processes do nothing, by themselves, to reduce the actual health and environmental risks of a facility. They do seem to reduce perceived risks and involuntarily assumed risks, thereby fostering a willingness at least to consider and negotiate proposals and ultimately increasing acceptance.

What has been termed here the voluntary approach goes beyond merely offering communities a choice of whether to opt in at the outset. It also offers them a greater degree of control over the siting process that follows and over the eventual operation of the facility. What Alun Richards labels a "co-management" model (see his chapter in this volume) was an inherent feature of both successful siting processes in western Canada. Acceptance by siting proponents of a community's right to negotiate a management and oversight role for itself also seems to be an essential step in reducing the perceived risks of an operating facility. Stringent environmental standards and engineering measures to reduce emissions and ensure a safe facility are vitally important but not sufficient. Risk mitigation needs to be accompanied by transparency. The operation not only must be safe; it must be seen by the local community to be meeting those standards and taking all appropriate measures. When communities understand that they can negotiate and secure such assurances, acceptance seems to follow more easily.

Another advantage of the voluntary approach is that the process seems to attract public support in the prospective, volunteer communities from local opinion leaders, who often have a community development agenda. Because local authorities must invite facility proponents to investigate their community, they are thus on record. The local leaders in Alberta and Manitoba—and, for that matter, in British

Columbia—not only supported the proposal but were generally active on the "yes" side of the facility referenda. A process where the community is not involved in the siting decision enables elected officials and local opinion leaders to avoid taking a firm stand and may even encourage them to use public resources in order to defeat the proposal. A benefit of involving local leaders is that it adds credibility to a group of "outsider" proponents.

The eventual failure of the B.C. siting attempt is evidence that the volunteer process alone is clearly not sufficient to ensure siting success. No community, not even a volunteer community, can offer assured, blank-check support for a hazardous waste facility or any other LULU. In the B.C. case, a collapse in credibility on the part of the proponents, unnecessary secrecy, and a fast-tracked process all led to evaporation of the trust and support that had initially developed. It suggests that no matter how voluntary a selection process and no matter how much willingness there is to negotiate with a community on the part of facility proponents, some communities may never accept a hazardous waste treatment plant. Voluntary siting is not a magic bullet. It is not capable of solutions in all cases.

APPLICABILITY OF THE VOLUNTARY MODEL

Is the experience of western Canada siting processes reproducible? Is it applicable elsewhere? The answer must be: to some extent. The three experiments in western Canada indicate that the facility was seen in volunteer communities as a means of local economic development or diversification. For this reason, the reproducibility of the process may be restricted to siting LULUs in certain economic areas, but not all. The process may not be as successful, for example, in commuter regions or other regions where local welfare is not perceived to be dependent on local economic activity. Caution is necessary here, though. The argument that hazardous waste facilities will encounter less opposition in nonagricultural areas may seem supported by the Alberta case, but the argument that hazardous waste facilities will generally be seen as incompatible in predominantly agricultural areas is not supported by the Manitoba case.

A second constraint on the reproducibility of volunteer siting success concerns community benefits. The relationship between the amount of economic stimulus provided by the facility and the population size of the host community is likely important. All the favorable referenda for waste facilities discussed here were in small communities with populations less than 10,000 people. The per capita economic benefit of a facility in such a community is much greater than that for large communities. The importance of perceived benefits obviously diminishes as the size of the host community increases. (On the other hand, the Manitoba facility could well have ended up in Winnipeg, a medium-sized city.)

A third consideration that may limit the applicability of the conclusions about voluntary processes is what might be called the Canada factor. All of these three cases, and both of the successful ones, took place in western Canada. Thus, it might be argued that what was important in these processes was not that they were voluntary but, rather, that they reflect something about small-town Canada, perhaps small-town western Canada. Is the secret to siting one of political culture rather than process? Is there perhaps an unusual degree of trust or of willingness to comply with perceived authority at work here?

While this explanation cannot be ruled out, the notion that Canadians in general are relatively more acceptant of such plants being sited in their backyards does not hold up under even limited scrutiny. Canadians may not have invented the NIMBY phenomenon, but they are no strangers to it (e.g., Armour 1984). Nor has it dissipated. A comprehensive hazardous waste treatment and disposal facility in Ontario, pursued since the early 1980s according to the infamous "Decide and Defend" (DAD) approach, was finally rejected in early 1995. The long-standing effort was officially abandoned by the provincial government without ever having gained support from the community around the selected site.[34] But no list of examples from Canada where top-down processes failed can prove the point that voluntary processes will necessarily work as well elsewhere as they did in the cases considered here. Only further analyses of other cases elsewhere will answer such questions. It might be noted, though, that some of the evidence about siting in the United States discussed in other chapters in this volume points to a degree of applicability of the western Canada experiences.

EQUITY CONSIDERATIONS

Some critics suggest that site selection, including voluntary site selection, is unfair because poorer communities are more likely to abandon their concerns for health and safety in return for the economic benefits associated with a hazardous waste treatment facility. In one of the earliest contributions to the now abundant literature on environmental equity and environmental racism, Julian Wolpert argues that "low income communities face a triple jeopardy: their communities are more susceptible in the selection of sites for major facilities (because land values are lower); their opposition will be relatively ineffective; and they will have little ability to exact concessions" (Wolpert 1976).

Morell and Magorian assert that a process of finding sites where "public opposition is lowest," which they term "the path of least resistance," should "be rejected on both strategic and equity grounds" (1982, 185). Critics of the processes in Alberta and British Columbia have argued that they did not involve any attempt to ensure that the costs of the facilities were equitably distributed.

Fairness and equity are undeniably important principles. A successful siting process should be fair and as democratic as possible, not just result in the construction of a needed facility. The question in these cases from western Canada is whether equity was in fact served. The evidence to support the argument that the Alberta or Manitoba volunteer siting process singled out poor or otherwise disadvantaged communities to host hazardous waste treatment facilities seems, in short, lacking.

First, arguments based on the relative economic disadvantage of host communities are, in these cases, factually flawed. The town of Swan Hills had an average household income of $44,023 in 1986, significantly higher than the Alberta norm ($36,796). Unemployment in Swan Hills in 1986 was 7%, one of the lowest in the province. Montcalm is much poorer than Swan Hills, but the community is not economically disadvantaged compared to other rural areas in Manitoba. While average per capita income in Montcalm's census district ($15,081) was lower than the 1988 provincial average ($18,341), the region fares better when the cities of Winnipeg and Brandon are excluded from the comparison. Montcalm's census district ranked eighth in per capita income out of the twenty-one rural Manitoba census districts.[35] Unemployment in Montcalm was 4.1%, compared with a 1988 provincial figure of 7.8%. The diversification of the Montcalm farm economy into specialized crops such as field peas, flaxseed, and canola, in addition to the more common wheat and barley, may have prevented a decline in farm incomes as severe as that faced in some other rural Manitoba communities.

It may well be true that the siting of other waste facilities in other times and other areas, and by other processes, has followed regressive patterns. The processes in Manitoba and Alberta, however, do not appear to have followed such a pattern. It is notable that many smaller, less well organized and less wealthy communities rejected the Manitoba and Alberta facilities. Not only does this history seem to demonstrate the neutral role of community income in these decisions about facility siting; it also contradicts the notion that low-income community opposition will always be less effective or ineffective. Process is not everything, but the voluntary process of siting means that community views will be relatively effective, rather than irrelevant, and that communities will have considerable ability to exact concessions, rather than little or none at all.

The second problem inherent in equity- and fairness-oriented criticisms of the voluntary site-selection process is mostly one of democratic principle. "Ultimately . . . siting facilities in places where people do not want them is, has to be, fundamentally wrong," Kent Portney correctly notes. "Unless people, especially the people most directly affected, accept these facilities, then the siting process and result must be considered less than optimum."[36] While it is not difficult to criticize many siting processes, these criticisms should not be applied willy-nilly to voluntary siting processes. To do so is to ignore the legitimacy of what is, in effect, a concurrence of opinion by three political levels representing the community. In each of the Canadian cases, the provincial government approved the siting through its cabinet-level control of the provincial waste corporation and its permitting activities. The local government approved through council resolutions supporting the siting process at various stages, without threat of being overridden if they opted for rejection. (In some cases, there was also negotiation of a site-operating agreement.) And the citizens of the host community approved the siting through a considered and popular referendum. The democratic nature of a voluntary approach cannot ensure equity and perfection; it does seem, however, to be as fair as possible, as well as successful.

NOTES

1. Whether by accident or by quietly pursued design, the system that is coming together in these provinces in the 1990s actually looks very much like some of the cooperative options outlined in the Reid-Crowther report. Only Alberta will have a fully integrated facility with an incinerator capable of destroying PCBs, while the other provinces have opted for developing more limited facilities for treating other wastes. The resistance evident in the 1980s to interprovincial shipment of wastes destined for treatment also seems to have dissipated. Alberta's Swan Hills plant, until 1995 closed to shipments of waste from outside the province, is now accepting wastes from other provinces. British Columbia is no longer planning its own comprehensive treatment and disposal facility. The Manitoba facility was developed as a specialized, rather than general, plant.

2. For a fuller discussion, see Rabe 1990, 1991, 1992; Castle 1992.

3. Kupchanko later summarized the approach as follows: "You take an area where the people want it, and then you engineer around any physical obstacles that present themselves." Finding an area "where the people want it," was, of course, more of a challenge than this comment implies. Eugene Kupchanko, interview by author, 1992.

4. These more one-on-one sessions, for one process participant at least, also allowed local citizens to raise their individual concerns in a more relaxed setting, especially "if you're the sort of person that's afraid to talk when there's a lot of people around." Because people were divided into small groups, "you could ask questions without feeling silly." Derek Kool, interview by author, 1992.

5. The town of Swan Hills had an average household income of $44,023 in 1986, significantly higher than the Alberta norm ($36,796). Unemployment in Swan Hills in 1986 was 7%, one of the lowest in the province. Source: Alberta Special Waste Management Corporation.

6. On the technical operation of the Swan Hills plant, see McQuaid-Cook and Simons 1989.

7. The most significant was a proposal by Enviro Waste Management Corporation to establish an incinerator for organic wastes. In 1986, Enviro Waste approached both the Kamloops native band (*Kamloops Daily News*, June 11, 1988) and the coastal community of Ocean Falls (B.C. Ministry of Environment, Lands and Parks files 10-3-14).

8. Report of the Hazardous Waste Management Advisory Committee, 1981.

9. Hazardous Waste Management Coalition, press release, February 2, 1987.

10. Confidential interview by Castle with a process participant.

11. On April 12, 1988, the Tumbler Ridge council voted to rescind its invitation to SWAC. An article in the *Peace River Block News* explains this reversal as a response to widespread local opposition to the facility (April 12, 1988). A meeting between BC Special Waste representatives and the town council in Princeton, B.C., was picketed by a crowd that chanted, "We don't want it!" (*Princeton News*, April 20, 1988). A May meeting in Cache Creek–Ashcroft was reported by the *Ashcroft–Cache Creek Journal* to have been "monopolized by a handful of opponents to the proposal" (May 10, 1988). The *Merritt Herald* reported that a 1,300-name petition against the proposed facility was handed to the local MLA in May (May 5, 1988).

12. Some local support is clear from coverage in the *Kamloops Daily News*, the *Ashcroft–Cache Creek Journal*, the *Princeton News*, and the *Merritt Herald*. B.C. Ministry of Environment, Lands and Parks files 10-3.

13. Quoted in the *Vancouver Sun*, May 17, 1988.

14. The relatively low majority in favor of the facility as well as the issue of the noninclusion of native voters (natives who have band elections on reserves are not included in municipal voting lists) were problematic for the facility proponents. If there were as many as 300 reserve natives and this group turned out to vote in the same proportion as other voters, the results could have been substantially different. If all natives voted against the facility, the majority in favor could conceivably have been as low as 52%. This inclusion of reserve natives, however, would not have been normal municipal polling procedure in British Columbia.

15. B.C. Environment files 10-3-8-1.

16. David Boyes, interview by Castle.

17. Confidential interview by Castle.

18. For a more detailed treatment of this case, on which the following discussion is based, see Castle 1993 and the chapter by Alun Richards in this volume.

19. Alun Richards, manager of external affairs for the Manitoba crown corporation, interview by first author, November 12, 1992.

20. On this arrangement, Wayne Neily of the Manitoba Environment Council noted, "Although we are conscious of maintaining our independence, sometimes we are on the same side as the government. . . . It wasn't so much that [MHWMC] was using us to sell their facility as much as we felt one was needed in the province. You could say just as easily that we were using them to get it built." Wayne Neily, interview by first author, September 23, 1992.

21. Total attendance at the open houses was 3,118. Of this total, 2,244 filed out detailed questionnaires; 1,536 supported further investigations in their community. See "Summary of Questionnaire Analysis" in Manitoba Hazardous Waste Management Corporation, 1992, 169.

22. Alun Richards, interview by first author, November 12, 1992.

23. These characterizations are based on a number of informal and confidential discussions in the R. M. of Montcalm on September 22, 1992.

24. Yves Sabourin, interview by Castle, September 22, 1992.

25. Minutes of the Rural Municipality of Montcalm, Council Meeting, November 27, 1989.

26. Denis Foidart, interview by Castle, September 22, 1992.

27. Florent Beaudette, interview by Castle, September 22, 1992.

28. Beaudette and Sabourin interviews by Castle.

29. Beaudette, interview by Castle.

30. Quoted in the *Winnipeg Free Press,* February 29 and March 1, 1992.

31. Ibid.

32. This summary is based on "The Agreement between Manitoba Hazardous Waste Management Corporation and the Rural Municipality of Montcalm," September 11, 1992; same as Montcalm By-Law Number 453/92.

33. See, for example, Davis and Lester 1988, 6, and the introductory chapter in this volume.

34. The Ontario government's decision not to proceed, announced on February 8, 1995, followed a lengthy environmental-review process in which the proposal of the Ontario Waste Management Corporation (OWMC), a crown corporation, was not approved in part for technical reasons. Questions were raised about whether there was a demonstrated need for such a large facility and whether the OWMC had shown its plans to be in every respect the best possible options. The vociferous and resilient opposition of the local community around the site that had been selected by OWMC, an opposition backed by broader environmental groups and concerns, was clearly a major factor in the cautious approach taken by successive governments and in the delays encountered. If not an important direct factor in the final negative decision, the opposition was at least indirectly critical, in the sense that it

delayed the siting until serious questions could be raised about its technical appropriateness, questions that were not widely asked when the process began.

35. Statistics Canada, *Income Estimates for Subprovincial Areas,* Ottawa, 1992, 13–216; Sova 1990, 17–101; Chem-Security (Alberta) Limited 1991, 2: 6–12, 6–13. "Rural" is defined here as a district not containing a town or city of more than 50,000.

36. Portney 1991, xiv. The undemocratic nature of siting decisions is also a focus for critics, such as Heiman 1990.

REFERENCES

Armour, Audrey, ed. 1984. The Not-In-My-Backyard Syndrome, Symposium Proceedings, Faculty of Environmental Studies, York University.

British Columbia, Special Waste Advisory Committee. 1988. *Development of the Special Waste Program for British Columbia—towards a Healthier Environment.* Report. Vancouver, May 31.

Castle, Geoffrey. 1992. *Siting the Alberta Special Waste Treatment Center.* Sustainable Development Research Institute Discussion Paper Series. Vancouver: SDRI.

———. 1993. Hazardous Waste Siting Facility in Manitoba: Case Study of a Success. *Journal of the Air and Waste Management Association* 43 (July): 1–7.

Chem-Security (Alberta) Limited. 1991. *Environmental Impact Assessment for Proposed Expansion of the Alberta Special Waste Treatment Center,* 2:6–12, 6–13.

Davis, Charles E., and James P. Lester, eds. 1988. *Dimensions of Hazardous Waste Politics and Policy.* New York: Greenwood Press.

Heiman, Michael. 1990. From "Not in My Backyard!" to "Not in Anybody's Backyard!": Grassroots Challenge to Hazardous Waste Facility Siting. *Journal of the American Planning Association* 56, no. 3 (Summer): 359–62

Kupchanko, Eugene. 1989. Case Study of a Successful Facility Siting. In *Standard Handbook of Hazardous Waste Treatment and Disposal,* edited by Harry Freeman, 3.39–3.49. New York: McGraw-Hill.

Manitoba Hazardous Waste Management Corporation. 1992. *License Application and Environmental Impact Assessment,* March.

McQuaid-Cook, Jennifer. 1987. Hazardous Waste Management System Development: Alberta, Canada. *Wastes Management,* April, 220–26.

McQuaid-Cook, Jennifer, and C. S. Simons. 1989. Development and Operation of a Waste Management System in Alberta, Canada. *Waste Management and Research* 7:219–27.

McQuaid-Cook, Jennifer, and Kenneth J. Simpson. 1986. Siting a Fully Integrated Waste Management Facility in Alberta. *Journal of the Air Pollution Control Association* 36 (September): 1031–36.

Morell, David and Christopher Magorian. 1982. *Siting Hazardous Waste Facilities.* Cambridge, Mass.: Ballinger.

Portney, Kent E. 1991. *Siting Hazardous Waste Treatment Facilities: The Nimby Syndrome.* Westport, Conn.: Auburn House.

Rabe, Barry. 1990. Toward Cooperation in North American Hazardous Waste Facility Siting: The Case of Alberta. *Canada-US Outlook* 1, no. 3/4 (April): 87–106.

———. 1991. Beyond the NIMBY Syndrome in Hazardous Waste Facility Siting: The Albertan Breakthrough and the Prospects for Cooperation in Canada and the United States. *Governance* 4, no. 2 (April): 184–206.

———. 1992. When Siting Works, Canada-Style. *Journal of Health Politics, Policy, and Law* 17, no. 1 (Spring): 119–42.

Reid-Crowther. 1980. *Hazardous Wastes in Northern and Western Canada.* 3 vols. Edmonton: Environment Canada.

Sova, Gordon, ed. 1990. *Corpus Almanac and Canadian Sourcebook.* Toronto: Southam.

Winnipeg Community Advisory Group on Hazardous Waste. 1992. *Draft Report.* City of Winnipeg, Manitoba, February.

Wolpert, Julian. 1976. Regressive Siting of Public Facilities. *Natural Resources Journal* 16 (January): 103–15.

Alternatives to NIMBY Gridlock: Voluntary Approaches to Radioactive Waste Facility Siting in Canada and the United States

BARRY G. RABE, WILLIAM C. GUNDERSON, PETER T. HARBAGE

Waste management and the siting of waste disposal facilities remain among the most controversial environmental issues in both Canada and the United States. Both nations continue to generate substantial quantities of solid, biomedical, hazardous, and radioactive wastes and have experienced considerable difficulty in finding politically and technically suitable areas for opening new treatment or disposal facilities. Indeed, in most areas of waste management in both nations, the siting process has regularly ground to a halt. Consistently across varied provinces and states, the announcement of a siting decision quickly meets an outraged public reaction, the so-called "Not in My Backyard" (NIMBY) syndrome. Time and again, proposed facilities are thwarted as either governmental or private-sector proponents back off in the face of such forceful political opposition.

The NIMBY syndrome has produced a mixed set of outcomes for effective waste management. On the one hand, many poorly conceived siting efforts have been rejected. In addition, the growing inability to open new waste treatment or disposal capacity has prodded governments and waste generators to place greater emphasis on waste reduction and recycling efforts. At the same time, however, even if the most ambitious current goals for reduction and recycling in Canada and the United States are realized, huge quantities of diverse types of waste require safe treatment and disposal methods. The inability to open new facilities means that new, safer technologies are not being employed, leaving the bulk of the waste management burden on older, less sophisticated technologies and, increasingly, waste storage at the site of generation. These ramifications of NIMBY politics are cause for considerable concern and suggest the need for at least some additional waste management capacity, particularly in those regions—or for those wastes—where current methods are suspect at best.

As a result, governments in both Canada and the United States need to revisit the issue of how they pursue siting. To date, both nations have tended to rely almost exclusively on top-down processes of site selection. Under these approaches, either governmental officials or private waste management firms use varying criteria to select their preferred sites. In both instances, the site evaluation process takes place in the absence of public consultation or any larger strategy to distribute on an equitable basis the burden for waste management across a province, state, or region. As a result, siting decisions, once revealed, seem coercive and unfair to proposed host communities. Consequently, communities feel they have been unjustly targeted to bear the sole burden for waste disposal, and they take effective collective action through NIMBY-type protest.

The most impressive exception to this pattern of siting gridlock stems from a series of cases in hazardous waste in the 1980s and early 1990s. The first, and perhaps most prominent, of these involved mid-1980s efforts by the province of Alberta to open a comprehensive facility for all hazardous waste generated in the province and to develop an integrated provincewide system of transportation, storage, reduction, and recycling. An extensive public-education process preceded any efforts to consider specific sites, and provincial authorities made clear that siting would only be considered among communities that formally volunteered after extensive study. This process, combined with a series of related innovations, culminated in the emergence of multiple volunteers and the eventual opening of a comprehensive waste management facility in Swan Hills in 1987 (Rabe 1991, 1994). This approach has been successfully replicated for hazardous waste in Manitoba and has begun to be adopted for hazardous waste in American states such as Minnesota and North Carolina, with generally encouraging results. Nonetheless, a voluntary approach to siting and waste management remains the exception for most types of wastes in most provinces and states.

A central concern of this chapter is the more general applicability of the Alberta approach to other areas of siting and waste management. This chapter examines the politics of attempting to site facilities for low-level radioactive waste in Canada and the United States. It reviews historic patterns of conflict—traditional top-down siting approaches that result in familiar gridlock—but also considers more recent efforts, in Ontario and Nebraska, to learn from the Alberta experience and devise a voluntary, comprehensive approach to the problem of siting and waste management. The following case analyses suggest that although this approach is no sure elixir for cooperation, it continues to appear far more promising than prevailing approaches and warrants further testing and development.

THE CASE OF LOW-LEVEL RADIOACTIVE WASTE

Few cases could provide as tough a test of the replicability of the Alberta approach as low-level radioactive wastes (LLRW). Whereas solid wastes pose certain environmental risks, neither they nor most hazardous or biomedical wastes pose a proven risk to both human health and the environment through contamination comparable to radioactive wastes. Indeed, in both Canada and the United States, the siting of facilities for either high-level or low-level radioactive wastes has been extremely controversial. The radioactive waste issue has probably triggered more political conflict over siting than hazardous waste. Consequently, neither Canada nor the United States has opened a new storage or disposal facility for either high- or low-level waste in more than twenty years (English 1992; Kemp 1992). Meanwhile, the waste continues to be generated, either being stored on site or being shipped to one of the few remaining waste disposal facilities that were opened before this aspect of waste disposal became so controversial.

This chapter reviews that familiar process of siting gridlock but devotes particular attention to whether siting agreements become more promising under a fundamentally different approach. Whereas high-level radioactive waste siting remains deadlocked in both nations, there are some promising possibilities for cooperation in low-level waste siting, particularly in the province of Ontario, which otherwise has experienced classic NIMBY in both hazardous waste and solid waste facility siting (Macdonald 1991, chap. 14). What makes the Ontario case especially interesting is that contrary to virtually all prior LLRW siting experience in Canada and the United States, multiple communities have voluntarily and actively been exploring the possibility of hosting a facility. Moreover, they have agreed to do so through a process that draws heavily on the very approach to siting initiated in Alberta in the 1980s. As we shall see, the Ontario case indicates that the voluntary approach to siting may indeed have replication prospects beyond the area of hazardous waste facility siting. This and important cases from the American siting experience will be examined in depth after a brief introduction to the basic history and evolution of LLRW policy in each nation.

EARLY STAGES OF LLRW MANAGEMENT IN CANADA AND THE UNITED STATES

The technical and political dimensions of LLRW management are quite similar in Canada and the United States, perhaps more so than they are for hazardous waste. Both federal governments played a fundamental role in endorsing and subsidizing the nuclear power industry in the decades following World War II. Despite tendencies toward regulatory decentralization in both nations, particularly Canada, both federal governments have assumed central responsibility for radioactive waste management. These governments continue to play a far more central role in virtually every aspect of radioactive waste regulation and management than they do for hazardous, biomedical, or solid wastes. Both governments, however, have delegated more authority to provinces or states in this area in recent years, although this trend has clearly advanced more rapidly in the United States than in Canada (Skogstad and Kopas 1992).

Canada and the United States also tend to speak much the same technical language concerning nuclear power and radioactive waste. The Canadian nuclear industry continues to promote actively its compact Candu 3 nuclear reactor in the United States (Urquhart 1990). Moreover, waste definitions and classification schemes are similar, and the technological options for waste reduction and disposal are largely identical. Perhaps most important, both nations have experienced strikingly similar patterns of conflict for radioactive waste facility siting. The seemingly intractable conflicts so evident in the American LLRW cases closely resemble the nation's pattern of hazardous waste facility siting. In turn, the Canadian approach to LLRW facility siting until 1986 was, in many respects, a carbon copy of the approach—leading to NIMBY results—that continues to be used in many individual states and multistate compacts.

The major sources of commercial LLRW in both nations include nuclear power plants, biomedical and industrial research, and nonmilitary governmental projects. The most common source, of course, remains nuclear power plants, many of which were opened in the 1960s and are beginning to approach the end of their functioning capacity. When such plants are closed, as appears inevitable for aging facilities in future decades, they no longer will generate new LLRW. At this stage, however, they must be decommissioned and decontaminated, which will create an entirely new—and potentially massive—type of LLRW disposal problem. In general, LLRW in both nations consists of materials contaminated with small amounts of radioactive substances, such as clothing, packaging, animal carcasses, medical fluids, and power reactor liquids. Such wastes need to be isolated from humans for 60–300 years. In contrast, high-level radioactive wastes include both the solid spent fuel from nuclear reactors and the concentrated fission products left in liquid form after spent fuel has been chemically reprocessed to retrieve plutonium and unfissioned uranium.

In the United States, the Pacific and Atlantic oceans served as repositories for LLRW during its first three decades of generation. Commercial land disposal of these wastes began in the early 1960s with the opening of six facilities, all of which used shallow land burial technology. This approach was fairly primitive, burying waste in shallow trenches and often storing waste in cardboard boxes until the burial was completed. Three of these facilities, however, were closed during the 1970s because of evidence of leakage and other forms of contamination. Of the three remaining facilities, the two located in Washington and South Carolina have begun to approach the end of their anticipated operational lives; the Beatty, Nevada, facility closed in January 1993. This facility shortage has triggered new legislation, the Low-Level Radioactive Waste Policy Act of 1980 and its 1985 amendments. These turned considerable siting authority over to the states and set up a complex system of incentives and timetables whereby states would assume responsibility for their wastes (English 1992). Most states or compacts, however, have employed traditional, top-down siting strategies, and not a single new facility has opened in the period following enactment of the new legislation.

In Canada, the Canadian Atomic Energy Act of 1946 established the basic framework for radioactive waste, a framework that endures to this day. It created the Atomic Energy Control Board (AECB), which retains authority for licensing and regulating the nuclear industry, including waste storage and disposal. It is distinct from Atomic Energy of Canada Limited (AECL), which is responsible for federal research and development related to nuclear energy, and Ontario Hydro, a major provincial public corporation that is responsible for management of that province's nuclear power plants. The AECB does share regulatory authority, in the sense that it typically confers in a working-group process with related federal and provincial agencies. Provincial involvement is a particular concern in that Canadian LLRW is overwhelmingly concentrated within its largest province. Sixteen of the nation's eighteen nuclear power plants are located in Ontario, and by all available measures, well over 90% of Canada's LLRW is generated within this single province, forcing the federal government to be sensitive to provincial concerns.

Contrary to the American approach of ocean dumping and later commercial-facility development, Canada has generally favored a quiet, de facto approach to LLRW, encouraging storage and disposal at or near the sites where the waste is generated. Most nuclear power plants keep such wastes on-site under interim-storage provisions, just as some American plants are beginning to do as disposal options narrow. In Canada, however, it is expected that many of these wastes will eventually be shipped to one or more permanent facilities. In turn, wastes generated from the use of radioisotopes sold by the AECL Commercial Products Division must be returned for burial at the AECL Chalk River Nuclear Laboratories in central Ontario. Finally, perhaps the most controversial disposal sites are located in and around the town of Port Hope; this has become a rough radioactive waste counterpart to the American Love Canal episode with hazardous wastes. Indeed, it was the controversy surrounding the Port Hope case that put the radioactive waste disposal issue on the Canadian political agenda more than fifteen years ago and triggered a search for more effective siting and disposal methods.

THE EVOLUTION OF CANADIAN LLRW MANAGEMENT

The problem of radioactive waste disposal at Port Hope actually precedes the formation of the AECB and the development of the nuclear power industry by more than a decade. Wastes generated by various domestic and military activities were deposited at several sites within the town of Port Hope from 1933 to 1948. Some of these wastes were transferred to nearby locations, including Welcome and Port Granby, in later years. It was not until 1975, however, that extensive radioactive contamination was discovered in Port Hope. This contamination, most of it in the form of radioactive soil, had been widely dispersed around the community. Radioactive waste was deposited in the Port Hope harbor, and a number of buildings in the town were constructed with contaminated material.

After the AECB completed a series of remedial actions in Port Hope and other contaminated areas, in 1980, it asked Eldorado Resources Limited, a federal crown corporation now known as CAMECO, to begin work on the design and siting of advanced LLRW disposal facilities.[1] Eldorado's approach to siting, however, in many ways resembled the pattern of Ontario's approach to hazardous waste facility siting and of troubled LLRW compact states such as Connecticut, New York, and Michigan. Without public consultation, Eldorado began examining alternative sites, using various technical criteria to assess the suitability of each potential area. In August, authorities announced that Eldorado "had taken options on two properties in the Port Hope area" and that they intended to complete surficial geological assessment on both sites before selecting one.

Typical of regulatory approaches to siting, the first opportunities for public participation were to be delayed until after these announcements, and confined to the federal environmental-impact assessment process. An assessment panel was appointed and received the formal site proposal and an environmental-impact statement. This material was also made available to the public, and local "issue identification" or "scoping" meetings were planned to acquire public input. The public reaction to the announcement of the two site candidates, however, was so strong that these meetings were delayed until after a final site had been selected. At this point, public opposition so intensified that the federal government suspended the process and dropped the two sites from further consideration.

The Emergence of the Voluntary Process

In response to this standoff, the minister of state for forestry and mines appointed an independent Siting Process Task Force in December 1986 to develop a less confrontational site selection process. The task force was commissioned to focus its efforts on the development of a process for siting a disposal facility in Ontario for the existing, ongoing, and historic wastes in the Port Hope area and, where advantageous, for the disposal of other existing and ongoing low-level wastes located in the province. The main objective of the process was to be the voluntary identification of one or more host communities, each with a suitable disposal technology.

The task force, in turn, suggested that a process driven by "social" rather than "technical" criteria be applied to resolve the LLRW disposal problem. In many respects, the successful experience of siting a comprehensive hazardous waste disposal facility in Alberta in the mid-1980s became a model for the task force as it rethought the process of LLRW disposal facility siting. This new process, proposed in a 1987 task force report entitled "Opting for Cooperation," drew heavily on the Alberta hazardous waste process and, in the words of one task force member, "built on the mistakes [Alberta] made." The task force was highly critical of the AECB for allowing technical criteria to drive the previous selection process and thereby failing to consult with affected communities until site selection was effectively completed. Not only had the AECB relied heavily on technical rather than social criteria in its siting endeavors, according to the task force; it also insisted that LLRW be disposed of through land burial rather than be stored or handled through other methods.

Given these concerns, the task force strongly recommended that a new siting process be established wherein potential volunteer communities would have the opportunity to learn about the problems of LLRW management as well as the possible payoffs if they decided to accept a site. It was the task force's position that volunteer communities should be free to opt out at any point in the exploratory phases of the process. Those deciding to proceed would become full partners in making decisions on all relevant matters. As a result, site selection and design, type of disposal technology, and impact management, along with other important features, were left open to negotiation rather than imposed from on high by authorities. Finally, not only would this cooperative-consultative process guarantee the health and safety of a volunteer community's citizens; equity payments would ensure that the community would be better off in economic terms after site construction was completed than it had been before. These provisions were not only a dramatic departure from prior LLRW siting efforts in Canada and the United States; they also featured many of the key components for successful siting, including voluntarism and comprehensiveness, so crucial to the Alberta and related cases in hazardous waste.

Implementing the Voluntary Process

The 1987 release of the task force report led to efforts to refine and ultimately implement the new siting process. On September 30, 1988, the minister of energy, mines, and resources appointed a new Siting Task Force (STF) to implement the first three phases of the voluntary approach to siting. The second "opting for cooperation" task force was asked to refine a five-phase siting process that considered the role of potential volunteer communities, a tentative discussion of disposal and storage options, the terms of reference for negotiation with local communities, and detailed cost estimates of implementation. Given the high interest generated by the process, the new STF was given twenty-three months to complete its work (Minister of Energy, Mines, and Resources 1990, 9).

Besides clarification of a few matters concerning site suitability and alternative technical options for potential host communities, some elements of the consultative process, particularly as they concern community liaison groups (CLG), had to be addressed. First of all, it was decided that all CLG meetings should be open to the public, despite the absence of American-like regulations that mandate such openness. Second, depending upon the response of those attending the meetings, a CLG could decide to opt out of the process at any time. Furthermore, a community could decide to receive newly generated, as well as historic, radioactive wastes even if the amounts would be relatively small and it would thereby be only a part of a larger waste management solution.[2]

While various guidelines and criteria were being established by the STF, consultations began with regions and communities. In November 1988, exploratory letters were sent to 850 communities. Four months later, follow-up letters were sent to invite two representatives from each community to attend one of eight regional information meetings. At these meetings, community representatives were informed of the need for improved management of LLRW, the process principles already established, and some of the safeguards available to those interested in pursuing the process. Community representatives, in turn, asked how communities would be consulted, how they would be defined in the first place (some contiguous communities were interested and wanted to proceed together), and how their CLGs, if formed, would measure community acceptance. They asked, furthermore, about the characteristics of, and the hazards associated with, LLRW, the dangers involved in transporting LLRW to their communities, the ongoing problems associated with managing such a facility, the type and number of jobs available to local citizens if a facility were built, and the impact on tourism.

After the regional consultations, twenty-six communities requested additional meetings and/or information. Between the regional meetings and the follow-up informational sessions held by the STF, five communities dropped out. An additional seven withdrew after the informational sessions, leaving fourteen to enter the next phase of the process. At this point, the remaining communities created their CLGs. Various formal and informal contacts, including ads in local newspapers, were used to create tentative lists of CLG members. The STF staff called those listed to see if they met the criteria. A final list was printed in newspapers to allow a two-week comment period regarding the representativeness of the list. By the end of November 1989, ten CLGs were formed. An additional one was completed in January 1990, followed by one more in March and the last two in April. By request of the local councils, the STF had taken responsibility for selecting CLG members.

As this process unfolded, the advantages of a voluntary over a top-down process became clear. In the voluntary process with which the STF has experimented, the typical kind of NIMBY confrontation has not occurred. To the contrary, the various CLGs networked—they shared information and ideas and learned from each other's experience, as exemplified by the January 1990 Thunder Bay meeting of chairs and facilitators of CLGs.

The results of extensive consultation—within communities, between them and the STF, and between the communities and outside experts they had selected—varied considerably. In Manitouwadge, Ear Falls, Red Lake, Atikokan, Mattice–Val Cote, and Upsala, both the CLGs and the local councils rejected further consideration of hosting a site. The CLGs of the United Townships of Rolph, Buchanan, Wylie, and McKay and of the Townships of Head, Clara, and Maria declined to continue. The CLGs of Deep River, Hornepayne, and Chalk River chose to withdraw, but their respective councils approved movement into the next phase. In contrast, the Elliot Lake and James Township CLGs wanted to continue, but their respective councils refused.

The concerns of the communities that opted out were familiar: 1) the potential health risks, and the risks to ground and surface water, were perceived to be too high, with a baseline health study considered absolutely essential before proceeding; 2) the potential negative impacts on economic development and tourism were believed to be too severe; 3) the potential transportation risks and effects were also thought to be too great; 4) finally, local citizens were not convinced of the long-term integrity of technology needed to manage LLRW.

Although these constituted the biggest concerns, others reflected a lack of trust in the federal government's willingness to pay for compensation and for a well-managed facility as well as a continuing distrust of the AECB. The AECB clearly lacked the public credibility of crown corporations developed for hazardous waste in Alberta and Manitoba. There was also expression of concern that were a community to accept LLRW, other sorts of waste, including high-level radioactive materials or hazardous wastes, would ultimately migrate there for disposal (Minister of Energy, Mines, and Resources 1990, 53–54). These fears were exacerbated by Ontario's extremely controversial efforts to locate sites for disposal of hazardous and solid wastes in the 1980s and 1990s (Macdonald 1991, chap. 14). In both of these cases, technical, top-down approaches to facility siting have led to the familiar outcome of gridlock. In hazardous-waste facility siting, for example, the province devoted more than twelve years and spent more than $100 million in attempting to impose a facility on a highly reluctant Ontario community.

Possible Site Volunteers

These withdrawals did not constitute the end of the voluntary siting process. Despite the concerns that prompted a number of CLGs or councils to opt out, several communities remained interested in the siting process and continued deliberations with STF officials through 1994. Moreover, one source community was interested in cleaning up LLRW located within its own boundaries.

Among the potential volunteer communities, the Deep River Council (a neighbor of the Chalk River facilities of AECL) chose to proceed after its ClG said no for four cluster communities. These communities are located in northeastern Ontario, with Ottawa 115 kilometers to the southeast, and have a combined population of approximately 3,800. The two major concerns of the citizenry involved the desire that the historic wastes of the Port Hope area be managed *there* before being moved anywhere and the fear that an accident or spill could occur if LLRW were transported through surrounding areas. The Deep River Council agreed to continue the process, subject to the following provisos: 1) given the risks and costs of transportation, siting a new facility where the waste is now located would be the best option; 2) any waste relocated to Deep River should be transported by rail, with the material stored in suitable containers; and 3) during or at the end of phase four, a referendum should be held concerning whether to continue the process, with any associated costs to be paid for by the government. Chalk River agreed to go along with the Deep River Council. Since there are no sites in Chalk River, it seems to want to be part of the process to protect its interests.

Technical studies are proceeding at Deep River, and a working group has been formed, consisting of CLG and STF members as well as representatives of relevant provincial and federal ministries, to develop criteria for matching sites with technologies. Much of their attention has focused on two preferred sites, one below and one above ground, which were announced on July 20, 1994. These sites are located on 10,000 acres in the southeast portion of Deep River. Geraldton, a community of 2,900 residents located the farthest west of all the potential sites, approximately 45 kilometers northeast of Thunder Bay, was intrigued by the process given its potential contributions to economic diversification, but its local council withdrew before a referendum that had been scheduled for November 1994. Two problems became interwoven to produce the withdrawal—annexation of land intended for a facility and political opposition. The Ontario Municipal Board ultimately approved the necessary township annexations, but only after administrative delays extended the process from the anticipated three months to one year. Given this delay, the CLG and council had only eight months to complete all relevant technical work prior to the November referendum, as an opposition group known as Stop the Ongoing Process (STOP) had successfully pressured to move the vote from March 1995 to the prior November. This meant that the technical work could not be completed in time for the referendum and the STF had to certify that this basic deadline would go unmet; the CLG asked the local council to opt out, which it did.

In addition to the potential volunteer communities, some source communities decided to proceed to the next phase as well. Port Hope, located 45 kilometers east of Toronto with a population of 10,300 residents, has been most active among communities in this category, although the local council has never passed a resolution expressing interest in participating as a volunteer community. Its citizens want, primarily, to find a permanent site for the wastes deposited within town boundaries and to secure restoration of its badly contaminated harbor in the process. Furthermore, they want to play a major role in the cleanup and hope to do this by having their CLG stay in operation until the process is over. To fulfill its role, the CLG seeks to be empowered to hire consultants and contract for studies as necessary. In short, the Port Hope community, through its CLG, hopes to participate in the establishment of criteria for selecting a site, in the cleanup of old sites, and in the decision regarding which wastes will be moved from the old to the new sites. In October 1992, the Port Hope council passed a resolution expressing a willingness to continue negotiations but emphasizing key conditions, such as components of the compensation package.

Construction of a new harbor for Port Hope, however, has become a central stumbling block in LLRW facility siting. The International Joint Commission has identified Port Hope Harbor as among the forty-three most environmentally hazardous sites bordering the Great Lakes and in need of immediate "remedial action," due to extreme contamination from LLRW and other toxic substances (Hartig and Zarull 1992). In response, the Port Hope CLG recommended that Port Hope Harbor be included under the criteria to be developed jointly by the Port Hope CLG and the Siting Task Force. It also recommended that the government of Canada consider creation of a new harbor in Port Hope as a demonstration of its commitment to the environmental integrity of the Great Lakes. Such linkages could formally integrate LLRW management with the larger issues of sediment contamination and areawide remediation, possibly leading to increased public support and greater integration of environmental management in this part of Ontario. The STF lacks authority, however, to make such commitments, confining Port Hope's likely involvement to management of its own abundant wastes. Greater integration of various aspects of the environmental regulatory process may be one of the greatest lessons from this process, as the federal government has committed to harbor restoration but has not formally linked it with the siting process.

Hope Township has also remained active, although only as a source community. Its CLG suggested the following as an interim response: 1) it wants control of how wastes are evacuated and transported from the Welcome LLRW site, as it is very concerned about radioactive and arsenic wastes; 2) since the township will be affected by decisions on what to do with its historic wastes, it should be part of the consultations that determine where the new site will be located; and 3) its CLG should continue to function until the process is complete. The Hope Township council went on record in support of the above and has continued to examine issues such as assessment of soil cleaning levels. Newcastle, a Toronto exurb with 34,100 residents, has also continued to be involved in the process. Like other source communities, its involvement has been motivated in large part by its concern over Port Granby wastes.

As this winnowing among possible candidates has continued, a new task force, created in 1991 and 1992, is expected to complete siting deliberations over four years. During this period, the potential volunteer community has begun to explore the technical, environmental, and social impacts of receiving wastes, while source communities have begun to examine the impact of removing wastes. Joint decision making has been emphasized, and operational planning has begun in the remaining volunteer community (Deep River) and the three remaining source communities (Port Hope, Hope Township, and Newcastle). Although it is possible that this process could collapse, many observers remain optimistic, as serious dialogue continues in multiple settings.

FEATURES COMMON WITH RELATED SITING AGREEMENTS

The signs of evolving cooperation in LLRW siting in Canada are remarkable, given the historical opposition to siting all sorts of LLRW disposal facilities in both Canada and the United States. Even more so than in hazardous waste policy, examples of LLRW siting successes are hard to find in either nation. The dismal American experience in LLRW siting is one that Canada has apparently learned from and chosen to avoid. Examining the sorry track record of many of its own provinces in hazardous and solid waste facility siting, and its highly contentious early approach to LLRW siting, Canada undertook a significant reform of its siting policy in 1986. Officials chose to model their new policy on the successful siting experiment completed in Alberta in 1984, since replicated in Manitoba in 1992 (Rabe 1994). As in these hazardous waste facility siting cases, several key features were present.

First, early and extensive public education and participation efforts were crucial to the siting processes in Alberta and Manitoba. Such a pattern closely resembles the Canadian LLRW siting effort. Public information and participation activities were not as numerous as in the western provinces, but were extensive and were based on the principles of voluntarism. All involved communities were given the option to withdraw at any point of the process. Moreover, they were also assured of direct input in the selection of technology at any facility that might be opened, a decision that is usually finalized by authorities in advance of any siting negotiations.

Second, the successful siting efforts in Alberta and Manitoba further indicate that citizens want reassurance that they will not become a magnet for wastes from other provinces or states if they agree to open a new disposal facility. Some formal agreement to share the burden for waste management in a fair way was crucial in both Alberta and Manitoba. In turn, Canada's LLRW siting efforts have promoted the notion of burden sharing in a variety of ways. Perhaps most important, siting officials have remained open on the ultimate number of places that will participate in waste management and on whether they will provide waste storage or more "permanent" disposal. It is possible, for example, that one major facility will emerge but that other communities will develop more modest facilities. Such options simply are foreclosed in the American states or compacts, where there is usually an intensive search for a single site to employ a single disposal technology to dispose of all the wastes that are eventually shipped its way.

The Canadian process has also benefited from a general sense of assurance that it will not become a magnet for wastes from other parts of Canada. Since Ontario generates the vast majority of Canada's LLRW, there is little threat of exploitation by neighboring provinces. Moreover, officials have stated that any new LLRW facility will not accept waste from the United States or other nations. A shift in this position could indeed undermine the cooperation that has emerged thus far.

The final aspect of burden sharing, combining siting efforts with expanded efforts to reduce the volume of waste being generated, has proven less prominent in LLRW deliberations in either Canada or the United States than in the hazardous waste facility siting agreements in Alberta and Manitoba. Both nations have made significant strides in waste reduction in recent years, but this has not formally been incorporated into siting deliberations. By contrast, several European nations have successfully made this linkage central to their siting policies (Kemp 1992).

Third, compensation and safety protection were of enormous importance in the successful hazardous waste facility siting cases. Canadian LLRW siting officials have proven extremely open to alternative siting technologies and varying types of compensation packages. In Port Hope, for example, the issue of facility siting has been directly linked with harbor dredging and cleanup, an action that, if completed, would offer considerable environmental and economic benefits for local residents.

Fourth, Alberta and Manitoba were extremely successful in devising new managerial partnerships to defuse concern over long-term commitments to the technical and financial feasibility of new facilities. As in hazardous waste, many private LLRW management corporations in both Canada and the United States have shoddy records when it comes to maintaining high levels of operational proficiency and assuring fiscal commitment to facility operation. In many siting controversies, the questionable integrity of private corporations that will be responsible for facility operation scuttles any possibility of agreement. As we shall see, slipshod public and private management helped undermine an American case in which, under a voluntary approach, an agreement may have been possible.

Both Alberta and Manitoba devised public-private partnerships designed to assuage those very reasonable concerns. Both established new crown corporations, public entities responsible for shared facility management with private firms that are overseen by provincial regulatory authorities. In Canadian LLRW, the governmental role will be similarly large and enduring in any new storage or disposal facilities. Indeed, the Canadian Siting Task Force has proven unusually successful at winning public trust and facilitating genuine dialogue over siting options in multiple communities. It has helped overcome understandable public reservations about both Eldorado and the AECB, a pair of public entities with long and somewhat controversial records.

These four broad features—public participation, burden sharing, compensation and protection, and a credible governmental role in siting and facility management—appear to have contributed to the evolution of cooperation in Canadian LLRW management, much as they have in prior hazardous waste facility siting agreements. We now turn to the American case, where traditional siting approaches have prevailed and faced predictable, NIMBY-type outcomes but some experimentation with a more voluntary approach offers some hope.

THE EVOLUTION OF AMERICAN LLRW MANAGEMENT

The 1980 Low-Level Radioactive Waste Policy Act and its 1985 amendments represent a significant departure from traditional American approaches to LLRW, which have been dominated by the federal government. They featured a series of federal mandates and incentives that encouraged individual states to form multistate compacts to manage LLRW (Derthick 1974; Brown 1986). This legislation retains federal authority for setting safety standards and orchestrates a system of timetables, incentives, and penalties related to compact creation. As is the case with many other interstate compacts, the radioactive waste compacts must be approved by Congress before becoming official.

This legislation, however, also delegates a significant degree of subnational authority. It leaves to individual states the decision about what other states (if any) will share compact responsibilities with them, how they will design and implement their siting processes, and what methods they will use for disposal. Among the powerful incentives that the federal government provides to states to enter into compacts rather than manage waste on a solo basis is that states belonging to a congressionally approved compact have the authority to exclude wastes from other states, whereas independent states cannot restrict such waste importation.

The compact approach has thus represented a unique federal government effort to delegate regulatory authority to individual states or clusters of them. Its enactment was driven in large part by the efforts of the governors of Nevada, South Carolina, and Washington. Their states hosted the only three waste disposal sites that continued to operate in the nation after three other sites closed in the late 1970s. The governors' threat to restrict access to these sites prodded Congress to accept the compact approach that was developed by the National Governors Association (Colglazer 1982; Kearney and Garey 1982).

The success of the compacts was to be measured by their ability to meet a series of federally established milestones that anticipated a fully functioning system of sites by 1993. These timetables were delayed and altered in the 1985 amendments when it became obvious that the initial legislation had been too ambitious. Select portions of this process were invalidated by a 1992 U.S. Supreme Court decision, which argued that it was unconstitutional to require states to "take title" to all wastes after specified deadlines (Kearney 1993). Much of the remaining structure, however, was not overturned, and at least in some cases, the siting process continues.

All of these complicated factors were designed to promote policy coordination and facilitate long-term support for waste disposal agreements. They operated on the assumption that states and compact regions would perceive long-run advantages from entering into compacts and accepting surcharge rebates and other compensatory benefits. In this process, it was assumed, they would successfully override political resistance to facility siting. This approach hinged, in many respects, on a presumption that economic compensation alone would serve as a reliable lubricant of siting conflict.

The compact approach has enjoyed one degree of success, in that forty-two states have entered into a total of nine congressionally approved compacts. The fifteen states belonging to the two compacts that feature disposal facilities in continued operation (in South Carolina and Washington) are set for the near future, with secure facility access and the ability to exclude out-of-compact wastes if they so choose. For the remaining states and territories, however, the compact approach has largely been a nightmare. No new siting agreements have been reached in any state or compact, and none are in sight in the near future. Consequently, many states, such as Michigan, are forced to store LLRW at more than fifty sites around the state; many of these sites were never intended for long-term storage or disposal, and they pose significant safety and security problems. Increasingly, LLRW that cannot be shipped to South Carolina or Washington must be stored in this fashion at the sites of nuclear power plants, hospitals, medical-research institutes, or other facilities that generate some amount of LLRW. South Carolina has allowed a small amount of LLRW imports from states demonstrating some continued commitment to resolving their own waste disposal problems, although it imposes increasingly steep surcharges. There is, however, no assurance as to how long this export option will remain available, as the state is free to close it at any time. In turn, those states perceived as not making a reasonable commitment to waste management have been completely cut off from further access to the facility.

The siting approach most commonly used in the states and compacts has eschewed broad public participation and a comprehensive waste management process in favor of a traditional top-down approach driven by technical criteria. In Connecticut, for example, the state's Hazardous Waste Management Service used such criteria to determine that three towns were its preferred candidates for a facility that would manage LLRW from Connecticut and New Jersey, its Northeast Compact partner. There had been no consultation with these communities and no warning that they would be finalists. As an elected official of one of these areas explained after learning the news: "I'll never forget it. I was in shock. I had to sit down for a minute."

Residents of the other two potential host communities responded similarly but then wasted little time before taking vigorous collective action against the proposal. Public meetings were finally held, but these became rallying events for opponents. Several interviewees recalled that at one of these meetings, an official of the management service said, in essence, that "the site will go in one of these areas, and you cannot stop it." After months of NIMBY opposition, however, the proposals were withdrawn, leaving the siting process in Connecticut and New Jersey up for grabs. LLRW is no small issue in these states, as combined they produce approximately 7% of the nation's total new LLRW generated each year.

Michigan exercised similar diplomatic and participatory skills in advancing its siting plans. The state grudgingly agreed to become the siting host for the seven-state Midwest Compact in 1986 and did receive considerable financial support through the rebates made available to the compact from the federal government. But like Connecticut, Michigan relied on technical criteria with no public participation in its process of selecting three sites. In October 1989, three rural areas scattered about the state were informed, by mail, that they were the finalists. None was at all aware that it was under consideration for managing LLRW from around the Midwest. As one town supervisor recalled: "I first found out when the press called to ask me how I felt about all of this. It was really upsetting to find out that way." From this point, the Michigan siting process almost perfectly paralleled the Connecticut experience. After nearly two years of contentious public meetings and NIMBY opposition, the state withdrew the three sites from further consideration. But as it failed to develop any alternative siting process, it was ousted from continued participation in the Midwest Compact, leaving it on its own to find a siting solution and cut off from any further access to the South Carolina facility.

Other compacts and states have struggled in somewhat similar fashion, although a few have tried to incorporate some of the conditions that have proven so crucial to the cooperation emerging over Ontario LLRW siting and in the earlier examples of successful hazardous waste facility siting. Among these, the Central Compact with its host state, Nebraska, remains the closest American counterpart to this alternative approach. It emphasized a number of the key themes discussed in the Ontario case and made far more progress toward a cooperative solution than other states and compacts, although its siting process faced some serious problems and its fate remains uncertain.

AN ATTEMPT AT VOLUNTARISM: THE CASE OF NEBRASKA

Between 1981 and 1987, Nebraska followed much the same process as had Connecticut and Michigan. The state was negotiating its status in a compact, and in March 1982, the Central Compact was formed. In addition to Nebraska, the compact included Arkansas, Kansas, Louisiana, and Oklahoma, and it collectively generates about 9% of the nation's LLRW each year. Then, beginning in 1986, the Nebraska Department of Environmental Control drafted several regulations regarding the construction of a LLRW facility. Several public hearings were held on the compact and the facility regulations, but few citizens attended the meetings. In June 1987, the compact selected a private firm, US Ecology, to develop, construct, and operate the Central Compact facility.

At this point, the Nebraska case began to diverge from the ones in Connecticut and Michigan. Nebraska was chosen as the compact's host state in 1988. This stimulated a citizens' group, Nebraskans for the Right to Vote, to gather the more than 150,000 signatures necessary to hold an initiative vote on whether the state should remain in the compact. In what was known as Initiative 402, 69% of participating voters supported continued compact involvement in November 1988. The initiative proved a major statewide issue, and several groups on both sides of the question organized and advanced their respective cases. The questions raised during the campaign focused not on site-specific issues but on larger concerns, such as whether Nebraska's waste generators would have to shut down if Nebraska rejected the possibility of hosting a site.

The initiative campaign overlapped with the initiation of US Ecology and Nebraska's public-participation efforts in April 1988. The Nebraska Citizens' Advisory Committee (NCAC) and the Nebraska League of Women Voters both played important roles in this process, supplementing related efforts of US Ecology. The committee was established by US Ecology and was designed to reflect diverse constituencies, as it included representatives of farming, ranching, business, environmental, and natural-resource interests. It was intended not only to gather public opinion but also to play a significant role in advising US Ecology on key siting and facility-design issues. Over seven months, the committee held six public meetings, at which the public was encouraged to give its input on any aspect of the siting program. At least one meeting was held in each major region of the state.

The League of Women Voters, in turn, sponsored two series of public workshops in 1988. The first series, held in May, involved US Ecology presentations of the siting criteria that were going to be used. Citizens were encouraged to give their input about the appropriateness of the proposed criteria. The second series featured a US Ecology presentation of the proposed facility design and, again, sought public suggestions. There were five workshops in each of these two series, generating literally thousands of comments and suggestions from all around the state. Both site criteria and facility-design plans were changed in response to concerns raised in these meetings.

Despite the significant involvement of the League of Women Voters, US Ecology played the most important role in the public participation process. As these other efforts were taking place, it mailed out invitations to each Nebraska municipality, county board, chamber of commerce, and natural resource district, asking if it was interested in the possibility of hosting a facility. Using the same voluntary philosophy successful in other siting cases, the firm made clear that it would not enter an area for site testing unless invited in by local authorities. Moreover, an initial positive response was not binding; communities were told that they could withdraw at any time they wished.

In response, fifty-two municipal governments and twenty counties expressed interest in exploring the possibility of hosting a facility. Within these fifty-two communities, 111 "potential siting areas" (PSAs) were identified by US Ecology. Such a response was astounding, given the pattern of near immediate rejection of siting proposals in virtually all other states and compacts. As the process continued, some of these candidates withdrew, for either political or geological reasons, but 27 PSAs in eleven different counties remained involved. Such a pattern was similar, in many respects, to the winnowing that occurred among potential volunteer sites in hazardous waste in Alberta and Manitoba and is occurring in LLRW in Ontario.

But at this point, the process began to sour. Rather than pursue continued public dialogue with all possible candidates, as in Ontario, the Central Compact moved very quickly to narrow the field and attempt to reach an agreement in short order. Their pace was driven in part by Nebraska law, which called for site announcement by January 1, 1989, one year ahead of the timetable in the federal legislation. The NCAC employed a "blind" format using technical information on remaining areas to select three of them for further examination and, in the process, eliminate all other possible candidates. The NCAC chose rural sites in Boyd, Nemaha, and Nuckolls counties, and US Ecology designated specific areas within these counties as its finalists on January 18, 1989. This selection was little more than three months after the statewide referendum, reflecting the extremely fast pace of site selection after the initial public participation efforts. Hence, there was little opportunity for the sorts of extended discussions and review of multiple candidates so crucial to the other siting agreements.

Generous compensation packages began to be offered at this point, as called for in the authorizing legislation, providing a steady flow of new, largely unrestricted income to each of the three county finalists. In 1989, for example, each of the three final counties received $100,000. US Ecology completed concurrent investigations at the three sites but focused most of its subsequent analysis on a Boyd County site located in McCully Township (pop. 160). This township is located near the village of Butte (pop. 540), upon which it is dependent for all educational and fire protection services. US Ecology was attracted to the Butte site for technical reasons and because it knew that it had strong support for the facility from the Butte Board of Commissioners, which had twice passed resolutions supportive of hosting the facility. US Ecology announced on December 31, 1989, that the Butte site had been selected as the preferred host area, closing its information office and related efforts in Nuckolls and Nemaha counties within a few months.

The selection of this site, however, was a serious blunder that undermined the public trust and dialogue that had been developed to that point. Despite the support of Butte elected officials and the neighboring communities of Anoka and Lynch, there were numerous warning signs that any effort to attempt to impose a siting decision in Boyd County would be ill fated. One day prior to the January 1989 announcement that a Boyd County site would be one of three finalists, the Boyd County Board of Commissioners had voted to rescind their invitation to US Ecology. In turn, the neighboring communities of Spencer (pop. 645, located 12 miles east of Butte) and Naper (pop. 171, located 12 miles west of Butte) had registered strong opposition in advance of the January announcement. Butte, Spencer, and Naper have a long-standing history of animosity toward one another, with a series of skirmishes over the placement of railroads, highways, the county fair, and the county seat. Some of these battles date back to the nineteenth century, but the tensions endure through the 1990s, only inflamed further by the LLRW proposal. Moreover, there is considerable tension among the communities over a proposed merger of the Butte and Spencer high schools, which would mean that one of these communities will soon lose its secondary school. There is also resentment over division of the compensation package, as Butte has received $300,000 in virtually unrestricted funds for each of the past several years while its neighbors have received little, if any, financial support. Finally, a strong grassroots organization, Save Boyd County, had already formed to oppose the facility before US Ecology had chosen to press ahead, and would later seek advice from national antinuclear groups on civil disobedience and harassment tactics. In short, Boyd County hardly seemed an auspicious place to pursue siting, especially when opposition was already mounting and other volunteer candidates had expressed interest in the possibility of hosting a facility.

The selection of Butte was viewed by many as a fundamental break of US Ecology's earlier promise to pursue siting only where a genuine commitment was evident. In fact, as late as December 1988, a US Ecology vice president had reassured the Boyd County commissioners that the firm would leave if at any future point local citizens wanted them to do so. Despite the strong opposition, a group of Butte citizens, People for Progress, have continued to rally support for the site and claim to have 500 supporters throughout the county. Since the siting process began, there have been two Village of Butte Board of Commissioners elections, and on both occasions candidates supportive of the site have been elected or re-elected. Butte's and McCully Township's elected county board representatives have continued to support the site. All subsequent US Ecology efforts to hold public meetings and discussions of the proposal, however, have proven extremely divisive, leaving the future of the siting proposal for Butte in near total disarray. The credibility of US Ecology has been badly damaged in Boyd County and the remainder of Nebraska by this episode, making it an unlikely candidate to lead any resumption of the voluntary siting process. "I don't think that US Ecology could hold another meeting in [Boyd County]," noted one siting opponent. "They would just get run out on a rail."

This failure should not, however, mar the considerable early achievements of the Nebraska siting process. Far more so than by any other compact, extensive efforts were made to involve the public. Much to its credit, this voluntary process attracted numerous potential participants. It is impossible to know whether selection of another finalist community, or keeping options open for numerous potential communities, might have led to a different outcome. Nonetheless, the compact's determination to proceed rapidly to a siting decision, as dictated by Nebraska legislation, and US Ecology's adamance about Butte in the face of numerous warning signs of conflict should not lead to rejection of the Nebraska case as typical of siting failure. At least in the stages leading up to site selection, the Nebraska case bears a good deal of resemblance to the Ontario LLRW case and those in hazardous waste discussed earlier. Moreover, there is a very real possibility of reviving the siting process, given continued support for a site by residents of Butte, Anoka, and Lynch.

CONCLUSIONS

The Ontario LLRW siting case indicates that many of the siting lessons drawn from hazardous waste facility siting can be applied with some success to the contentious issue of radioactive waste disposal. In turn, the American experience in LLRW generally confirms the overall pattern of failed siting in hazardous waste, although the Nebraska case suggests a possibility of implementing some version of the voluntary approach in the United States. In recent months, some states and compacts have begun to explore this approach, most notably Connecticut, which has chosen to revise its siting efforts in this way. However, it remains unclear whether such states will merely extend a general invitation for volunteers or, instead, attempt to develop the more comprehensive process necessary to enhance the prospects of extended dialogue over siting options and, in turn, increase the likelihood of siting agreements. Far from foolproof, the voluntary approach, in all its facets, does offer a fresh alternative to prevailing approaches to controversial facilities. Besides waste management facilities, both Canada and the United States have experienced increasing conflict over a wide array of siting efforts, including proposed facilities for drug and alcohol rehabilitation, nursing homes and hospices, public housing, and air and rail transport. National and subnational governments in both nations have generally groped for strategies that can foster serious public dialogue over siting options and lead to cooperation in final siting decisions. Both governmental and private-sector officials in a wide array of policy areas plagued by NIMBY-type reaction may well benefit from careful consideration of these more successful efforts in waste facility siting.

NOTES

1. CAMECO is now a semiprivate firm because of a joint venture with a Saskatchewan-based private firm.

2. In addition to these and related guidelines, it was felt that tentative site elimination criteria had to be established. In level one, if just *one* criteria was met, the site would be eliminated; in level two, one of the criteria could be present and the site would not be immediately eliminated. Individual communities could add further criteria if they wished. Five level-one criteria were established: 1) resources identified by the Ministry of Natural Resources as requiring protection; 2) federal or provincial parks, existing or proposed; 3) natural areas (wildlife, wetlands); 4) unstable lands (flood, fault, earthquakes); and 5) archeological, historic, or cultural resources. After modest revisions suggested by CLGs in January 1990, the fourteen communities meeting with the STF agreed to the reasonableness of these criteria by February and March. They were, however, given the option of defining and ranking them through their CLGs in ways that they deemed appropriate to local circumstances.

REFERENCES

Brown, Paul M. 1986. Environment Canada and the Pursuit of Administrative Decentralization. *Canadian Public Administration* 29:218–36.

Colglazer, William E., Jr., ed. 1982. *The Politics of Nuclear Waste.* New York: Pergamon.

Derthick, Martha. 1974. *Between State and Nation: Regional Organizations of the United States.* Washington, D. C.: Brookings Institution.

English, Mary R. 1992. *Siting Low-Level Radioactive Waste Disposal Facilities: The Public Policy Dilemma.* New York: Quorum Books.

Hartig, John H., and Michael A. Zarull, eds. 1992. *Under RAPs: Toward Grassroots Ecological Democracy in the Great Lakes Basin.* Ann Arbor: University of Michigan Press.

Kearney, Richard C. 1993. Low-Level Radioactive Waste Management: Environmental Policy, Federalism, and New York. *Publius* 23 (Summer): 57–73.

Kearney, Richard C., and Robert B. Garey. 1982. American Federalism and the Management of Radioactive Wastes. *Public Administration Review* 42 (January–February): 12–24.

Kemp, Ray. 1992. *The Politics of Radioactive Waste Disposal.* New York: Manchester University Press.

MacDonald, Doug. 1991. *The Politics of Pollution: Why Canadians Are Failing Their Environment.* Toronto: McClelland and Stewart.

Minister of Energy, Mines, and Resources, Government of Canada. 1990. *Opting for Cooperation: The First Phases.* Ottawa: Ministry of Energy, Mines, and Resources.

Rabe, Barry G. 1991. Beyond the NIMBY Syndrome in Hazardous Waste Facility Siting: The Albertan Breakthrough and the Prospects for Cooperation in Canada and the United States. *Governance* 4 (April): 184–206.

———. 1994. *Beyond NIMBY: Hazardous Waste Siting in Canada and the United States.* Washington, D.C.: Brookings Institution.

Skogstad, Grace, and Paul Kopas. 1992. Environmental Policy in a Federal System: Ottawa and the Provinces. In *Canadian Environmental Policy: Ecosystems, Politics and Process,* ed. Robert Boardman. Toronto: Oxford University Press, 43–59.

Urquhart, John. 1990. "Canadian Nuclear Industry Expects Orders to Rise, Plans Compact Reactor for O.S." *Wall Street Journal,* February 10, B6.

Democratic Dialogue and Acceptable Risks: The Politics of High-Level Nuclear Waste Disposal in the United States

Michael E. Kraft

Disposal of high-level radioactive waste (HLRW) is the ultimate siting controversy. No nation has yet built a permanent repository for the wastes associated with nuclear power and strategic weapons production, and in most the near-term prospects are not encouraging. The obstacles to repository siting are similar to those that have long afflicted the building of hazardous waste storage, treatment, and disposal facilities. They include technical and management challenges, but above all they are social and political. Simply put, people are enormously fearful of the health and environmental risks they associate with the wastes and of the prospect of nuclear accidents and mismanagement, and they are intensely opposed to having unacceptable risks imposed on their communities. Their opposition is difficult to overcome in a democratic society, particularly when they distrust project promoters and the decision-making processes they follow.

Effective siting programs seem to require a form of citizen involvement that can build trust and create a publicly acceptable process for evaluating and distributing the risks associated with the technology—in this case, chiefly commercial nuclear power. Yet designing and implementing such programs are by no means simple matters, given current levels of citizen distrust of government and skepticism within government and industry about public participation on technical issues.

Public perceptions of technological risks such as nuclear waste are often dismissed as ill informed by scientific experts and industry and government officials. The implication is that public views are unworthy of serious consideration (Hadden 1991; Shrader-Frechette 1991). Agency programs of citizen participation are affected by judgments about the public's interest, knowledge, and capabilities. Perhaps as a consequence, such programs may be sharply circumscribed and extend few genuine opportunities for democratic dialogue that may facilitate solutions. Token hearings and comment periods are offered, but little thought is given to the design of effective forums for policy legitimation on such complex and conflict-ridden issues. Without such forums, agencies may cut themselves off from useful feedback and criticism, weakening policy implementation and increasing public fear and distrust. Severely eroded trust and confidence in an agency raises the probability of policy failure. Can such failure be avoided, and if so, how?

In this chapter, I focus on high-level nuclear waste disposal in the United States as a case study in democratic approaches to risk management with implications for other siting controversies. I draw from some of the most detailed and rigorous studies available of public reactions to nuclear waste issues (Dunlap, Kraft, and Rosa 1993) to describe the public's views of repository siting, the policy consequences, and alternatives for the future. A major premise of this analysis is that highly negative public perceptions of nuclear waste do not preclude democratic solutions to repository siting. They do, however, call for unconventional approaches to public participation and a serious and sustained effort to rebuild public trust in government agencies responsible for the waste programs.

THE POLITICAL CONTEXT OF PUBLIC OPPOSITION

Communities across the United States and in other nations have shown a remarkable ability to frustrate the location of diverse facilities that residents believe will create adverse impacts on their health and well-being, property values, or environmental quality. Whether we call the public's action a NIMBY (Not in My Backyard) response or something else, the political dynamics and policy consequences are familiar. However worthwhile the facility might be for the state or nation, it is rational for a given community or neighborhood to object, especially because highly vocal and persistent opposition has a low cost and a high rate of success. The short-term effect of this opposition is political gridlock, forcing project proponents to look elsewhere for a more receptive community or to reformulate their proposals until they meet with community acceptance.

The negative response by communities is partially explained by the distribution of costs, risks, and benefits associated with the proposed facility. The costs and risks are geographically concentrated, while the benefits accrue to a larger, more dispersed population. It makes sense locally to just say no. Yet there is more than a rational calculus of costs and benefits at work. Particularly in cases involving hazardous and nuclear wastes, experience suggests that the public's reaction may be largely unaffected by provision of scientific studies of risk, assurances of safety offered by siting proponents, pledges to uphold environmental and other laws, and even offers of compensation that seemingly balance the local costs and risks (Brion 1991; Portney 1991). Underlying the public demands for exceptional levels of site remediation as well as the strong resistance to new facilities is a well-documented fear of chemical and nuclear wastes (Slovic 1987). Distrust of project proponents compounds the problem by constraining an open dialogue on the issues.

Conventional descriptions of NIMBY-style reactions also fail to do justice to policy history, personal experience, and the attitudes, perceptions, and motivations that spur citizens to adopt an oppositional stance in these cases. Absent from such accounts is any recognition that contextual variables matter in shaping the way in which the public perceives a project and hence its response to it, or that those variables affect the feasibility of policy responses that might be taken to break a political stalemate (Kraft and Clary 1991).

Case studies of community opposition and systematic survey research suggest a short list of variables that seem to affect the type and extent of public reaction to hazardous and nuclear waste facilities (Krimsky and Plough 1988; Piller 1991; Goldsteen and Schorr 1991; Hadden 1991; Dunlap, Kraft, and Rosa 1993). These include

- the magnitude of environmental and health risks posed by the facility—including the reliability of risk estimates and the degree of consensus among technical experts;
- public perception of environmental and health risks and other facility impacts on the community;
- the record of performance and the reputation of government agencies or other project proponents, particularly their perceived credibility and competence;
- the extent and nature of media coverage and the salience of the issues;
- the extent, form, and timing of opportunities for public involvement in decision making;
- the political skills and leadership abilities of community activists and project proponents.

Not only assessments of community opposition to facility siting and its effects but also the development of effective conflict resolution strategies would benefit from exploration of such propositions, whether through analytic case studies, survey data, or other methods.

The case of HLRW disposal in the United States offers a suitable means to examine some of these questions. For this purpose, I provide a description of the Nuclear Waste Policy Act (NWPA) of 1982 and its implementation to date. I then focus on public views of radioactive waste and the basis of public opposition to the implementing agency, the Department of Energy (DOE).[1] The U.S. experience with high-level waste illustrates why the standard approaches to citizen involvement and risk evaluation often fail and suggests opportunities for promoting democratic dialogue and ensuring greater success in the future in this and related siting controversies.

U.S. NUCLEAR WASTE POLICY: FROM POOR DESIGN TO FLAWED IMPLEMENTATION

HLRW is derived principally from civilian nuclear power plants in the form of spent fuel. A substantial volume comes from military operations, particularly the production of plutonium for nuclear weapons. HLRW is an inevitable by-product of the generation of electricity from nuclear fission. It consists mainly of irradiated thirteen-foot fuel rods bundled together into fuel assemblies. Each reactor generates an average of thirty tons of spent fuel each year. The fuel rods contain uranium oxide pellets with largely decayed or depleted uranium U-235, nonfissionable U-238, and a mixture of radioactive fission by-products of widely varying levels and types of radioactivity and half-lives. The combination is highly dangerous to living organisms and thus must be isolated from the biosphere and human contact for thousands of years.

HLRW should be distinguished from low-level radioactive waste (LLRW). The radioisotopes found in LLRW are far less potent than those in high-level waste; they also have relatively short half-lives. They are nevertheless potentially injurious to living organisms. They need to be contained for anywhere from several days to 500 years. LLRW includes irradiated tools and other equipment, clothing, ion exchanger resins, medical and pharmaceutical waste, and assorted residues from nuclear research and defense activities.

HLRW differs from other hazardous wastes in the degree of risk posed (both actual and perceived) and in the technical standards set for repository siting, which in turn create ample scientific uncertainty to fuel public and state opposition to a repository. Driving the process is the recognition that civilian nuclear power has no future without an acceptable waste disposal policy. Hence, electric utilities and the nuclear power industry are intensely interested in the development of waste repositories. Opponents of nuclear power are equally likely to question the safety of proposed facilities.

Waste Storage and Disposal

Since the 1950s, spent fuel from commercial nuclear plants has been stored in water-filled basins at each reactor site. Such storage was to be short term to allow heat to dissipate, after which it was to be reprocessed to separate usable plutonium and uranium from fission by-products. The former was to be reused and the latter disposed of permanently. But fuel reprocessing in the United States never developed as planned, and government support for it was ended in the late 1970s under President Jimmy Carter. Public officials worried that the plutonium might find its way to other countries and lead to the proliferation of nuclear weapons (Carter 1987). In the United States, all spent fuel is considered to be high-level nuclear waste and eventually is to be disposed of in geologic repositories, a method of permanent disposal almost universally seen as preferable to the alternatives.

Although continued water storage at reactor sites presents no insurmountable technical or safety problems that have not already been dealt with, it is far more expensive than other forms of storage. Thus, it is not a practical option for the long term. As power plants run out of storage space, which is increasingly occurring in the 1990s, and as reactors are decommissioned, alternative storage and disposal strategies will be needed. These include reliance on either another form of temporary storage—for example, dry storage in steel and concrete casks on-site or shipment of waste to a federally approved Monitored Retrievable Storage facility—or disposal in geologic repositories. Electric utilities in Virginia and South Carolina already use dry-cask storage, and others anticipate use of this method in light of the expected delay in opening of a repository. In the end, protection of public health and the biosphere, as well as security of the waste itself and its plutonium content, will require disposal in geologic repositories.

Estimates of the total amount of high-level nuclear waste vary, but the stock of spent fuel rods (the major component of civilian high-level waste) in the United States in 1993 totaled about 25,000 metric tons; about 2,000 metric tons are added each year (U.S. NWTRB 1993). High-level waste currently is stored in thirty-three states at over seventy sites and at three DOE nuclear reservations in South Carolina, Washington, and Idaho (U.S. GAO 1992). DOE, which has managerial authority for radioactive waste from civilian nuclear plants, estimates that the total quantity of waste will reach some 40,000 metric tons by 2000 and 80,000 metric tons by 2030 if no additional nuclear generating plants are ordered; the amount would be greater if additional nuclear plants are constructed. Large quantities of waste are accumulating in other nations reliant on nuclear energy, with the annual global production estimated at 8,000 tons a year. Should present and planned nuclear plants continue to operate, the world would require one radioactive waste repository (holding 70,000 tons) *every nine years* (Häfele 1990, 97). That scenario indicates the value of examining U.S. experience to learn from this nation's mistakes.

Policy Goals and Means

After decades of complacency over waste issues, followed by a comprehensive study by the Carter administration's Interagency Review Group, the U.S. Congress in 1982 enacted the Nuclear Waste Policy Act (NWPA). It established an elaborate national program for the permanent disposal of high-level commercial and military radioactive waste in mined geologic repositories, to be managed by the newly created Office of Civilian Radioactive Waste Management within DOE.[2] The U.S. Nuclear Regulatory Commission (NRC) and the federal Environmental Protection Agency (EPA) set regulatory standards for waste disposal, with DOE serving as the lead implementing agency.

The original 1982 act mandated that DOE survey multiple potential sites nationwide, using a rigorous set of technical guidelines, and submit environmental assessments of these sites for review and comment through a public hearing process. Congress expected DOE to consult closely throughout the process with states or Indian tribes that might be affected by the repository's location. Indeed, DOE was to enter into binding, written agreements on screening and characterization studies with the states the president recommended as candidate sites. In addition, it was to provide funds to states and Indian tribes to assist them in conducting their own reviews of the sites and to hire consultants.

This combination of careful scientific site review and extensive "consultation and cooperation" with affected parties was a critical feature of the original policy design. It was expected to build public confidence in the siting process and thus promote acceptance of the final decision on a repository location. Such acceptance by the public was widely understood to be essential for program success, although the degree to which it was appreciated by members of Congress and DOE officials is unclear.

NWPA called for DOE to recommend final sites to the president by 1985 and 1989 for a first and second repository, respectively, and the first repository was to begin accepting spent reactor fuel by 1998. That early date reflected optimism that the process as designed would work. NWPA also provided that a state may halt development of a repository after the president recommends it by formally notifying Congress of its disapproval, subject to a two-house override (Carter 1987).

The act seemed to be a compromise between rapid siting and public involvement, one that could unite divergent values, but in retrospect it fell short. By including conflicting policy goals and failing to specify how they were to be reconciled, DOE was handicapped from the beginning. In particular, Congress did not clarify the precise role to be played by the states and the public, and it saddled DOE with a wildly unrealistic siting schedule that undercut its efforts to promote public involvement and acceptance of the repositories (Kraft 1991, 1992).[3]

Public Participation and Opposition to Repository Siting

The opportunities for public participation in the repository-siting process were impressive on paper. Yet the form and timing of most of this citizen involvement as well as DOE's overall implementation of the policy were unlikely to help the federal government build the requisite public confidence in the process. One dubious assumption was that holding hearings near each candidate site after completion of initial assessments would foster public acceptance. The reality of public opposition became evident when DOE reached the stage of site-specific evaluations and the siting process came to resemble an old standard: "Decide, Announce, and Defend."

In the spring of 1983, the department identified nine candidate sites in six states in the West and South for the first of the two repositories authorized by NWPA. Texas, Utah, and Mississippi each had two sites. Nevada, Washington State, and Louisiana had one each. Under NWPA, DOE was to conduct environmental assessments for each of these nine sites. DOE released those assessments in December 1984 and narrowed the list of potential sites to five locations: Yucca Mountain, Nevada; Hanford, Washington; Deaf Smith County, Texas; Davis Canyon, Utah; and Richton Dome, Mississippi. Over 20,000 public comments were received on the assessments, an unmistakable sign of the saliency of the nuclear waste issue in these states (Clary and Kraft 1989).

By February 1985, Secretary of Energy John Herrington tentatively identified three sites as the only ones slated to undergo characterization, or a fuller study of technical suitability. The three were Yucca Mountain, Deaf Smith County, and Hanford. By this time, Nevada, Washington State, and Texas, as well as several environmental groups, had filed suit against DOE, challenging the legality of the designation process. The secretary persisted with his recommendation of these three sites in an official announcement on May 27, 1986, even though departmental analysis had ranked the candidate sites differently. The failure to select the more highly rated Richton Dome was seen by some as evidence that the secretary considered the potential for political opposition at each site. So, too, was a shared characteristic of the final three sites. Each had a history of association with nuclear technologies, which conceivably would make the surrounding communities more receptive than others to the repository. For over forty years, Hanford was the location of a major nuclear-weapons production facility that employed thousands of local residents supportive of nuclear technology. Yucca Mountain was adjacent to the Nevada Test Site, and Deaf Smith County was home to the Pantex assembly plant.

Disagreement with DOE within these three states and in the other semifinalist western states centered on the environmental assessments conducted under NWPA. The states and environmental groups found the assessments wanting (Clary and Kraft 1989). Criticism came from many other quarters, however, including the National Academy of Sciences Board on Radioactive Waste Management and the NRC. These arguments against the assessment process focused on omissions and other deficiencies in the data used, the methodology for site evaluation and comparison, and DOE bias in the evaluation process. For example, the National Academy's Board on Radioactive Waste Management found the initial methodology of comparative site assessment "unsatisfactory, inadequate, undocumented, and biased" and called for significant changes, particularly in the use of independent experts. DOE did revise its methodology to some extent, but it rejected a major role for outside experts, leaving its work subject to much the same criticism—that it was subjective and biased, and masked genuine uncertainties about long-term risks (Colglazier and Langum 1988, 327–28). In brief, the states and others would not accept DOE's site analyses as scientifically valid. This response would be repeated frequently over the next decade.

Controversies similar to those in the West emerged during the second round siting process in the East. In January 1986, DOE announced that 12 locations in seven states in the upper Midwest and East had been selected from its original list of 235 potential sites in seventeen states. Once again DOE defended its choices in terms of the official siting guidelines. The potential states had been scored and compared across nine attributes and were screened for disqualifying conditions such as high population density and proximity to national and state parks. That process led to a proposal to study further the final sites in Wisconsin, Minnesota, Maine, New Hampshire, Virginia, Georgia, and North Carolina. As in the West, however, the ostensibly scientific selection process depended on the credibility of the supporting studies. Residents of the finalist states were unimpressed. DOE prepared for public review a draft Area Recommendation Report, a preliminary environmental assessment. During three months in early 1986, DOE conducted thirty-nine briefings and thirty-eight public hearings in fifteen states on the report. Some 18,000 people attended these sessions, and another 3,200 individuals and organizations provided written comments; a total of some 60,000 comments (as DOE classifies them) were received (U.S. DOE 1987, 2). Few citizens liked what they saw in the report, and even fewer were prepared to support DOE's repository-siting plan (Kraft and Clary 1991, 1993).

Intense public opposition and fears of political repercussions in an election year led the Reagan White House in late May 1986 to halt the siting process in the East. Chances for successful implementation of NWPA diminished further when the western states objected strenuously to what they viewed as inequitable treatment that left them vulnerable to imposition of a repository. As a result of these and other criticisms and the continued pressure from a nuclear industry eager to have an operational repository, Congress was forced to revise NWPA in 1987.

Congressional "Redirection" of the Waste Program

Facing sharply conflicting proposals to declare a siting moratorium and study the issues further (favored by Representative Morris Udall, D.-Arizona), or to press ahead with a "sequential" siting process that would examine only a single site, Yucca Mountain, Nevada, Congress chose the latter (Carter 1989). Senator J. Bennett Johnston (D.-Louisiana) was the chief congressional advocate of the sequential process that would largely maintain the momentum of the siting process. He was persistent and aggressive in persuading his colleagues in Congress to accept that course of action as preferable to the alternatives (Kraft 1992). The Nuclear Waste Policy Amendments Act of December 1987 canceled authorization for the second repository, directed DOE to evaluate the Yucca Mountain site, and stipulated that the department cease consideration of the other western sites in Texas and Washington State. DOE was to report to Congress between 2007 and 2010 on the need for a second repository. The amendments also authorized a Monitored Retrievable Storage (MRS) facility (for temporary storage of waste), linking its construction to completion of the repository itself.

Congressional sponsors, particularly Senator Johnston, defended the 1987 act as a cheaper and potentially more rapid route to repository siting than the 1982 act—since only a single site would have to be studied in depth. It also was attractive politically because the other forty-nine states were disinclined to criticize DOE or the program and thus risk reopening the issue of repository location. But by authorizing a single repository in a specific location, Congress eliminated two of the most positive features of the original act: the distribution of risk between the East and the West (implicit in the two-repository specification) and a site selection process that was, at least plausibly, systematic and objective, as opposed to openly political.

The Yucca Mountain Strategy: Public Opposition Persists

Much of the optimistic prognosis for the 1987 act depended on the willingness of Nevada to cooperate with DOE. As it turned out, six years of bitter relations between the state and DOE followed, which many attributed to Nevada's "wounded psyche" at being singled out for the repository. Few state officials believed that cooperating with DOE was in their political self-interest, which suggests that the logic behind Congress's grant of modest financial compensation as a supposed inducement for cooperation by the host state was flawed. Indeed, state officials, particularly the governor, have led the opposition, and understandably so. Surveys indicate that state residents oppose the repository by about four to one (Flynn and Slovic 1993).

Economic incentives, at least in the form approved by Congress, did little to induce state cooperation with the federal government. They also were offered late in the siting process, after the mood in Nevada had turned negative. Thus, compensation was unlikely to help build public and state support for the repository. The 1987 act authorized payment of $10 million a year to the host state during site characterization and licensing and $20 million a year during operation of the repository. The latter was far less than Senator Johnston's original proposal of $100 million per year. These amounts were also contingent on Nevada's abandoning its right of disapproval, an unappealing offer to the state. In a separate form of compensation, Nevada received about $15 million per year for its oversight activities and for support of affected counties near the site to oversee DOE activities at Yucca Mountain and to mitigate the effects of the site characterization or potential development of the repository. Nevada used these funds to commission elaborate socioeconomic studies and technical assessments that allowed it to confront DOE on virtually every conceivable repository issue. In response, Congress sharply reduced funding in 1993.

Controversy in Nevada has centered on conflicting estimates of the risks of repository construction at Yucca Mountain, located about 100 miles northwest of Las Vegas, and the adequacy of DOE's 6,300-page site characterization plan (see table 1). The scale of the characterization and the uniqueness of the repository itself inevitably invited skepticism toward DOE site evaluation efforts, contributing to state opposition. Key disputes have focused on scientific findings related to the site's technical suitability for containing the waste over the exceptionally long times required under federal standards. Nevada has used significant uncertainty over the risks posed by the facility to regularly dispute DOE decisions on the acceptability of the Yucca Mountain site. Moreover, project opponents in the state argue that

TABLE 1. Characteristics of the High-Level Nuclear Waste Repository at Yucca Mountain

Location:	100 miles northwest of Las Vegas
Size:	100 miles of tunnels excavated in volcanic rock 1,000 feet below the surface of the mountain. Underground mined area will total 1,400 acres. Aboveground area will be 150 to 400 acres, with a 3-mile control perimeter.
Amount of waste in repository:	70,000 metric tons
Sources of waste:	Civilian nuclear power plants and defense waste. Defense waste should total 8,000 to 9,000 metric tons.
Period of containment of radioactivity:	10,000 years under 1985 EPA standards
Cost of characterization:	$6.3 billion (1991 est.)
Length of characterization:	6 years (DOE) or 11 to 19 years (GAO 1993 est.)
Number of people involved in characterization:	2,000 (1,700 of whom are nongovernmental contractors)
Transportation to site:	Waste canisters placed within large metal transportation casks certified by NRC and shipped by trucks and rail to the site. Estimated shipments: 28,000 by highway and 10,000 by rail.
Total cost of repository:	$26 billion (1990 est.) for one; $34 billion for two
Management of repository:	Office of Civilian Radioactive Waste, U.S. Department of Energy

DOE's scientific investigation of the site cannot be unbiased given political and organizational pressures on the department to find the site to be acceptable. Nevada has argued that a finding of acceptability is, in effect, predetermined and that the rules have been altered to favor such a conclusion. Lending credence to some of those suspicions, in 1989 DOE abandoned its initial two-year study of Yucca Mountain after concluding its technical quality was unacceptably low. The combination of scientific uncertainty and lack of confidence in DOE-sponsored investigations makes it unlikely that any amount of study or reassurance by DOE scientists can convince opponents of the repository's safety. This is particularly so now that trust in DOE has all but disappeared.

Relations between Nevada and DOE deteriorated steadily between 1987 and 1989, when the state legislature, in effect, vetoed placement of the repository at Yucca Mountain and refused to issue environmental permits required for site characterization to proceed. In 1990, the Ninth Circuit Court of Appeals ruled against Nevada and backed DOE's assertion that the state veto was premature and illegal. The Supreme Court declined to review the case. Nevada issued the permits, but the state has continued its concerted campaign against the Yucca Mountain repository, contributing in some ways to a delay in site characterization. That stance has led a good many observers, including a National Research Council panel, to conclude that the repository program was unlikely to succeed because of public fears, state opposition, and deficiencies in program design and management. These deficiencies include a high degree of inflexibility in the work schedule, rigid technical specifications, and unreasonable regulatory standards (National Research Council 1990; Flynn et al. 1992).

In many respects, the repository itself is not an especially difficult engineering challenge, even if uncertainties over containment of radioactivity remain. The repository is designed to contain over 100 miles of tunnels excavated in volcanic rock 1,000 feet below the surface of the mountain but above the water table. Waste canisters, which are to be removed from special transport casks that are used to get them to the site, would be emplaced in the repository and monitored for up to 100 years. After that time, the repository would be backfilled and sealed permanently, and marked to warn future generations. The canisters are expected to contain the waste for 300 to 1,000 years. Containment of wastes would then depend on the stability of the volcanic rock. The most important environmental risk is possible leakage of radionuclides into groundwater and thus into the outside environment. Other risks are associated with transporting such a large quantity of waste to Yucca Mountain; one estimate puts the total at 28,000 shipments by highway and 10,000 by rail.

If the basic engineering of the repository and the handling of waste containers are unexceptional, the costs of both site characterization and repository construction have become important issues. They raise questions about the economic justification for a geologic repository in comparison with alternatives such as aboveground storage facilities (Keeney and von Winterfeldt 1993). Originally estimated in 1982 to cost some $200 million, the cost of characterization for this one site was projected by DOE in 1991 at $6.3 billion. In May 1993, the GAO estimated that given the pace of work at the time, characterization would likely take at least 5 to 13 years longer than the original 6-year estimate and cost more than DOE calculated (U.S. GAO 1993). In late 1988, DOE believed that site characterization would be completed and a license submitted to the NRC by 1995 and that the repository would be ready to accept waste by 2003. One year later, the secretary of energy revised the schedule, estimating that operations would begin in 2010. The GAO has estimated that the repository will not be available until 2015 at the earliest, and possibly not until 2025. In 1990, DOE projected the full cost of the repository program to be about $26 billion (in 1988 dollars) for a single repository and $34 billion if two were constructed (U.S. GAO 1992, 8, 20). NWPA mandated that a Nuclear Waste Fund, supported by fees imposed on producers of electricity, cover the costs. Utilities contribute about $600 million per year to the fund, with consumers ultimately footing the bill. With the enormous investment they have made to date, utilities are anxious to see Yucca Mountain stay on track for the earliest possible opening. In the meantime, they are actively seeking short-term solutions that will relieve them of the burden, economic and political, of continuing to store spent fuel rods at reactor sites.

The scope and cost of site characterization reflect both the status of the project as the first of its kind and the demanding standards such a repository would have to meet under EPA and NRC rules. The guidelines call for isolation of the waste from the biosphere for 10,000 years, twice the length of recorded human history. In the Energy Policy Act of 1992, Congress required the EPA to establish new standards for environmental protection at the Yucca Mountain site to replace standards the EPA adopted in 1985, portions of which were struck down by a federal appeals court in 1987. Congress also charged the National Academy of Sciences (NAS) with the task of advising EPA on the standards. Opponents of Yucca Mountain, including Senator Richard Bryan (D.-Nevada), argued vociferously against the new standards as yet another attempt to weaken environmental and health safeguards at the site and ease construction of the repository. The NAS Committee on Technical Bases for Yucca Mountain Standards was to consider the 10,000-year standard, the possibility of dry-cask storage for 100 years, and assumptions made about institutional control of the repository site over time, among other issues.

In the same 1992 act, Congress asked the NRC to modify the licensing criteria for Yucca Mountain on the assumption that DOE will continue to monitor the facility indefinitely after closure. The act required that DOE serve as a permanent custodian for the repository by establishing a "continuous institutional presence" that would prevent any future human intrusion into the site and thus any exposure to radiation. Previously, the NRC assumed that DOE would monitor the site for no more than 100 years.

Taken together, the 1992 congressional specifications were a clear attempt by supporters of the Yucca Mountain strategy to further limit opposition to the repository unless extraordinary evidence of environmental risks could be assembled. It is less evident that such efforts finally will lead to successful repository siting. DOE still faces a considerable hurdle in obtaining a license from NRC to operate the repository—which is required for repository construction to begin. And the department continues to face intense public and state opposition that may yet derail the siting process.

Another policy design element that is proving to be less helpful to the siting process than supporters imagined it would be concerns outside technical reviews. Congress expected confidence in the waste program to be bolstered through NWPA's provisions for external oversight. There were to be regular reviews by the NRC and studies by the NAS and by a Nuclear Waste Technical Review Board appointed by the president. While these reviews and studies are broadly supportive of the policy goals, they have frequently underscored both weaknesses in DOE's management of the program and technical uncertainties that have contributed to loss of public confidence in DOE and in the repository program itself. DOE's own actions and its organizational culture have reinforced these tendencies. As discussed below, DOE's low credibility is a major reason for continuing public opposition to the repository. These conditions indicate that the most intractable siting issues are more political and institutional than technical.

Policy Learning at DOE?

In recent years, scholars have suggested the importance of organizational or policy learning as a critical ingredient in improving implementation (Mazmanian and Sabatier 1983; Sabatier 1987). Programs that have proved unsuccessful may be resurrected and redirected, and eventually achieve their intended goals. Whether this occurs depends greatly on the actions of policy entrepreneurs, within and outside government, who are attentive to new studies and information as well as to opportunities for policy change. In the present case, such policy learning has been limited by DOE's seeming incapacity to change its way of doing business. The department has long suffered from a variety of what might be termed organizational pathologies that date back to its predecessor agencies: the Energy Research and Development Administration and especially the Atomic Energy Commission. In its various incarnations, DOE has tended to promote nuclear technologies uncritically and minimize or misrepresent the associated uncertainties and risks of those technologies. It has long operated in relative secrecy, immune from public scrutiny—a legacy of the Cold War environment in which its nuclear-weapons production work took place (Gerber 1992; Rosa and Freudenburg 1993).[4] In addition, DOE has given little credence to social-science knowledge in assessing nuclear waste programs (Bradbury 1989). It has been far more likely to focus on scientific and engineering knowledge. As a result, it may have deprived itself of precisely the kind of information it needed to improve program effectiveness.

It is difficult to judge how much change has occurred recently in DOE's approach to the repository program. Over the past several years, it has promised to give more weight to so-called "institutional" issues, such as state and public relations (Isaacs 1989), and President George Bush's secretary of energy, James Watkins, strongly prodded the department to deal more openly with health, safety, and environmental issues that it had long ignored. Yet DOE's critics doubt that much organizational change has yet occurred or is likely in the near term (Cook, Emel, and Kasperson 1990; Rezendes 1991). Even Admiral Watkins left office convinced that the department's organizational culture remained a significant barrier to its implementation of remediation and waste disposal programs.

Some developments, however, are moderately encouraging. Following a workshop held in November 1990 on problems in the repository program, DOE pledged to consider public trust and confidence in program decisions, assign equal importance to institutional and technical activities, and allow affected governments and interested parties to be involved in decision making. DOE wrote those commitments into its revised mission plan for the repository program (U.S. GAO 1992, 47). Questions remain, however, about DOE's ability to meet these "strategic principles" to the extent necessary for securing public approval. In this regard, three other actions are worth noting: the Nevada Initiative, the secretary of energy's task force, and creation of the Office of the Nuclear Waste Negotiator.

The Nevada Initiative

One indicator of a persistent tendency of the nuclear industry (and DOE) to discount public and state opposition and to seek solutions through other kinds of interventions became evident in the fall of 1991. The industry launched a large-scale public-relations campaign dubbed the Nevada Initiative. It spent a reported $800,000 in 1991 alone and laid plans to spend $9 to 10 million over three years. The American Nuclear Energy Council (ANEC), an industry trade association, asserted that the campaign was intended merely to correct misinformation about nuclear waste and the Yucca Mountain site and to educate the public on the issues. Confidential documents leaked by an antinuclear group, however, made clear that ANEC and the advertising and political consulting firm it hired had a much broader and more explicitly political agenda. They hoped to counter public opposition through carefully crafted advertisements touting the safety of nuclear power and the repository, reinforced through endorsements by DOE scientists. Their plan predicted that a majority of Nevada's citizens could be persuaded to support the site. ANEC and its consultants also envisioned that such artificially inflated public support would force state officials to cooperate more with DOE—for example, in issuing work permits needed for site assessment (Dunlap, Kraft, and Rosa 1993, 312–15; Flynn, Slovic, and Mertz 1993, 499–500).

Such beliefs about the potential effects of a public information campaign are at odds with decades of social-scientific research that indicates public attitudes toward nuclear power and radioactive waste are far too negative and deeply held for such a campaign to affect them (Flynn, Slovic, and Mertz 1993, 501). The impacts of the ANEC public-relations effort were what one would expect given the highly adversarial climate in which consideration of the Yucca Mountain site has taken place. A statewide telephone survey of Nevada residents, conducted a few weeks after the start of the Nevada Initiative, indicated that despite good coverage (some 72% of Nevada residents reported having seen or heard the advertisements), the campaign did not change many minds. More than half of the respondents said the advertisements did not alter their negative opinion of the repository, and the extent of repository opposition among state residents remained unchanged. About half of the respondents who had seen the advertisements said they did not believe them, and two-thirds reported negative comments about them. Fewer than 20% of respondents offered positive or moderately positive comments. As might be expected, citizens who already were predisposed to favor the repository were likely to offer positive comments about the ads, and those opposed to the repository were likely to offer negative comments. For these reasons, the information campaign seems to have had only a minor effect on public views of the proposed repository (Flynn, Slovic, and Mertz 1993).

The political-strategy documents leaked in late 1991 had an important effect on subsequent public-education efforts in the state. Following a period of public outrage and ridicule and of intense criticism from state officials who resented ANEC and DOE attempts to pressure them in this manner, ANEC canceled the initiative in the spring of 1992. A survey conducted in June 1992 indicated that opposition to the repository remained constant at about 75%. Yet only 3% of respondents said they were more trustful of the repository program, whereas nearly 41% were less trustful (Flynn, Slovic, and Mertz 1993, 500). Such evidence hints at some obvious limits to public-education programs that are designed and implemented in a manner similar to the 1991 Nevada Initiative. That lesson was not lost on ANEC. In the fall of 1992, it began a new and more subdued information campaign with a series of low-key TV spots that avoided the excesses of the 1991 effort. Whether it will be any more successful remains to be seen.

The Secretary of Energy's Task Force

In what seemed to be a more promising venture, in April 1991 DOE created a high-level Task Force on Radioactive Waste Management under the auspices of the secretary of energy's advisory board. Its purpose was to study how DOE might "strengthen public trust and confidence" in the repository program. The task force met seven times over twenty-one months and heard from nearly 100 spokespersons from various organizations and state and local governments. It also commissioned independent studies and contracted with NAS and the National Academy of Public Administration to assist in "designing and leading trust-evoking organizations."

The task force issued its final report in November 1993 (U.S. Secretary of Energy, 1993). It concluded there "is widespread lack of trust in DOE's radioactive waste management activities," and it characterized this distrust as "not irrational" and "[not] a manifestation of NIMBY." It predicted that the distrust "will continue for a long time" and that organizational restructuring and other incremental changes are "not likely to affect the level of public trust and confidence" in DOE. Rather, it said, "what is required is a full commitment from senior managers of the Department and its contractors coupled with improved institutional capacity to design, implement, and evaluate measures for strengthening trustworthiness." The task force found no consistent and effective approach to restore trust and confidence within the waste management office in DOE, but it offered detailed recommendations for what the department should do to improve the situation. These focused on open, frequent, and sustained interaction with affected parties and on increased DOE competence and management capacity. The task force concluded that if adopted, such actions could "provide a sufficient basis for DOE to show it is *worthy of trust*."

It is not yet clear what effects the report will have on nuclear waste policy and DOE decision making under Hazel O'Leary, secretary of energy in the Clinton administration, and over a longer term. At the request of Nevada officials, in June 1993 O'Leary endorsed an independent review that was to focus on financial aspects of DOE's Yucca Mountain project. She also requested a review of major published reports that were written between 1989 and 1993, that were critical of DOE management of the waste program (Thurber 1994). In a series of actions in late 1993 and early 1994 related to secret Cold War–era radiation experiments that began receiving extensive coverage by the nation's news media, O'Leary promised more open decision making at DOE on a range of issues, including environmental-restoration efforts at DOE facilities and the waste disposal program itself. She acknowledged that one purpose of such efforts was to improve the public's trust and confidence in a department long accustomed to a culture of secrecy and often seemingly oblivious to the views of its critics (Schneider 1994).

Although on record as favoring an acceleration of the repository program, O'Leary quickly learned how intractable the program had become. One example was a Keystone Center stakeholders meeting she arranged in 1993, with government officials, nuclear power industry and electric utility executives, environmentalists, and Native American representatives. The issue proved too contentious for negotiation. The same thing occurred at a public meeting DOE arranged in Las Vegas on August 10, 1993, to develop a consultative process for the Yucca Mountain project. The workshop resulted in no immediate agreement, but recommendations were transmitted to O'Leary to consider in her ongoing review of the waste disposal program (Kriz 1993).

The Office of Nuclear Waste Negotiator

In another important development, in August 1990 the Senate confirmed President Bush's nomination of David H. Leroy, former lieutenant governor of Idaho, as the nation's first nuclear waste negotiator. The amendments act of 1987 created the independent Office of the Nuclear Waste Negotiator in the executive branch. The act authorized the negotiator to identify states or Indian tribes willing to host either the temporary MRS facility or the permanent repository, and to negotiate with governors and governing bodies of the tribes to determine the conditions under which they would assume the role of host. DOE had hoped that the MRS facility would allow it to meet the 1990 statutory deadline under NWPA for acceptance of waste from nuclear power plants. In 1992, Congress extended authorization for the negotiator's office through January 1995.

Some progress has occurred in persuading Indian tribes and other governmental entities to take an interest in the MRS facility (though not in the permanent repository). They receive up to $100,000 for applying to study the potential for the facility. If they decide to continue the study, they are eligible for up to several million dollars in federal grants to assist them in making a final environmental assessment. In some states, however, applying for even the initial study grant was politically dangerous. In Grant County, North Dakota, in 1991, voters recalled all of the county commissioners after they applied for and received two study grants. Even with such political constraints, by fall 1993 the negotiator's office had received about thirty applications, chiefly from Indian tribes. In August 1993, the Mescalero Apache tribe in southern New Mexico and the Skull Valley Band of

Goshute Indians in Utah were the first of the tribes, counties, and other governmental jurisdictions to move to the final stage of negotiation by requesting third study grants of $2.8 million each.

Leroy (whose resignation was accepted by President Bill Clinton in June 1993) and others have defended the negotiations as a fair and open dialogue with the American public consistent with a voluntary siting process. Yet critics have voiced concern about several elements of this process (Leroy and Nadler 1993; Lippman 1992). Chief among these concerns is a belief that emphasis on tribal lands may conflict with norms of environmental equity or justice. Issues of environmental justice have assumed increasing importance in the siting of hazardous waste facilities in the United States after several scholarly studies and an EPA report (U.S. EPA 1992) that found such sites were located disproportionately in poor and minority communities.

The circumstances in the case of high-level nuclear waste are different, particularly in the voluntary nature of agreements to study tribal lands for possible sites. The repositories are not being imposed on any community or tribe, and no jurisdiction is obliged to accept them. In a paper prepared for an October 1993 International Nuclear Congress and reflecting on his experience as negotiator, Leroy noted that to guide its efforts, the negotiator's office adopted ten principles that reflect such a philosophy: the process must be voluntary; participation does not constitute a commitment to proceed any further; any dialogue is terminable by the volunteer for any reason; study grants and other financial assistance will be provided to jurisdictions that wish to evaluate facilities technically; discussions will focus primarily on safety and secondarily on benefits; the volunteer can make its own choices about technology, ownership, and control; all issues will be considered relevant; a prospective host is entitled to receive benefits; the process must be open to broad public participation and is to be conducted in full view of citizens and the media; and the views of all stakeholders—opponents as well as proponents—must be considered. Such principles share much in common with the well-known "facility siting credo" that characterizes a broadly participatory and voluntary siting process.[5] Leroy noted that such a "voluntary process of siting these facilities has become a worldwide norm and a worldwide necessity."

Assuming that such principles do indeed characterize the negotiator's relationship with possible host communities, issues of environmental equity may be less important in this instance than might initially be supposed. While it is possible that offers of generous benefit packages (especially in poor communities) unduly constrain voluntary acceptance of risk, such a bias cannot necessarily be presumed. Ultimately, whether a siting decision reflects the community's collective interest depends on the extent of democratic dialogue and public participation in the decision-making process. There is no reason to suppose that Indian tribes are any less capable of making such choices than other governmental entities, since much evidence now suggests citizens possess the requisite capacity to judge acceptable risks (discussed later). According to press accounts, little vocal opposition to the interim-storage facility existed within the Mescalero tribe, at least in 1993, and the elected tribal council was greatly impressed with the economic benefits of hosting the facility (Wald 1993b).

In November 1993, the U.S. Senate confirmed President Bill Clinton's nominee for negotiator, former congressman Richard Stallings. Stallings promised to continue negotiations with the Mescalero Apaches and Skull Valley Band of Goshute Indians even in the face of continuing opposition by the states' governors and congressional delegations (Kriz 1993). In New Mexico, the governor, both houses of the state legislature, both U.S. senators, and all three U.S. representatives have been adamantly opposed to interim storage on the reservation. Reflecting the extent of their opposition, Senator Jeff Bingaman (D.-New Mexico) fought successfully in Congress to delete financing for the Mescaleros' study grant from the fiscal 1994 budget (Wald 1993b). Even with such political opposition outside the tribes, it is not clear whether either state has the legal authority to prevent a sovereign tribe from accepting the repository. The dispute may require resolution in the federal courts. According to DOE, to accept a repository, a sovereign tribe would need the approval only of Congress and the president, not the states.

Perhaps further complicating the issue of environmental equity are recent estimates of the amounts the tribes might receive for agreeing to host an MRS facility. Because of the extremely high cost of on-site waste storage at nuclear power plants (especially in water pools), some nuclear industry and financial experts estimate that the tribes might negotiate a fee as high as $50 million per year on a twenty-year contract (Wald 1993a).[6] The impact on individuals could be substantial. For example, for the 3,400-member Mescalero tribe, such an amount would come to about $15,000 a year per person if paid directly to individuals.

Despite these recent signs of interest in new approaches at DOE, many critics doubt the department will succeed in the face of profound public fear of nuclear waste and continuing opposition to the repository program (e.g., Slovic, Layman, and Flynn 1991; Flynn et al. 1992; Flynn and Slovic 1993). Even the nuclear industry has begun looking for alternatives as some power plants begin to run out of storage capacity and hopes fade for construction of a permanent repository by DOE's new deadline of 2010. Aside from creation of an MRS facility, dry-cask storage is receiving increased attention, and utilities in Michigan and Wisconsin have already petitioned the NRC to allow them to begin preparation for cask storage on site. National environmental groups have favored on-site storage, in part because of the risks of transporting the waste (Kriz 1993). Others, including the secretary of energy, have preferred a centralized storage site. Continuing doubts about the viability of the Yucca Mountain strategy and the rise of local disputes over dry-cask storage led Senator Paul Wellstone and nine other senators in March 1994 to urge President Bill Clinton to create an independent commission to examine the full range of issues related to U.S. nuclear waste disposal policy.

Debate over the future of the Yucca Mountain site and over the U.S. nuclear waste program itself continued through 1995, and it was especially evident in the Republican-controlled 104th Congress. Congress remained divided on the preferred policy direction, with several key committee chairs in the House of Representatives declaring that the Yucca Mountain site was "no longer viable" as a permanent site because of its cost and the continuing political and technical controversies. Yet intensive nuclear industry lobbying sought to counteract that opposition and to build support for construction of a centralized interim-storage facility at the Yucca Mountain site that could begin accepting waste by 1998, as initially stipulated by NWPA. The secretary of energy also announced in late 1994 a new public participation policy that emphasized a new departmental culture of openness and accountability to the public. DOE continued to work on that plan in 1995.

PUBLIC REACTIONS TO NUCLEAR WASTE REVISITED

The case of high-level nuclear waste disposal illustrates the general argument made in the introduction to this chapter that public reaction to facility siting proposals depends on crucial contextual variables. Among the most important in this instance are the unique qualities of a geologic repository and conflicts over estimates of ultra-long-term environmental risks; the record of performance and reputation of DOE as the lead agency, and especially its low credibility among stakeholders; and DOE's institutional capabilities, particularly its very limited abilities historically to involve pertinent stakeholders in the siting process. Many conclusions are possible in a case like this, but one stands out from the others: a different approach to public and state involvement in the process at the earliest stages might have yielded a more positive outcome.

NWPA explicitly recognized the importance of public participation in nuclear waste policy, reflecting a broad commitment to democratic processes in such siting decisions. Many leading members of Congress and other knowledgeable observers remain convinced that successful implementation is possible only with public support for the siting process and public confidence in the implementing agency. Even staunch defenders of the U.S. nuclear waste repository program have acknowledged that the siting strategies to date have proved inadequate because they did not identify or deal well enough with public fears of radioactive waste and they were unable to build sufficient confidence in the siting process (Isaacs 1989; Leroy and Nadler 1993; U.S. NWTRB 1993; U.S. Secretary of Energy 1993).

The American public remains deeply concerned about health and environmental risks associated with radioactive waste and is distrustful of DOE's efforts to build a repository. There is little reason to expect these public perceptions and attitudes to change any time soon. These conditions suggest the importance of improving our understanding of public concerns, their origins, and their persistence. To address such questions, Dunlap, Kraft, and Rosa (1993) assembled some of the most detailed studies available by sociologists, psychologists, political scientists, and economists on the public's multifaceted views of nuclear waste and repository siting issues. A synopsis of this work hints at what might be done to create a more effective siting process for nuclear waste without dwelling on specific policy recommendations. The conclusions should pertain as well to parallel cases in the management of risky technologies.

What is most remarkable is that the findings are consistent even though they emerge from a diverse set of studies that employ different sampling and measuring instruments within different geographic areas at different stages in the repository siting program, from the mid-1980s to the early 1990s. This is especially so for the two major findings: radioactive waste is perceived to be a particularly dangerous threat, and the public has little trust and confidence in DOE's ability to protect it from this threat. Both the high level of perceived risk and the low level of trust in DOE are significant factors underlying the degree of public opposition to the repository program. The convergence of conclusions across eight empirical studies suggests a high degree of validity.

All studies centered on the core question of the psychological, social, and political acceptability of a high-level nuclear waste repository (HLNWR) in the United States. The geographic focus varied from the nation as a whole to small counties and local communities near proposed sites, especially Yucca Mountain. In all cases but one (which used content analysis of public hearing transcripts), samples were drawn in the conventional manner for survey research, which relies primarily on telephone interviews. Some questions were structured and others were open-ended, which provides ensurance that question wording did not shape people's responses.[7]

The principal findings of these studies include the following (Dunlap, Kraft, and Rosa 1993, 296–303):

- Given an opportunity to vote on having an HLNWR in their state, an overwhelming number of citizens *would vote against it.*
- Compared to other environmental hazards, high-level nuclear waste is believed to be one of the *most dangerous.*
- Compared to any other industrial or municipal facility, an HLNWR is the *least desirable,* often by wide margins.
- Images of an HLNWR are so negative that they reveal *deep dread,* revulsion, and anger.
- Perceptions of the risks of an HLNWR are closely tied to *trust* in government and trust in agencies responsible for repository management, especially DOE.
- There is widespread belief in the *likelihood of accidents* at a repository or in the transportation of waste to it, and great concern over such accidents.
- Economic concerns appear in some analyses but are often overwhelmed by concerns over *safety and health,* including that of future generations.
- *Knowledge* is not a consistent factor in citizen opposition to a repository; it is correlated sometimes positively, more often negatively, and sometimes not at all with repository attitudes.
- If a repository were to be sited at Yucca Mountain, Nevada, the city of Las Vegas and the state would likely *suffer substantial economic losses* due to a significant convention decline.
- *Past experience* with other hazardous facilities or with other federal facilities is of only minor significance.
- That *opposition* to an HLNWR *is deep and widespread* is revealed in the fact that almost without exception, perceptions and attitudes toward a repository are unaffected by standard demographic variables.

Especially noteworthy is that perceptions of risk are markedly influenced by the level of trust in institutions responsible for managing the risks (see also Flynn and Slovic 1993; Pijawka and Mushkatel 1991; and Slovic, Flynn, and Layman 1991). There is a strong inverse relationship between trust in government and risk perceptions and attitudes (low trust is associated with perceptions of high risk). Such a finding reaffirms the importance of contextual features in shaping perceptions, and it underscores a common finding in the risk literature. Formal risk analysis, based on a "technical" view of risk and widely used by the nuclear community, is often at odds with "cultural" definitions of risk employed by ordinary citizens (Hadden 1991; Krimsky and Plough 1988). The former relies on statistical measures of the probability of events and their consequences, but it often neglects issues of great concern to the public, such as economic impacts, or minimizes the public's fear of low-probability events, such as accidents. The public's concerns are not superficial but are deeply felt and tied to images of dread related to nuclear power. These characteristics, along with distrust of DOE, suggest the public's perceptions and views will be difficult to change in the short term.

Empirical findings like these also illustrate the importance of social and historical context in shaping public acceptability of the repository. Public attitudes toward radioactive waste are affected by changing perceptions of nuclear power and its safety, which have shifted dramatically (and negatively) since the Three Mile Island and Chernobyl accidents (Rosa and Freudenburg 1993). Moreover, it is difficult to separate current perceptions of the risks associated with nuclear waste from the broader and more diffuse perceptions of government in general and of those agencies charged with disposing of nuclear wastes. Trust and confidence in government have plummeted over the past several decades and were only slightly, and temporarily, reversed early in Ronald Reagan's presidency (Lipset and Schneider 1987). At all levels of government, one is apt to find a "decline of deference" among the public for institutions and professions once respected and trusted (Laird 1989). As O'Hare and Sanderson put it, facility opposition can be traced to the "pessimistic expectations that citizens, industry, and government all hold of each other and themselves" (1993, 375). At the same time, public fear of technological risks has increased, seemingly unrelated to any measurable change in the risks to which people are exposed (Slovic 1987). The combination is deadly for a program such as nuclear waste management.

The implications of these results is that technological fixes alone will not repair the many torn seams in the NWPA. Opposition to a repository is too strong and the public is too distrustful for DOE to have much success in altering public views. A pertinent conclusion was drawn in the summary chapter for the collected studies:

> In this volume citizens from every region of the nation, from every social strata, from every political persuasion have, through the instruments of social science, expressed their views on this question. Against the rigorous evidence provided by them, an unequivocal conclusion emerges: A HLNWR is unacceptable to a majority of Americans for the foreseeable future. Majorities of citizens from Nevada, Texas, Washington state, Arizona, California, and the nation as a whole, while expressing a wide

> variety of views on the siting of a HLNWR, are unanimous about their rejection of such a facility. For people everywhere and in every walk of life a repository is simply, and unequivocally, unacceptable; it is far too risky and those responsible for its management are not to be trusted. (Dunlap, Kraft, and Rosa 1993, 317)

The results from Nevada and Washington State, which are consistent with the general conclusions, are especially noteworthy because these are states with extended experience with, and support for, nuclear technologies.[8]

Another finding is important as well. Individual citizens have demonstrated a willingness and capacity to weigh the importance of social issues and political concerns related to an HLNWR. This was particularly striking in the content analysis of public hearings in the finalist states in the Midwest and East, where the "attentive public" testifying at hearings proved to be fairly knowledgeable about nuclear waste issues, able to discuss specific repository impacts in their areas (e.g., on water quality, tourism, and public health), and prepared to offer alternatives to DOE proposals. This and other analyses indicated opposition to repository siting was strongly related to perceived DOE credibility, a relationship largely unaffected by other variables. The findings are contrary to the dismal image of public knowledge and capabilities found in conventional descriptions of the NIMBY syndrome—and very likely still commonplace among DOE officials (Kraft and Clary 1991, 1993). Such discoveries, however, are encouraging for those who believe enhanced democratic dialogue on the management of risky technologies is indeed possible. They also are consistent with many recent studies indicating the potential for citizen politics in the United States and elsewhere (Berry, Portney, and Thomson 1993; Hill 1992; Ingram and Smith 1993).

As the secretary of energy's task force noted, trust once lost is not easily regained. Yet many of its recommendations to DOE consist of efforts to regain trust—and to avoid further erosion of public support (U.S. Secretary of Energy 1993, 36–46; see also Slovic 1993). It is the enormous difficulty of reestablishing that trust that leads many independent analysts to the conclusion that the only feasible short-term solution to the repository stalemate is an extended moratorium on geologic disposal, perhaps for as long as 100 years (Shrader-Frechette 1993; Flynn et al. 1992). The availability of dry-cask storage on site makes such a moratorium on the repository program possible. The passage of 50 to 100 years should increase the quality of scientific knowledge about long-term consequences of repository siting and may suggest alternatives to permanent geologic disposal. It should also allow sufficient time for the institutional and political changes so necessary for a successful radioactive waste program.

CONCLUSIONS: THE PROMISE OF DEMOCRATIC DIALOGUE

Despite the policy failure this case represents, the experience has suggested promising approaches to siting radioactive waste repositories as well as other facilities posing similar challenges. They are based on the premise that the obstacle to successful siting may not be too much democracy or citizen participation but, rather, too little—or at least the wrong kind. Involving the public, even in well-designed programs, does not guarantee successful siting of facilities. If handled poorly, it could even increase public fear and opposition. The history of nuclear waste disposal, however, indicates that not providing a suitable role for the public leads more assuredly to program failure.

What form of citizen involvement and what other institutional arrangements contribute to success or diminish the chances of failure? Our knowledge of such strategies is limited, but there are lessons one can draw from studies of hazardous waste siting within the United States and cross-nationally. Both empirical and theoretical studies suggest the importance of building agency credibility through genuine commitment by policy makers to public participation, the designing of methods for meaningful information exchange and public dialogue on the issues, the promotion of equity in risk distribution, the sharing of project benefits (in part through compensation to communities bearing the risks), and the adoption of mitigation and control procedures that include strict safety standards and significant citizen roles in facility oversight and operating decisions. Extensive citizen involvement and adherence to principles of voluntary acceptance of facilities lie at the heart of such processes and are captured in the facility siting credo (see, e.g., Elliott 1984; Hadden 1991; Kunreuther et al. 1990; Kunreuther, Fitzgerald, and Aarts 1993; Lynn 1987; Mazmanian and Morell 1994; National Research Council 1989; Rabe 1991). Of special importance is the development of institutions and processes that not only allow laypersons to have full access to scientific and technical information on which safety decisions are based but also promote two-way communication and learning. Establishment of such participatory programs seems to be a prerequisite for building public trust and a sense of citizen control over the risks at issue.

There is some evidence that this perspective is gaining support among those involved with nuclear waste disposal policy in the United States, particularly in the experience of the nuclear waste negotiator. The first negotiator, David Leroy, expressed great faith that a voluntary siting process with "proper public involvement processes" could lead to a repository. The heritage of public fear and distrust of government agencies, however, must be addressed in any such processes to gain public acceptance. Leroy concluded that "open, broad, voluntary public participation by local citizens and local politicians in the siting of controversial facilities is the best way to win that support" (Leroy 1993). The social-science literature reinforces such a judgment.

Assessment of alternative approaches that promise to meet expectations for the democratic dialogue that lies at the heart of such processes would go beyond the scope of this chapter. But the subject is beginning to attract attention. Mary English and colleagues, for example, recently have surveyed a range of "mechanisms for stakeholder involvement" in decision making on remediation of contaminated sites, particularly those associated with DOE and Department of Defense environmental-restoration programs. Their inventory of small-group approaches (advisory committees, task forces, citizen juries, etc.), direct democracy (town meetings, ballot initiatives), surveys, technology-based approaches (e.g., interactive teleconferences or computer-based polling), and multiparty mediation is comprehensive and instructive (English et al. 1993). Daniel Fiorino (1990) offers a comparable, if briefer, survey of institutional mechanisms for citizen participation in environmental risk decisions, as does Susan Hadden (1991) for hazardous waste issues. Table 2 lists the most commonly suggested mechanisms, both conventional and participatory.

Others will need to extend these analyses to empirical study of the use of such approaches, which has been done selectively for citizen panels and other forms of participation in different policy areas (e.g., Kathlene and Martin 1991). Only in that way will we learn what the effects of such public involvement on decision making actually are, and the advantages and disadvantages of relying on such varied mechanisms. Multiple criteria may be suggested for evaluating the suitability of different forms of public participation and democratic dialogue on nuclear waste disposal or comparable siting issues: (1) representativeness, (2) political equity, (3) capacity to increase citizen knowledge and enhance decision-making abilities, (4) extent of citizen influence on decision making, (5) opportunities for dialogue and deliberation among citizens and between citizens and other decision makers, (6) capacity to build institutional trust and confidence, and (7) adaptability or flexibility under changing conditions. Such criteria may be helpful in judging the suitability of public-involvement mechanisms, whether considered theoretically or in empirical studies.

From a comparative perspective, the fragmented and adversarial U.S. political system may be less amenable than other systems to creating and fostering such democratic dialogue. For that reason, cross-national studies would be especially instructive. It appears, for example, that Canada's approach to HLRW disposal is far more attentive to issues of ethics and stakeholder involvement than has been the case in the United States (Hardy Stevenson and Associates 1991; Allan 1993). Similarly, Sweden and France have successfully sited

TABLE 2. Forms of Public Participation and Democratic Dialogue

CONVENTIONAL MECHANISMS	Participation in elections, lobbying, and litigation Public notice and comment periods for regulations Public hearings and meetings
PARTICIPATORY MECHANISMS	
Small Group	Citizen review panels and juries Task forces and workshops Advisory committees and boards, including community safety and oversight boards Technology tribunals Independent experts employed by citizen groups
Direct Democracy	Town and neighborhood association meetings Ballot initiatives and referenda Computer-based polling and teleconferences
Public/Citizen Surveys	Conventional public-opinion surveys Focus groups
Multiparty Mediation	Mediation between citizens and government/industry Negotiated rule making

Sources: Fiorino (1990); Hadden (1991); English et al. (1993).

LLRW repositories and have faced less public opposition to high-level geologic repository siting than has the United States, although it is premature to accord them success in the latter endeavor. Sweden has used negotiation with local opponents and national groups to reach consensus, and France has relied more on state authority and compensation (Cook, Emel, and Kasperson 1991; see also Kemp 1992).[9] In general, the U.S. political process has provided more opportunities for citizen groups to shape environmental policy than is the case in Europe and Japan. Yet it is not clear if the openness of the U.S. policy-making process implies greater potential for meaningful citizen participation in a dialogue on acceptable risk.

If new participation programs emphasizing democratic dialogue offer promise for siting risky facilities, that alone would be reason to celebrate. The case of high-level nuclear waste suggests another benefit. Public involvement may improve the quality of technical decisions by providing essential oversight and review of decisions that might otherwise be left to technical specialists with narrow training and experience. Under the latter conditions, it is all too easy to overlook analytic and bureaucratic deficiencies and to ignore important political, social, and economic issues because no one is present to press them. The history of a long-troubled nuclear power industry, the neglect of health and safety issues at federal nuclear weapons production facilities, and continuing doubts about DOE's implementation of NWPA all speak to the dangers of such dependency and the long-term cost it imposes on the political system.

NOTES

1. Fuller accounts of the formulation and implementation of U.S. nuclear waste policy may be found in Carter (1987), Colglazier and Langum (1988), Clary and Kraft (1989), Jacob (1990), Cook, Emel, and Kasperson (1990), and Dunlap, Kraft, and Rosa (1993, esp. chap. 1).

2. A review of the technology of waste management and of alternatives to geologic disposal, a history of waste management in the United States, an overview of the DOE program established by the 1982 legislation, and the statute itself can be found in a thorough report by the U.S. Office of Technology Assessment (OTA) report (U.S. OTA 1985).

3. In light of subsequent implementation problems, it is noteworthy that OTA observed as early as 1982 that "the greatest single obstacle that a successful waste management program must overcome is the severe erosion of public confidence in the Federal government that past problems have created" (U.S. OTA 1982, 10). The same argument was advanced in OTA's comprehensive report on nuclear waste disposal in 1985 (U.S. OTA 1985).

4. DOE's record of closed decision making, misrepresentation of nuclear risks, and carelessness in the handling of radioactive wastes is nowhere better described than in Michele Stenehjem Gerber's (1992) history of the Hanford nuclear site in Washington State. Cleanup activities at Hanford continue to reveal a pattern, over the past four decades, of extraordinarily poor management of the wastes and a striking disregard for public health and safety.

5. The facility siting credo includes the following elements: instituting a broad-based participatory process, seeking consensus, working to develop trust, seeking acceptable sites through a volunteer process, considering a competitive siting process, setting realistic timetables, keeping multiple options open at all times, achieving agreement that the status quo is unacceptable, choosing the solution that best addresses the problem, guaranteeing that stringent safety standards will be met, fully addressing all negative aspects of the facility, making the host community better off, using contingent agreements, and working for geographic fairness. See Kunreuther, Fitzgerald, and Aarts (1993).

6. The costs of on-site storage can be especially high for decommissioned plants, which bring in no revenue. The Edison Electric Institute has estimated, for example, that to keep the closed Rancho Seco plant in California active just to run the water pool for waste storage can run between $7 million and $21 million per year (Wald 1993a). The cost of decommissioning Yankee Rowe in Massachusetts is estimated at $247 million, of which $60 million is to build casks to store the waste. The number of retired plants is expected to rise sharply over the next decade.

7. Survey research is limited in describing views of the politically attentive public, since it is such a small percentage of the total sample. Perhaps most important to the argument in this chapter, we cannot be sure how active participation by the public and meaningful dialogue among the parties in conflict might alter perceptions of, and attitudes toward, the repository. This was one reason for relying on content analysis of hearing records to supplement the survey data, even if public hearings are but one form of citizen involvement and not a very useful one for encouraging two-way communication and accommodation of interests.

8. The most recent surveys of Nevada residents indicate no significant reduction in public opposition to the repository. In spring 1993, over 70% of Nevadans remained staunchly opposed to the waste facility, and nearly as many objected to any negotiations between Nevada and the federal government over compensation to the state. Trust in DOE remained low, with 66% of respondents indicating they did not trust DOE "to provide prompt and full disclosure of any serious problems with the Yucca Mountain project" (Flynn, Mertz, and Slovic 1993).

9. Two nations, France and Great Britain, reprocess nuclear fuel for nonmilitary purposes, with the large $4.2 billion British facility at Sellafield approved only in late 1993. Such reprocessing should reduce the amount of high-level waste headed toward repositories. As noted above, the United States ended its reprocessing program under the Carter administration in the late 1970s.

REFERENCES

Allan, C. J. 1993. Building Confidence in Deep Geological Disposal of Nuclear Fuel Waste: Canada's Approach. Paper presented at the International Nuclear Congress, October 3–6, Toronto.

Berry, Jeffrey M., Kent E. Portney, and Ken Thomson. 1993. *The Rebirth of Urban Democracy*. Washington, D.C.: Brookings Institution.

Bradbury, Judith A. 1989. The Use of Social Science Knowledge in Implementing the Nuclear Waste Policy Act. Ph.D. diss., University of Pittsburgh, April.

Brion, Denis J. 1991. *Essential Industry and the NIMBY Phenomenon*. New York: Quorum Books.

Carter, Luther J. 1987. *Nuclear Imperatives and Public Trust: Dealing with Radioactive Waste*. Washington, D.C.: Resources for the Future.

———. 1989. Nuclear Waste Policy and Politics. *Forum for Applied Research and Public Policy* 4 (Fall): 5–18.

Clary, Bruce B., and Michael E. Kraft. 1989. Environmental Assessment, Science, and Policy Failure: The Politics of Nuclear Waste Disposal. In *Policy through Impact Assessment: Institutionalizing Analysis as a Political Strategy*, edited by Robert V. Bartlett. Westport, Conn.: Greenwood Press.

Colglazier, E. W., and R. B. Langum. 1988. Policy Conflicts in the Process of Siting Nuclear Waste Repositories. *Annual Review of Energy* 13:317–57.

Cook, Brian J., Jacque L. Emel, and Roger E. Kasperson. 1990. Organizing and Managing Radioactive Waste Disposal as an Experiment. *Journal of Policy Analysis and Management* 9:339–66.

———. 1991. A Problem of Politics or Technique? Insights from Waste-Management Strategies in Sweden and France. *Policy Studies Review* 10 (Winter): 103–13.

Dunlap, Riley E., Michael E. Kraft, and Eugene A. Rosa, eds. 1993. *Public Reactions to Nuclear Waste: Citizens' Views of Repository Siting*. Durham, N.C.: Duke University Press.

Elliott, Michael L. P. 1984. Improving Community Acceptance of Hazardous Waste Facilities through Alternative Systems for Mitigating and Managing Risk. *Hazardous Waste* 1:397–410.

English, Mary, Amy K. Gibson, David L. Feldman, and Bruce E. Tonn. 1993. Stakeholder Involvement: Open Processes for Reaching Decisions about the Future Uses of Contaminated Sites. Knoxville, Tenn.: Waste Management Research and Education Institute, University of Tennessee, September.

Fiorino, Daniel J. 1990. Citizen Participation and Environmental Risk: A Survey of Institutional Mechanisms. *Science, Technology, and Human Values* 15 (Spring): 226–43.

Flynn, James, Roger Kasperson, Howard Kunreuther, and Paul Slovic. 1992. Time to Rethink Nuclear Waste Storage. *Issues in Science and Technology* 8 (Summer): 42–48.

Flynn, James, C. K. Mertz, and Paul Slovic. 1993. *Spring 1993 Nevada State Telephone Survey: Key Findings.* Carson City: Nevada Agency for Nuclear Projects.

Flynn, James, and Paul Slovic. 1993. Nuclear Wastes and Public Trust. *Forum for Applied Research and Public Policy* 8 (Spring): 92–100.

Flynn, James, Paul Slovic, and C. K. Mertz, 1993. The Nevada Initiative: A Risk Communication Fiasco. *Risk Analysis* 13 (October): 497–502.

Gerber, Michele Stenehjem. 1992. *On the Home Front: The Cold War Legacy of the Hanford Nuclear Site*. Lincoln: University of Nebraska Press.

Goldsteen, Raymond L., and John K. Schorr. 1991. *Demanding Democracy after Three Mile Island*. Gainesville: University of Florida Press.

Hadden, Susan G. 1991. Public Perception of Hazardous Waste. *Risk Analysis* 11 (March): 47–57.

Häfele, Wolf. 1990. Energy from Nuclear Power. In *Energy for Planet Earth*, edited by Scientific American, 95–105. New York: W. H. Freeman.

Hardy Stevenson and Associates. 1991. Moral and Ethical Issues Related to the Nuclear Fuel Waste Disposal Concept. Toronto: Hardy Stevenson and Associates, April.

Hill, Stuart. 1992. *Democratic Values and Technological Choices*. Stanford, Calif.: Stanford University Press.

Ingram, Helen, and Steven Rathgeb Smith, eds. 1993. *Public Policy for Democracy*. Washington, D.C.: Brookings Institution.

Isaacs, Thomas H. 1989. DOE Goals: Excellence, Openness. *Forum for Applied Research* 4 (Fall): 36–38.

Jacob, Gerald. 1990. *Site Unseen: The Politics of Siting a Nuclear Waste Repository*. Pittsburgh: University of Pittsburgh Press.

Kathlene, Lyn, and John A. Martin. 1991. Enhancing Citizen Participation: Panel Designs, Perspectives, and Policy Formation. *Journal of Policy Analysis and Management* 10 (Winter): 46–63.

Keeney, Ralph L., and Detlof von Winterfeldt. 1993. *An Analysis of Strategies to Manage Nuclear Waste from Power Plants*. Decision and Risk Analysis Report 93–1. Los Angeles: Institute of Safety and Systems Management, University of Southern California, January.

Kemp, Ray. 1992. *The Politics of Radioactive Waste Disposal*. New York: St. Martin's Press.

Kraft, Michael E. 1991. Public and State Responses to High-Level Nuclear Waste Disposal: Learning from Policy Failure. *Policy Studies Review* 10 (Winter): 152–66.

———. 1992. Technology, Analysis, and Policy Leadership: Congress and Radioactive Waste. In *Science, Technology, and Politics: Policy Analysis in Congress*, edited by Gary C. Bryner. Boulder: Westview Press.

Kraft, Michael E., and Bruce B. Clary. 1991. Citizen Participation and the NIMBY Syndrome: Public Response to Radioactive Waste Disposal. *Western Political Quarterly* 44 (June): 299–328.

———. 1993. Public Testimony in Nuclear Waste Repository Hearings: A Content Analysis. In *Public Reactions to Nuclear Waste: Citizens' Views of Repository Siting*, edited by Riley E. Dunlap, Michael E. Kraft, and Eugene A. Rosa, 89–114. Durham, N.C.: Duke University Press.

Krimsky, Sheldon, and Alonzo Plough. 1988. *Environmental Hazards: Communicating Risks as a Social Process*. Dover, Mass.: Auburn House.

Kriz, Margaret. 1993. Wasting Away. *National Journal*, October 9, 2430–33.

Kunreuther, Howard, Douglas Easterling, William Desvousges, and Paul Slovic. 1990. Public Attitudes toward Siting a High Level Nuclear Waste Repository in Nevada. *Risk Analysis* 10 (December): 469–84.

Kunreuther, Howard, Kevin Fitzgerald, and Thomas D. Aarts. 1993. Siting Noxious Facilities: A Test of the Facility Siting Credo. *Risk Analysis* 13 (June): 301–18.

Laird, Frank N. 1989. The Decline of Deference: The Political Context of Risk Communication. *Risk Analysis* 9 (December): 543–50.

Leroy, David H. 1993. NIMBY and the Nuclear Negotiator in the United States. Paper presented at the International Nuclear Congress, October 3–6, in Toronto.

Leroy, David H., and T. Scott Nadler, 1993. Negotiate Way out of Siting Dilemmas. *Forum for Applied Research and Public Policy* 8 (Spring): 102–7.

Lippman, Thomas W. 1992. Sure, It's Radioactive, but It'll Mean Money. *Washington Post National Weekly Edition*, July 13–19, 33.

Lipset, Seymour Martin, and William Schneider. 1987. *The Confidence Gap: Business, Labor, and Government in the Public Mind.* Rev. ed. Baltimore: Johns Hopkins University Press.

Lynn, Frances M. 1987. Citizen Involvement in Hazardous Waste Sites: Two North Carolina Success Stories. *Environmental Impact Assessment Review* 7:347–61.

Mazmanian, Daniel A., and David Morell. 1994. The "NIMBY" Syndrome: Facility Siting and the Failure of Democratic Discourse. In *Environmental Policy in the 1990s,* edited by Norman J. Vig and Michael E. Kraft. 2nd ed. Washington, D.C.: CQ Press.

Mazmanian, Daniel A., and Paul A. Sabatier. 1983. *Implementation and Public Policy.* Glenview, Ill.: Scott, Foresman.

National Research Council. 1989. *Improving Risk Communication.* Washington, D.C.: National Academy Press.

———. 1990. *Rethinking High-Level Radioactive Waste Disposal: A Position Statement of the Board on Radioactive Waste Management.* Washington, D.C.: National Academy Press, July.

O'Hare, Michael, and Debra Sanderson. 1993. Facility Siting and Compensation: Lessons from the Massachusetts Experience. *Journal of Policy Analysis and Management* 12 (Spring): 364–76.

Pijawka, D. David, and Alvin H. Mushkatel. 1991. Public Opposition to the Siting of the High-Level Nuclear Waste Repository: The Importance of Trust. *Policy Studies Review* 10 (Winter): 180–94.

Piller, Charles. 1991. *The Fail-Safe Society: Community Defiance and the End of American Technological Optimism.* Berkeley: University of California Press.

Portney, Kent E. 1991. *Siting Hazardous Waste Treatment Facilities: The NIMBY Syndrome.* New York: Auburn House.

Rabe, Barry G. 1991. Beyond the NIMBY Syndrome in Hazardous Waste Facility Siting: The Albertan Breakthrough and the Prospects for Cooperation in Canada and the United States. *Governance* 4, no. 2: 184–206.

Rezendes, Victor S. 1991. Long Road to Recovery Begins at DOE Plants. *Forum for Applied Research and Public Policy* 6 (Spring): 19–26.

Rosa, Eugene A., and William R. Freudenburg. 1993. The Historical Development of Public Reactions to Nuclear Power: Implications for Nuclear Waste Policy. In *Public Reactions to Nuclear Waste: Citizens' Views of Repository Siting,* edited by Riley E. Dunlap, Michael E. Kraft, and Eugene A. Rosa, 32–63. Durham, N.C.: Duke University Press.

Sabatier, Paul A. 1987. Knowledge, Policy-Oriented Learning, and Policy Change. *Knowledge: Creation, Diffusion, Utilization* 8:649–92.

Schneider, Keith. 1994. Disclosing Radiation Tests Puts Official in Limelight. *New York Times,* January 6, A1, A14.

Shrader-Frechette, K. S. 1991. *Risk and Rationality: Philosophical Foundations for Populist Reforms.* Berkeley: University of California Press.

———. 1993. *Burying Uncertainty: Risk and the Case against Geological Disposal of Nuclear Waste.* Berkeley: University of California Press.

Slovic, Paul. 1987. Perception of Risk. *Science* 236 (April 17): 280–85.

———. 1993. Perceived Risk, Trust, and Democracy. *Risk Analysis* 13 (December): 675–82.

Slovic, Paul, James H. Flynn, and Mark Layman. 1991. Perceived Risk, Trust, and the Politics of Nuclear Waste. *Science* 254 (December 13): 1603–7.

Slovic, Paul, Mark Layman, and James Flynn. 1991. Risk Perception, Trust, and Nuclear Waste: Lessons from Yucca Mountain. *Environment* 33:6–11, 28–30.

Thurber, James A. 1994. *Report on Selected Published Works and Written Comments Regarding the Office of Civilian Radioactive Waste Management Program, 1989–1992*. Washington, D.C.: Center for Congressional and Presidential Studies, School of Public Affairs, American University.

U.S. Department of Energy (U.S. DOE). 1987. Office of Civilian Radioactive Waste Management. *Annual Report to Congress*. Washington, D.C.: Department of Energy, April.

U.S. Environmental Protection Agency (U.S. EPA). 1992. *Environmental Equity: Reducing Risk for All Communities*. Washington, D.C.: Environmental Protection Agency.

U.S. General Accounting Office (U.S. GAO). 1992. *Nuclear Waste: DOE's Repository Site Investigations, a Long and Difficult Task*. GAO/RCED–92–73. Washington, D.C.: General Accounting Office, May.

———. 1993. *Yucca Mountain Project behind Schedule and Facing Major Scientific Uncertainties*. GAO/RCED–93–124. Washington, D.C.: General Accounting Office, May.

U.S. Nuclear Waste Technical Review Board (U.S. NWTRB). 1993. *NWTRB Special Report to Congress and the Secretary of Energy*. Washington, D.C.: NWTRB, March.

U.S. Office of Technology Assessment (U.S. OTA). 1982. *Managing Commercial High-Level Radioactive Wastes: Summary*. Washington, D.C.: Government Printing Office.

———. 1985. *Managing the Nation's Commercial High-Level Radioactive Waste*. Washington, D.C.: Government Printing Office.

U.S. Secretary of Energy. 1993. *Earning Public Trust and Confidence: Requisites for Managing Radioactive Waste*. Proposed final report of the Secretary of Energy Advisory Board Task Force on Radioactive Waste Management. Washington, D.C.: Department of Energy, June.

Wald, Matthew L. 1993a. Nuclear Industry Seeks Interim Site to Receive Waste. *New York Times*, August 27, A1, A11.

———. 1993b. Nuclear Storage Divides Apaches and Neighbors. *New York Times*, November 11, A8.

Hazardous Waste Management and Facility Siting in California

DAVID MORELL

INTRODUCTION

Hazardous waste management is a major environmental issue across the United States; controversy has been particularly apparent in California, the country's most populous state.[1] Here, as elsewhere, new hazardous waste management facilities have been needed both to help clean up leaking sites from the past and to manage industry's new hazardous wastes. New laws banning further land disposal of most untreated hazardous wastes, first in California and then nationally, made siting of new treatment facilities even more important. California industry generates immense quantities of hazardous wastes every year, well in excess of the state's treatment or disposal capacity. Therefore, many hazardous wastes are exported to treatment facilities and landfills in states such as Alabama, Louisiana, Texas, and Utah.

Initiatives to site the needed new facilities, however, have been fraught with difficulty. Communities outraged at the possibility of having a facility built locally have mounted effective resistance to the siting proposals.

"Not in My Backyard" (NIMBY) attitudes prevented siting of the new hazardous waste facilities needed to move toward developing an effective solution. Low public confidence in the ability of either the private or the public sector to manage hazardous wastes in a safe, environmentally sound manner reinforced NIMBY, as localities refused to allow new or expanded treatment or disposal facilities to be sited. The specter of endless cost overruns associated with siting delays caused many project developers to retreat and inhibited entirely the emergence of other proposals.

By the early 1980s, California was experiencing numerous controversies over hazardous waste management and facility siting. Only one licensed hazardous waste landfill remained in southern California at the end of 1980, compared with five facilities at the beginning of that year. Debates raged over ways to clean up federal and state

Superfund sites and over bureaucratic inefficiency in the state's hazardous waste regulatory programs.

A variety of solutions were proposed to meet California's need for new hazardous waste facilities in the face of competing business, local NIMBY, and other interests. Some people, mainly in business, called for state preemption over local disapproval of proposed facilities. In contrast, community groups called for complete local veto power over proposed new facilities. Legislators sought a compromise that would establish a balanced process to streamline and remove some of the complexity of the facility siting process. The solution had to resolve the dilemma of the state's need to site a number of additional treatment facilities, such as incinerators, in the face of intense local-government and public opposition to having a "dump" of any kind built locally. This legislation (AB 2948) was finally passed in 1986 (California Health and Safety Code 1986).

This act, intended as the foundation of an innovative new approach to California's hazardous waste problem, has failed in many respects. Analysis of this failure can help guide future hazardous waste management efforts in more productive directions. This chapter first discusses creation of AB 2948 and describes its new process for hazardous waste management planning and facility siting. It then examines the failures of the act in order to determine their causes. Three specific case studies then illuminate how the process failed in the efforts by Rhone-Poulenc, Dow Chemical, and Chemical Waste Management to site hazardous waste incinerators. These cases are essential in providing real-world detail as the basis for exploring what worked, and especially what did not, in the innovative AB 2948 process. Finally, the overall effort is explored from the differing perspectives on the "need" for new facilities as perceived by participants in the hazardous waste facility planning and siting process.

THE CALIFORNIA HAZARDOUS WASTE MANAGEMENT ACT

In 1982, the legislature formed a new bipartisan Hazardous Waste Management Council to address the need for effective and equitable hazardous waste facility siting. This fourteen-member group comprised key state agency directors, state legislators, elected county and city officials, and representatives from industry, environmental, and citizen groups.

Over the following two years, the Hazardous Waste Management Council met regularly to explore options and devise recommended solutions. Its recommendations formed the basis for AB 2948, which the council formally presented to the legislature in 1984. This bill easily passed both the assembly and the senate that year, but was vetoed by the governor. The same ritual of passage and veto was repeated in 1985.

Governor George Deukmejian, a pro-business conservative who took office in January 1983, twice vetoed the bill despite the presence of his own appointed state agency heads on the council in 1983–84 and their agreement (at least formally) with the council's final recommendations. His opposition apparently derived from several sources: his personal skepticism about the value of public sector hazardous waste planning; his doubts about the wisdom of enforcing regulatory bans on land disposal of untreated wastes; his belief that the private hazardous waste management business would be overly constrained by this new bill's approach to government planning and dispute resolution; and especially his unwillingness to accept innovations that had emerged from his predecessor, Governor Jerry Brown. Despite all the evidence of facility-siting failure to that point, in California and across the United States, Deukmejian apparently believed that private-sector siting, if left alone, could successfully overcome local NIMBY resistance. Deukmejian was finally convinced to sign the again passed legislation in 1986, just in advance of his successful re-election bid.

The rationale behind the act was to "ensure the availability of safe, effective and economical facilities for the management of hazardous waste beginning with local planning and siting in cooperation with state government, the public, and hazardous waste industry" (Lennard and Labadie 1988). A balanced process called on all concerned to participate in decisions on the siting and permitting of new hazardous waste management facilities.

The balance set out in AB 2948 was designed explicitly to avoid both state preemption and local veto. Localities (city or county, depending on the specific location of the proposed new facility) were to reach their own decisions on proposed new off-site commercial hazardous waste management facilities. These local decisions were expected to reflect the area's new County Hazardous Waste Management Plan (CHWMP). The county plan was to provide for effective management of all the hazardous wastes expected to be generated in a county, either through accommodating the siting of needed new or expanded facilities or through agreements with other counties. The CHWMP had to be approved both locally and by the state hazardous waste management regulatory agency. The local decision (yea or nay) could be reviewed and overturned by a new state appeals board, whose primary criterion was the consistency between the proposed facility and the CHWMP. This process focused on the siting of new facilities ("land use" decisions); it did not restrict any flows of hazardous wastes (which generators could continue to send wherever economics, corporate preferences, and other factors so dictated).

Equity-Based Siting as the Central Principle

The principles of need, equity, and "fair share" were the centerpieces of California's new approach to this complex and controversial issue. AB 2948 called on small counties, as well as larger, industrialized ones, to assume responsibility for the development of sufficient facilities to treat and manage the hazardous wastes being generated locally. Clearly, however, not all of California's fifty-eight counties needed a full set of large facilities to manage all types of possible hazardous wastes. Counties were to cooperate through a fair-share concept so that the full range of facilities would emerge regionally and statewide. Each CHWMP would define what types of facilities that county needed and where they could be located environmentally.

The intent was to build a positive dynamic in favor of siting facilities needed locally, as the counter to inherent NIMBY tendencies. In this way, California's total needs for hazardous waste management were to be attained through the collective efforts of all fifty-eight counties, each taking affirmative responsibility to accept facilities consistent with its own share of total statewide hazardous waste treatment and disposal needs.

By focusing on local need, the process attempted to overcome NIMBY by allowing local residents to see how facilities proposed for their country were consistent with the wastes being generated there. Smaller counties that did not need a full-scale facility could negotiate agreements with other counties to mange their fair share of the group's wastes. Larger counties would thus be required to accept larger facilities. While some smaller counties might choose to accept a larger facility, for its employment and other economic benefits, such a facility could not be forced on them against their will. This new hazardous waste management theory encouraged responsibility and accountability instead of the old-fashioned alternatives of preemption and coercion.

There remained, however, an important and large new "stick" to enforce fair share's impressive "carrots": the appeals board mechanism. AB 2948's new project-specific appeals board provided a way for companies trying to site facilities to appeal a local siting rejection (and for aggrieved local citizens to appeal a siting approval). If the appeals board found the proposed facility consistent with an approved county plan, the local decision could be overridden. The appeals mechanism thus enhanced the likelihood of county plans being implemented. In the absence of an approved CHWMP, a facility could be sited if it was consistent with the area's pre-1983 General Plan and patterns of industrial zoning. This encouraged counties to prepare good CHWMPs. The basic idea was to hold each county accountable for meeting the commitments included in its plan.

County Hazardous Waste Management Plans

The new siting process revolved around each county's hazardous waste management plan. Preparing these CHWMPs was not mandatory; each county could decide whether to develop one. Nevertheless, all fifty-eight California counties decided to write their own plans. AB 2948 provided a golden opportunity for the counties to create strong plans that would lay the basis for resolving a major policy problem yet effectively prevent state preemption and avoid inequitable facility siting.

The main incentive for counties to write a plan was the linkage between the CHWMP and the new appeals board. With a good CHWMP in place, AB 2948 prohibited the board from forcing hazardous waste facilities (e.g., large ones) on communities (e.g., small ones) that did not want or need them. While the plan had to accommodate sufficient facilities to meet local needs (or the plan would not be approved), larger/more facilities could not be forced on an unwilling locality. The absence of a plan, however, would allow siting of these facilities in any area zoned for industry.

The plans thus provided local and county governments with explicit powers of self-determination. While the state retained responsibility to regulate hazardous waste management, county and local governments, through their powers over land use, were allowed the primary role in planning the kinds of hazardous waste facilities to be developed there.

AB 2948 stated that the counties had to plan for effective management of the hazardous wastes produced in their county. They were responsible for approving the siting of facilities to manage these wastes, though not the wastes of other counties. This protected smaller rural counties from having to plan to accept facilities that would manage the massive amount of wastes generated in the larger industrial counties (such as Los Angeles).

The legislation (California Health and Safety Code 1992) stated clearly that the purpose of the CHWMPs was to examine

- the types and amounts of hazardous wastes being generated in the county and estimates of future hazardous waste production (in the year 2000);
- existing hazardous waste facilities that handle, treat, dispose of, and recycle the hazardous wastes produced in the county;

- the potential in the county of recycling and reducing the volume of hazardous waste at the source of generation;
- the need to manage the small volume of hazardous waste produced by businesses and households;
- the need for additional hazardous waste facilities to manage properly the volumes of hazardous waste currently produced or expected to be produced during the planning period;
- specific siting criteria to place needed facilities in the right locations and identify suitable sites within the county;
- the goals, objectives, and policies for the siting of hazardous waste facilities and the general management of hazardous waste through the year 2000;
- development of a schedule that describes how and when the plan will be implemented.

For a CHWMP to be approved, both the country's board of supervisors and the city councils of two-thirds of the cities in that county (containing at least half the population of the county) had to adopt that plan. Once approved locally, the CHWMP had to be submitted for approval to the state Department of Health Services (DHS), the state agency responsible for regulating hazardous wastes.

The act allocated money for counties to develop their hazardous waste management plans. This funding was another incentive to get the counties involved. State money, however, was not sufficient for the counties to prepare fully their CHWMPs, and state funds were not available to implement county plans after their approval; local funding was needed.

The State Appeals Board

The appeals board provided the latent power behind implementation of this balanced process. This board comes together upon appeal of a permit decision. It has the power to override any local disapproval or approval of a proposed hazardous waste facility. An override requires a vote of at least four of the board's seven members:

- the state director of DHS, the hazardous waste regulatory lead agency (this role was later transferred to the Department of Toxic Substances Control [DTSC]);
- the chair of the state Air Resources Board;
- the chair of the state Water Resources Control Board;
- one county supervisor serving at large statewide;
- one county supervisor from the affected county's board of supervisors;

- one city council member serving at large statewide; and
- a council member from the city most directly affected.

Some decisions require five of the seven votes. Clearly, any override requires the approval of at least one and perhaps two elected officials, as well as of the three state agency leads.

For private developers attempting to site a facility, the appeals board offers the opportunity to overcome local rejection of their project. The local land use decision may be overturned and the facility approved if the appeals board finds the proposed project consistent with the approved CHWMP or, if an approved plan does not exist, with the area's pre-1983 General Plan.

County Advisory Committees

The act mandated formation of county advisory committees to involve local communities more fully in planning for effective management of hazardous wastes. This mechanism was intended to blunt the NIMBY situation in which facilities are opposed once they are proposed for specific locations.

These committees, put together by each county, were to be composed of at least seven members representing industry, environmental organizations, the cities, and the general public. At the county's choice, other members with expertise in hazardous waste management planning could be appointed to the committee.

The responsibilities of these committees included advising the county staff, the board of supervisors, and the cities on issues related to the proposed new CHWMP; holding informal meetings to provide information and to involve the public; and helping shepherd the plan through the local approval process. These committees were an important component of AB 2948's overall approach to siting.

Local Assessment Committees

AB 2948 established unprecedented power sharing at the local level over the process of approving proposed new hazardous waste management facilities. By formalizing local authority over siting, the act revised the process by which hazardous waste facilities were approved.

A local assessment committee (LAC) was now to be established for each off-site facility proposal. Formed when a permit application is submitted to a local government, the LAC was directed to "negotiate with the applicant on the terms, provisions and conditions for project approval which would protect the public health, safety and welfare, and the environment of the County and would promote the fiscal welfare of the county through special benefits and compensations," as stated in the act. The legislation made the LAC responsible for "representing the interest of County residents and adjacent communities in negotiations with the project applicant."

The act also included an incentive for counties to site facilities in their jurisdictions because it allowed local jurisdictions to tax up to 10% of the gross revenues of a hazardous waste treatment or disposal facility located in their area.

IMPLEMENTATION ACTIONS AFTER PASSAGE OF AB 2948

As mentioned earlier, all fifty-eight counties decided to prepare draft CHWMPs, most meeting the deadline of March 31, 1988. The plans provided the first comprehensive picture of county-by-county generation of hazardous wastes. The counties were also able to develop "a greater understanding of the relationship of local generators and the disposal methods that they use and their effects on the local economy" (Penn 1988).

The counties were enthusiastic about having a plan to handle their hazardous wastes that would avoid the fear of state preemption or of being forced to accept a disproportionate siting burden. Counties expressed readiness to rely on AB 2948's fair-share provisions to protect them from having to accept unwillingly facilities that exceeded defined local needs. This was especially true for rural, low-population counties that had long feared having a facility stuck in their extensive open space and cheaper available land.

The legislation and the planning process worked out in the bill were, however, gravely undercut when DHS in 1987 released its guidelines, which set out a vastly different interpretation of the law's central equity-based fair-share provisions.

Failures in the Process

DHS was formally responsible for approving all the CHWMPs. In its guidelines, DHS required all counties to include in their CHWMPs siting criteria for all kinds and sizes of hazardous waste management facilities, regardless of the size of the county and its own local waste management needs.

Local governments and DHS came into increasing conflict over what the act said and what was required by the guidelines. As DHS insisted on its position, local incentives to accept their fair share of facilities for hazardous waste management disappeared.

AB 2948 is based on a central premise: if each county took steps to identify its own needs and thereby assumed responsibility to accept siting of proposed facilities needed to meet those needs, the state's overall hazardous waste management requirements would be adequately covered. Many observers were hopeful that the NIMBY attitudes in many localities might decrease if all counties took responsibility for effective management of their own share of the state's total.

The law had authorized DHS to develop state guidelines that each county would be compelled to follow in order to receive DHS approval of its CHWMP. Such state approval was required to give these county plans official standing in any subsequent decisions of the appeals board. DHS used its mandatory guidelines, however, to redefine the goals of the act in accord with its own views. Despite the central role of fair share and intercounty agreements in the statute, the regulatory agency refused to recognize either principle as acceptable.

DHS argued that a private, open-market system would develop facilities when and where the market was conducive, regardless of equity considerations or formal CHWMPs. DHS chose to ignore the overwhelming evidence that the market approach typically focused on attempts to site large facilities in remote areas that themselves had little industry and thus created little waste. These proposed facilities usually encountered intense NIMBY opposition and were rejected in the siting process. On the other hand, land in industrial, urban areas was often more expensive and less available than in rural areas, which had more open space and cheaper land for facility development. Rural areas might be more desperate for economic development of any kind, and their residents might seem less organized and politically weaker. For these reasons, hazardous waste management firms often preferred to site hazardous waste facilities in these remote regions.

Another reason for the breakdown in the AB 2948 process could be attributed to DHS's belief that its responsibility to protect public health gave it the authority to impose requirements on CHWMPs different from the explicit provisions of AB 2948. Part of the conflict was rooted in the different mandates each side was following. The counties were working solely under AB 2948; DHS, on the other hand, was also responding both to another state law requiring preparation of a state hazardous waste plan (AB 650) and to federal requirements that California prepare a statewide Capacity Assurance Plan.

The formal reason DHS gave for its requirement that each county's CHWMP contain siting criteria for a full range of facilities was to be able to assure the U.S. Environmental Protection Agency, as required by federal law, that California had sufficient capacity to deal with all of its hazardous wastes over the next twenty years. By deleting the fair-share aspect of AB 2948, the state could argue that certain areas in California would clearly be available for the siting of hazardous waste treatment or disposal facilities, regardless of local opposition. Once again the state was trying to preempt NIMBY instead of finding a way to work with localities to accommodate their interests and meet their local needs.

According to Tony Eulo of the California Local Government Commission, a close observer of this local/state conflict, the state saw fair share as a hindrance to inducing potential siting proposals from the hazardous waste management industry. Basically, the state was totally opposed to placing any limit on the types or sizes of possible new hazardous waste facilities. Because fair share would grant each county the authority to identify the size and type of facility needed in its region, the principle was viewed by the state agency as a major roadblock to the overall statewide hazardous waste management and siting process.

DTSC's Valerie Heusinkzeld argues that it simply is not economically profitable for the private sector to site a multitude of small facilities on a regional (county) scale. In her view, fair share-proponents were naive to think that such an encompassing statewide problem could be managed on a county-by-county or even a regional level.

David Bauer, formerly with IT Corporation, a major hazardous waste disposal and treatment firm, also feels that the fair-share concept would impose many economic impediments. He asserts that focusing on hazardous waste management on a regional or county level precludes the economies of scale needed for private industry to site and operate these facilities. Because of high overhead costs and the lengthy permitting process, industries would not be willing to take the risk of siting smaller facilities, Bauer argues. There is no guarantee that the waste stream from that region will remain constant. Bauer notes that the CHWMPs do not take into consideration the shifting economic conditions of a market system and that economic conditions are too volatile for industries to depend upon wastes only from certain industries in a fixed region.

Since DHS's mission is to protect public health and regulate hazardous wastes, it chose to interpret AB 2948 (once passed) as a preemption tool, thus making its job simpler. DHS stated, in a November 3, 1987, letter to Santa Cruz County supervisor Gary Patton, that its mission is "to facilitate the establishment of hazardous waste treatment facilities to implement the statutorily mandated curtailment of land disposal . . . to help people of the state to escape or avoid threats to their health and environment." If they could not win public cooperation, they would force it by granting approval only to those CHWMPs that contained a full range of facility-siting criteria.

Those who had originally supported AB 2948, however, certainly did not interpret it as having given the state preemption powers. In his December 23, 1987, response to DHS, Patton stated:

> I understand that your job, from its statewide perspective, is made much easier when you can require local jurisdictions to dispose of wastes generated elsewhere in the State. We do not agree that this type of State-mandated disposal plan—if not consented to by the affected County—is called for in AB 2948. We regard the legislation as an attempt to structure a compromise between those who would completely preempt and supersede local authority, and those who suggest that local jurisdictions have no obligation whatsoever to plan for hazardous waste treatment and disposal.

The proponents of AB 2948 argue that without the ability of every county to make provisions in its own plan to meet its share of the state's hazardous waste management needs, NIMBY attitudes toward siting would continue to prevail over each facility-siting proposal.

Karen Peters from Orange County's Hazardous Materials Program notes that DHS did not understand the counties' mechanics and political framework. She asserts that the state was shortsighted in its decisions because it was unwilling to look at hazardous waste management on a localized level; that DHS was unwilling to accept the fact that communities needed assurance they would not be targeted as dumping grounds for wastes from the entire region or state; and that the state agency was not willing to grant the counties this guarantee, despite passage of state legislation that had done so.

Sacramento County supervisor Grantland Johnson focused on the correlation between the interests of big business and the decisions made by DHS.

> Instead of honestly addressing the public's concerns, DHS concentrates on developing strategies to override them. . . . This ensures that the gridlock that has characterized the siting of hazardous waste facilities would continue. . . . Every county is committed to reducing the generation of hazardous wastes and siting needed hazardous waste management facilities. The DHS, on the other hand, is fighting to maintain the ineffective and inequitable process that has created our toxics problem in the first place. (*Sacramento Bee*, December 18, 1988)

Suzanne Larson of the Association of Bay Area Governments (ABAG) feels that DHS eliminated fair share from the siting process at the technical-staff level; that because DHS had never been involved in local land use issues, it had no understanding of the zeal and dedication to maintain land use authority at the local level; and that DHS simply had no idea of the threat its guidelines posed for local governments.

Clearly, most cities and counties would never have supported AB 2948 in the legislature if they had known that this DHS interpretation would arise; and without their support, AB 2948 would never have passed into law. To the counties, the DHS guidelines represented exactly the same state preemption that AB 2948 had explicitly been designed to avoid. Fair share under the act called on counties to accept only those facilities that met their own needs. The state could not force on a county a facility that exceeded those needs, as defined in its CHWMP. Small counties might need a transfer or storage facility; bigger, more industrialized counties might need an incinerator. This is what the concept of fair share is all about. Each county was asked in AB 2948 to prepare siting criteria related to the amount of waste its own industries generate.

Had the fair-share language remained in effect in the DHS guidelines as in the act, CHWMPs for rural and smaller counties that do not generate large volumes of hazardous wastes would not have had to provide for a large treatment facility. A large incinerator or treatment plant could not be sited in their county because it would be inconsistent with the local CHWMP. The CHWMP thus was meant to protect the smaller counties in such a situation. In contrast, if every CHWMP followed the DHS guidelines and included provision for, say, a commercial off-site incinerator, a private incineration company could appeal a local land use decision that barred such development anywhere.

Under fair share, Modoc County's plan would not include an incinerator, since Modoc does not generate enough incineratable waste to require an incinerator. Thus, an incinerator could not be sited there against Modoc's wishes because of the conflict with the CHWMP. In contrast, in the absence of fair share, Modoc could be required to site an incinerator to handle wastes from as far away as Los Angeles.

This would inspire local opposition, since Modoc residents would be asked to carry a disproportionate share of the toxics burden. In Contra Costa County or Los Angeles County, where large quantities of waste are generated and people benefit from employment opportunities in the industries that generate these wastes, local residents might be more understanding of the need to accept siting of an incinerator to manage those wastes. Thus, costs and benefits would be distributed locally and more "fairly," as AB 2948 had intended.

In summary, the DHS guidelines violated the fair share principle and took away the locals' voice. By requiring all counties to develop criteria for a large off-site commercial incinerator, among other facilities, DHS presumably made it easier for a private company to appeal local land use rejection of such a facility. A small county could, therefore, be forced to site an incinerator that would treat wastes from larger, more industrialized counties. With the guidelines as official state policy, the state could utilize the act as a preemption tool to simplify its responsibility for managing hazardous wastes and protecting public health.

The interpretation put forward by DHS in its guidelines greatly weakened the siting process. Counties had accepted the AB 2948 process and had agreed to develop plans to manage the hazardous wastes generated in their own county and to accept local siting of facilities that would treat their share of California's hazardous wastes. Many had agreed to work with neighboring counties in broader fair-share compacts. Under DHS's mandate, however, each would now have to include all kinds and sizes of facilities in its plans. And the state appeals board could then force the siting of a treatment plant designed essentially to serve generators from far away. Yet in the absence of an approved plan, AB 2948 might well allow hazardous waste operators to site a facility in any area zoned for industry.

From the counties' perspective, their whole agreement with the state over the content of the act had been betrayed. The formal winner was DHS; it could assume successful siting of all the facilities wanted by any developer and fulfill its responsibility to prepare a federal Capacity Assurance Plan for California at the expense of the fair-share principle adopted by the counties.

Of course, by changing the fundamental dynamics of the siting process, no one won. DHS's revisions made the process of siting hazardous waste facilities in California just as inequitable as it had been before passage of the act. Formalism prevailed, but no new facilities have been sited.

The Counties' Approach: Regional Agreements

What DHS soon realized was that its approach to the act would force the agency to reject almost all the counties' draft CHWMPs, since the local plans almost all contained explicit fair-share provisions. In light of this potential problem, DHS proposed a modest compromise that would allow limited fair-share and regional-agreement principles under certain circumstances. The counties, already ahead of DHS, had moved on their own to supplement the absence of the fair-share principle in the DHS guidelines with intercounty agreements and bilateral negotiations with other counties. The act permitted counties to include in their CHWMPs descriptions and details of intercounty compacts that they had formed to manage more efficiently on a regional level the hazardous wastes generated. DHS's compromise reflected this reality.

The concept of forming equity-based regional agreements, known as YIMBY (Yes in Many Backyards), was first developed by eight southern California counties in a Southern California Hazardous

Waste Management Project (SCHWMP) prior to passage of AB 2948. The idea behind YIMBY was that each county would assume responsibility for its fair share of the region's waste problem but not have to manage itself all its own wastes (Mazmanian and Morell 1992). SCHWMP participants acknowledged that it was neither feasible nor necessary for each county in the region to site the full set of incinerators, recycling plants, residuals repository landfills, neutralization facilities, and other kinds of plants to treat all of their wastes.

The seven counties formed intercounty compacts to ensure that a certain level of equitable facility siting would occur. Under such agreements, one county would agree to accept one type of facility larger than its own needs, while another county would reciprocate with an agreement to site a different type of larger-than-local facility. "As long as aggregate treatment capacity meets the region's hazardous waste treatment needs, the system will be in accord with the fair share principle while providing adequate economies of scale and efficiency criteria" (Mazmanian and Morell 1992).

A regional compact would protect all the counties that concurred with the agreement. In an intercounty agreement, a smaller, less industrialized county could stipulate that it would accept siting of a hazardous waste facility as long as the county was compensated for receiving such a project and on the proviso that the county need not accept responsibility for managing more hazardous wastes than had been defined as equitable in the agreement.

In this way, smaller counties would be protected from having large facilities forced upon them by the appeals board, as could occur if they struck out on their own with or without an approved CHWMP. The approach taken by DHS also compelled large counties to negotiate intercounty agreements so that their CHWMPs could be approved. The large counties did not want hazardous waste facilities sited in their industrial zones, which would occur if they failed to prepare an approved CHWMP and the state appeals board wanted to approve siting a hazardous waste facility in their county.

Status of the Siting Process

A resolution between the counties and DHS over fair share has still not been reached. By 1993, seven years after AB 2948 was finally approved and eleven years after formation of the council that had designed the act, the process had basically fallen apart. As shown in table 1, only twenty plans had been approved. In 1991, the Toxic Substances Control Division (TSCD) within DHS was transferred to the California Environmental Protection Agency (Cal-EPA), and its name was changed to the Department of Toxic Substances Control

TABLE 1. Approval Status of County Hazardous Waste Management Plans

Approved	*Disapproved*	*Resubmitted, Pending Review*	*Withdrawn after Draft Plan Submitted*
Alpine	Alameda	Alameda	Del Norte
Calaveras	Amador	Butte	Humboldt
Contra Costa	Butte	Fresno	Mendocino
El Dorado	Colusa	Inyo	Santa Cruz
Glenn	Fresno	Marin	
Imperial	Inyo	San Francisco	Total = 4
Kern	Lake	San Mateo	
Kings	Lassen	Santa Clara	
Los Angeles	Madera	Sonoma	
Modoc	Marin	Stanislaus	
Monterey	Mariposa		
Orange	Merced	Total = 10	
Riverside	Mono		
San Benito	Napa		
San Bernardino	Nevada		
San Diego	Placer		
Santa Barbara	Plumas		
Solano	Sacramento		
Tulare	San Luis Obispo		
Ventura	San Mateo		
	San Francisco		
Total = 20	Santa Clara		
	Shasta		
	Sierra		
	Siskiyou		
	Sonoma		
	Stanislaus		
	Sutter		
	Tehema		
	Trinity		
	Tuolumne		
	Yolo		
	Yuba		
	Total = 33		

Source: Department of Toxic Substances Control, Office of Statewide Planning, Planning and Environmental Analysis Section (November 1992).

(DTSC). California's major budget crisis seemed to put further plan approval on hold. Many counties were still trying to get DTSC to review their plans; DTSC, however, no longer had the statutory authority to review the plans, nor the $2–$3 million it would cost to do so (Keene 1993).

Many counties have not given up their efforts to gain approval of their CHWMPs, even though the deadlines listed in the act passed long ago. Approval is a high priority for the counties, as it would lessen their vulnerability to unwanted facilities being sited under the appeal process. In 1993, the League of California Cities and the County Supervisors Association of California sponsored Senate Bill 786, which would extend the deadlines for DTSC to review and approve the remaining CHWMPs. The two organizations have supported this legislation with the hope that it would enable more county plans to be approved, avoiding the situation in which the appeals board could complete siting in any area zoned for industry. SB 786 includes the following provisions and objectives (Keene 1993):

- remove the requirement that CHWMPs be reviewed and approved by DTSC; assist as many counties/cities as possible in plan approval;
- retain requirements for LACs but emphasize waste reduction instead of disposal;
- clarify that cities/counties may amend their plans to remove objectionable anti-fair-share language and incorporate the CHWMP into existing county/city General Plans; and
- retain the appeal process.

In summary, once DHS destroyed the focal point of AB 2948 (the fair-share principle) the legislation never had an opportunity to improve hazardous waste management and the facility-siting process in California. Despite its passage into law, AB 2948's innovative approach to dispute resolution in this difficult policy arena was never really given an opportunity to be tested. Siting of hazardous waste treatment facilities remains as arduous as it was before the act. The difficulties inherent in the process are illustrated by three case studies of recent efforts to site hazardous waste incinerators in post–AB 2948 California.

DOW CHEMICAL COMPANY CASE STUDY (CONTRA COSTA COUNTY)

The Dow Chemical Company case provides an excellent example of the NIMBY phenomenon that state and local governments face in siting incinerators, a phenomenon that the AB 2948 process was meant to help resolve. Dow's facility in Pittsburg, California, manufactures agricultural products, caustic soda, chlorine, latex, solvents, and other chemicals. In 1986, this facility disposed of 36,000 tones of hazardous wastes through solar evaporation in surface impoundments on-site; it also sent 5,700 tons of hazardous wastes off-site for treatment and disposal at Dow plants in Louisiana (CHWMP 1989).

Dow first proposed to construct a new rotary-kiln hazardous waste incinerator at its Pittsburg plant in 1985 (before passage of AB 2948). This incinerator was estimated at the time to cost between $11 and $13 million (Cuff 1988a). Ninety-five percent of the waste to be burned in the incinerator would come from Dow's Pittsburg plant, with the remaining 5% coming from Dow's Walnut Creek and Fresno research laboratories (CHWMP 1989). Thus this was primarily an on-site unit; and it was entirely non-commercial. No non-Dow wastes would be accepted.

The new incinerator was designed to burn 7,000 tons of Dow hazardous wastes a year: tars, polymers, contaminated gloves and work suits, laboratory research materials, and other wastes. This figure reflects the enormous waste reduction projected by Dow's engineers: some 80%! The incinerator's waste ash would be trucked to an off-site hazardous waste landfill in California for final disposal. Use of this new incinerator would mean a 95% reduction in the total number of miles that Dow hazardous waste trucks would travel, with the commensurately lower probability of accidents and related hazardous spills (Cuff 1988a). Since Dow already treated and disposed of a large portion of its wastes on-site, operating under a permit from DHS, a new on-site incinerator seemed like a reasonable next step. Because surface impoundments were being closed nationwide, the shift to incineration was necessary if wastes from this facility were to continue to be managed on-site.

To understand what occurred in the attempts to site this hazardous waste incinerator, it is necessary to understand the role of environmental-impact assessment in the California siting process. When a company files its permit applications, it also files a Notice of Intent under the California Environmental Quality Act (CEQA). This begins the siting approval process. Under CEQA requirements, a lead agency is appointed: usually the city council of the community where the proposed hazardous waste facility is to be located, or the county board of supervisors if the facility lies outside city limits. The lead agency completes an environmental assessment and, if needed, develops an environmental-impact report (EIR) as outlines in CEQA. The draft report is made available for public review and comment. The EIR is then modified and a final report is produced. While CEQA review is under way, the new AB 2948 process (after 1986) called for an LAC to make a recommendation to the city council (or county board of supervisors), which then decides whether to certify the EIR as adequate. The city council also determines if amendments need to be made to local zoning.

Other regulatory agencies involved in the permitting process review the EIR and decide to certify or reject it. Their formal regulatory permit reviews come later. If the EIR is certified, the permitting agencies then decide whether the permit applications are complete. At this point, permit approvals or denials can be made by state and regional agencies, as well as by the local agency. Local permit decisions may then be appealed, if necessary, through the new process established by AB 2948.

In 1988, the LAC concluded that operation of Dow's incinerator would increase the statistical risk of cancer in the neighboring area by three cases in ten million (Cuff 1988a). Residents of Pittsburg and the neighboring town of Antioch were adamantly opposed to *any* increase in the area's cancer rate. They expressed fears that increases in the cancer rate would be even greater, since the pollutants emitted by Dow's hazardous waste burning included benzene, carbon tetrachloride, and dioxin.

Dow felt that it had proposed a good facility in a good location. The company was already successfully and safely operating a large chemical plant at this site. The proposed on-site incinerator would have modern, state-of-the-art air pollution control technology and would be located more than 2,000 feet from any home and 5,000 feet from any school or prison (as dictated in siting criteria outlines in Contra Costa County's General Plan). Its facility was designed to handle only Dow wastes. The land for the proposed new facility was already in industrial use, and zoned accordingly.

To ease the opposition anticipated from some members of the public, Dow offered public tours of an incinerator model and presented a slide show to interested citizens. In April 1988, Dow sponsored a "toxic waste forum" designed to inform the public about the risks of transporting wastes and the benefits of on-site incineration. The keynote speaker at this forum came from DHS's TSCD. The *Contra Costa Times* ran a few articles on the EIR and announced scheduled public hearings. Attendance at the first public hearing in May 1988 was poor, suggesting public disinterest in the proposed on-site noncommercial incinerator.

At a June 1988 public hearing on Contra Costa County's CHWMP, the Dow siting proposal gained more attention. Four environmental groups angrily expressed their belief that incineration was not the right answer to the county's hazardous waste management needs.

Three of the groups were local: West Contra Costa Toxics Coalition, Contra Costans against Toxic Pollution, and Citizens for a Better Environment (CBE). Greenpeace was also active at the meeting, representing statewide and national interests (Duff 1988b). The groups argued that source reduction had been overlooked in the county's planning process as the preferred way to deal with the area's toxics crisis. CBE proposed a new tax on toxic waste generation, plus a requirement that all businesses prepare and disclose their waste reduction plans. While most others noted that Dow had already been reducing its waste stream in significant annual increments, the activists advocated "zero waste generation." They felt it would be better to force industry to continuously reduce its wastes than adopt a CHWMP that would only "band-aid" the situation by allowing a new incinerator and planning to accept even more such facilities in the future as more need arose.

County Hazardous Materials Commission members explained at this meeting that they had proposed (and Dow had agreed to) restrictions on the volume of wastes to be handled in Dow's new incinerator and had set aggressive countywide goals for source reduction and recycling, hoping to reduce overall waste generation in 2000 by 40% from 1985 levels. (Dow's own reductions were much greater than this rate.) Since Contra Costa County in 1986 generated more hazardous waste per capita than any other county in California, its CHWMP would have to deal with both source reduction and incineration as well as with other methods of treatment and disposal (Cuff 1988b).

Dow's incinerator proposal was mentioned only briefly in the county's final CHWMP published in 1989, which simply noted that a permit application had been filed. Since Dow's incinerator was designed solely to handle Dow's own wastes, it did not factor greatly in the county's waste management plans, which focused on off-site facilities. Dow is listed among the county's largest hazardous waste generators and is mentioned several times in the CHWMP in discussions of total quantities generated and the overall need for new facilities. The CHWMP also points to Dow as the company that had achieved the largest amount of waste reduction from 1984 to 1987 (68%) (Contra Costa 1989).

In 1990, opponents of this incinerator, still few in number, asked the Pittsburg city council to reject the permits that Dow needed. Dow established a community advisory panel to help improve relations between the company and residents near the plant. Though environmentalists were skeptical that this was just another industry ploy, the panel was successful in dealing with the impact of a toxic spill at the plant in the fall of 1991. The panel quickly called a meeting with representatives of nearby residents to tell them about the spill before the media had a chance to do so. This sort of activity helped to build effective communication between the company and the surrounding community.

In May 1992, EPA and DTSC (successor to DHS) announced tentative plans to grant the federal Resource Conservation and Recovery Act (RCRA) permit needed to build the incinerator. Since all other local, county, and state permits were in order (Dow Representative 1993), this was the final step needed for approval before the city could grant necessary building permits. The regulatory agencies made plans to hold informational meetings followed by formal public hearings on the permits and to accept written comments, which would be taken into account in the final-approval process. These sessions took place without undue incident.

As the comment period drew to a close, however, with very few residents expressing interest in Dow's project, a group called CAP-IT (Citizens against Pollutants and Industrial Toxins) suddenly mounted a campaign against the incinerator. The reality that the incinerator had been tentatively approved and that the public had done nothing to stop it inspired the group's actions. Most residents had not previously been aware of the proposed facility nor of the incinerator's negative consequences (as CAP-IT saw them) for their community. Though Dow had a community outreach program, people had either not understood what was being said or had not cared to listen to the company. CAP-IT asserted that there was a need for much greater action and citizen education to allow the concerned public to protest.

Possibly, in retrospect, Dow should have done more earlier to educate local citizens to its own point of view. Dow should not have assumed that public apathy up to that point in the approval process represented latent public support.

There was also speculation that the CAP-IT group was angry at having lost its recent battle against the proposed siting in east Contra Costa County of a large new municipal solid waste landfill (Dow Representative 1993). While the new solid waste landfill was unrelated to Dow's proposed hazardous waste incinerator, CAP-IT activists felt that "all the wastes in the state were being dumped on East County."

At this crucial moment, overwhelming local response materialized, led by CAP-IT. Angry citizens showed up at a city council meeting. Others began distributing fliers and going door-to-door informing residents that a dangerous incinerator was about to be built in their backyard. A week earlier, the DTSC public meeting had been quiet; now, after all this organizing activity, an emotional crowd jammed the city council room demanding to be heard. They charged that they were getting stuck with the incinerator because they were low-income minorities living in an industrial area. "Environmental racism" became the topic of conversation (Been 1994a; Boyle 1993; Bullard 1990); soon residents were requesting extension of the formal regulatory comment period. In July 1992, responding to the sudden public interest, DTSC extended the comment period for two months and scheduled three additional public hearings.

In August 1992, faced with opposition of growing intensity, Dow withdrew its permit application and terminated plans for its incinerator. CAP-IT was overjoyed. They credited the withdrawal to their timely and effective campaign to inform citizens of the dangers of hazardous waste incineration in their community. Dan Offield, CAP-IT's leader and founder, was relieved that there "was still hope for the public protest system" (Richards 1992b).

From Dow's perspective, the incinerator proposal failed for several reasons, including economics and public opposition. Because of inflation, the cost of building the incinerator by 1992 had already almost doubled to $20 million. Dow's site manager envisioned the permit approval process lasting at least another year after public comment was reopened, causing costs to escalate further. Given the change in economics, Dow felt it would be a waste of the firm's (and the regulators') time to continue with the proposed project (Richards 1992a). Last, Dow's "Waste Reduction Always Pays" program had already reduced hazardous waste generation significantly from 1985 levels. Dow stated that it supported the permitting and public-comment process involved in siting hazardous waste management facilities but its project was now uneconomical (Cal-EPA 1992a). Today Dow still exports a large portion of its remaining California hazardous wastes to Louisiana, since no facilities have been sited in California to treat these materials.

RHONE-POULENC BASIC CHEMICAL COMPANY CASE STUDY (CONTRA COSTA COUNTY)

Developments in this case provide further insights into the facility siting process operating under AB 2948. In January 1985 (before AB 2948 became law), Stauffer Chemical Company applied to EPA and DHS for a hazardous waste incineration permit to modernize the rotary-kiln incinerator at its sulfuric-acid recycling plant in Martinez. The proposed focus here on commercial treatment of off-site wastes strongly distinguishes the Martinez siting proposal from Dow's proposed on-site treatment unit. Size was also a major difference: Stauffer's incinerator was twenty times as large as Dow's. In fact, the new incinerator's capacity would have far exceeded Contra Costa County's total needs.

As the permitting process began, Martinez was designated lead agency under CEQA. In this capacity, the city authorized several studies and reports to determine projected impacts from the incineration project. The county formed an LAC as required under AB 2948. Martinez, in March 1987, prepared and reviewed an initial environmental study—a precursor to the draft EIR. Rhone-Poulenc Basic Chemical Company (RPBC) acquired Stauffer in 1989 and continued the efforts to begin operating a commercial incineration facility at the plant.

A risk assessment completed in March 1989 elicited strong reactions from the different groups involved in the siting process. Citizens for a Safe Environment (CSE), a local environmental group, felt the assessment did not address all of the environmental risks. The group believed the RPBC incinerator would add to the heavy air pollution from oil refineries and chemical plants that already existed in the area. CSE also felt the assessment did not adequately consider the risk of accidents at the plant or spills on the highway and that it did not look at issues of human or equipment failure at the plant.

As part of the permitting process, several more reports were developed to assess the economic impacts and feasibility of the incineration project. A financial-impact report requested by the LAC was completed in February 1990. In May 1990, a baseline monitoring report detailed RPBC's emissions from the plant's existing operations. Information from this report was used to revise the risk assessment, an updated version of which was issued in February 1991. Further amendments were made in September 1991 in response to agency and toxicologist comments that Martinez had required on the draft report.

In mid-1990, the city council adopted a CSE-sponsored citizen petition to limit storage and disposal within the city limits solely to those hazardous wastes generated within Martinez. Since RPBC's incinerator was designed to treat far more wastes than what the city alone produced, its design far exceeded these limits. RPBC argued that the city's plan was not "consistent with the county's hazardous waste management plan," and that city "siting criteria and zoning ordinances" would "not apply to the company's application for a conditional use permit." RPBC asked Martinez to relinquish its role as lead agency in the environmental-assessment process to either the state Office of Planning and Research (OPR) or Cal-EPA. OPR responded to RPBC's request, stating that while it did not have the legal authority to assume the role of lead agency in the CEQA process, OPR was available to help mediate the growing dispute between RPBC and Martinez.

Beginning in July 1990, members of the LAC began to question the need for more hazardous waste incineration in Contra Costa County or even at the state level. The LAC suggested a market needs analysis report to provide information on the origins of the wastes to be treated at RPBC and on the amount of profit RPBC could anticipate from its expanded rotary-kiln incinerator. Martinez required RPBC to prepare this report. In December 1990, this initial study was prepared; revisions were added based on comments received from the LAC.

In August 1990, RPBC informed the LAC that the company was facing indictment in Los Angeles on misdemeanor and felony charges for violations of the California Health and Safety Code by its facility located there. RPBC attempted to reassure the LAC that these charges constituted a "misunderstanding." The charges were of great concern to members of the LAC and residents of Martinez, however, reinforcing their negative attitude toward the proposed incinerator and RPBC.

In December of that year, the LAC was still not convinced of the need for, or safety of, RPBC's incineration project. Controversy intensified in February 1991 hearings, focused on the revised risk assessment. In June 1991, RPBC updated its original permit applications and applied for a conditional land use permit (and its related zoning ordinance text amendment) from Martinez and a permit to operate from the Bay Area Air Quality Management District (BAAQMD).

Two LAC members toured the company's incinerator in Texas in December 1991. The visitors concluded that this incinerator caused "more negative than positive impacts" on the surrounding community and that "the petro-chemical industry is not to be trusted." The LAC also voiced concern in January 1992 over RPBC's proposed use of a process to blend hazardous wastes prior to incineration, as the company was doing at its out-of-state incinerators. The committee felt the blending process posed an added risk.

In April 1992, when the draft EIR was finally released, a dispute over payment of fees developed. RPBC noted it had financed the draft environmental study with over $700,000. Martinez claimed, however, that RPBC had not met the city's March 2, 1992, deadline to pay an additional $450,000 technical-assistance fee charged by the city council. At this point, RPBC openly expressed its frustration with the LAC and Martinez. RPBC termed Martinez "inept and uncooperative" and declared that it would no longer fund the LAC. When the company did not respond to the new payment deadline set for June 1, Martinez declared the incineration application invalid.

Negotiations ensued, but they came to a halt when a serious chemical spill and fire occurred at RPBC's Martinez sulfuric-acid recycling plant on June 22, 1992, killing one man and seriously injuring another. This accident, one of the worst industrial accidents ever in the Bay Area, galvanized the local community in intense opposition to the incineration project.

In August 1992, RPBC formally withdrew its plans for this off-site commercial hazardous waste incinerator. The company had spent upward of $4 million in its attempt to obtain the necessary permits, yet the permitting process had only reached as far as completion of the draft EIR. Though the company stated that negative business projections "were the driving force behind [its] decision," it did admit that "the . . . accident and the visibility it created, concentrate[d] the attitudes of the public to our facility here."

Roles and Perspective of Different Participants

This siting case had several key players. The main group that supported the incineration project was, of course, the company proposing it: RPBC. Various parties fought against the incinerator: CSE, local residents, the city council, and even the LAC.

As project proponent, RPBC was the hazardous waste incinerator's main supporter. The company's stated position in newspaper editorials was that it wanted to "help solve society's problems, . . . reduce the amount of toxics" in the Bay Area, "limit transportation hazards," and "prevent unnecessary landfills." RPBC believed that its project would not have any negative health effects on the community.

CSE was formed by Martinez residents in response to RPBC's proposed incineration project. Its members, concerned about the environmental risks of such a project, fought against the incinerator with protests, participation at LAC meetings, and involvement in the siting process. CSE was concerned that this large incinerator would encourage toxic waste producers to continue their "dangerous practices" by giving them ready access to a way to dispose of their wastes. The group advocated source reduction instead as the best way to decrease the problem of hazardous wastes. Throughout the process, this group asked RPBC to withdraw its permit applications.

CSE reflected the attitudes of numerous Martinez residents, many of whom came to the LAC hearings to voice their concerns about the project. These residents felt that they were already subjected to excessive hazardous waste pollution from all the industry that existed in their area. To add a large incinerator in their community was simply too much for them.

Martinez played one of the largest roles in the process. This city was lead agency in the CEQA process and held the authority to grant or deny local approval of the proposed hazardous waste incinerator. The city's opinions of the project and its analyses of the numerous reports had a distinct slant against the incinerator. Having such a large hazardous waste facility in the city was not seen as an attractive prospect.

The U.S. Environmental Protection Agency and Cal-EPA (including DHS/DTSC) had to decide if RPBC would receive the federal and state hazardous waste incinerator permits for which the company had applied. Both of these agencies refrained from taking an active role at the CEQA stage. Had the siting process continued further, these regulatory agencies would have become much more involved. Both of them view incineration as a viable and necessary method for treatment of certain hazardous wastes. BAAQMD, responsible for approving RPBC's air quality permit, also took no active role in the siting process. The Regional Water Quality Control Board would have had to modify permits held by RPBC to accommodate the incinerator. This agency served only as an advisor in the process, reviewing and commenting on the EIR.

Contra Costa County, too, did not seem to take a direct role in the siting process, although it was indirectly involved in the deliberations through the LAC. Membership on the LAC consisted of two environmental representatives, both from Martinez; three community representatives, one from Concord and two from Martinez; and two business representatives, one from Pleasant Hill and the second position vacant in the later months of the siting process. Community representatives served as both chair and vice chair of the LAC. In its dialogue with RPBC, the LAC seemed to adopt a consistently negative attitude about the proposed project.

Analysis of the Rhone-Poulenc Case

The Rhone-Poulenc siting case provides another opportunity to study how the facility siting process worked under AB 2948. This case allows us to examine the siting process outlined in a CHWMP.

The process started when RPBC first submitted its applications for incineration permits. The LAC was created presumably to help facilitate negotiations between the city and RPBC over the terms and conditions of these permits. RPBC and the different agencies moved along through the process until the 1992 fire at the company's plant, when the process stopped (still at the draft-EIR stage).

Mistrust intensified as disagreements arose over the facts in the different reports, the charges against RPBC in Los Angeles, and confusion over the roles and responsibilities of Martinez, the LAC, and CSE. Frustration among the participants grew, and interactions between them became more strained. This was exemplified in RPBC's request that Martinez pull out as the lead agency and let a state agency take over and, in February 1991, by the doubts expressed by the LAC over the value of ongoing talks with RPBC.

At several points, the LAC, Martinez, and RPBC seemed to be reaching a stalemate. No one was working to establish the basis for effective compromise over conditions or terms, acceptable to all parties, that would allow for eventual approval of all the necessary permits. The LAC's position as a negotiator was increasingly compromised, as it joined those working to oppose the incinerator. Most of the LAC's own fact finding and its analyses of other's reports supported the position that an incinerator—especially such a large one—should not be sited in the community.

Under AB 2948, the LAC's basic responsibility was not to determine if the facility should be approved or denied. The Martinez city council would make that decision locally (along with the several required state and federal environmental regulatory-agency permit approvals). AB 2948 directs the LAC to "negotiate with the applicant on the terms, provisions and conditions for project approval which would protect the public health, safety and welfare, and the environment of the County and would promote the fiscal welfare of the County through special benefits and compensation." In this case, the LAC became simply another group that RPBC had to try to persuade—in the end, unsuccessfully—to accept its project.

Understandably, the LAC was charged with representing local residents. But how is "local" to be defined in this context: The immediate vicinity? Martinez? Contra Costa County as a whole? While residents living near the RPBC plant obviously did not want the incinerator sited, the situation was never clarified regarding the interests of the county as a whole. The market needs analysis might have provided useful perspective, but not when used only to prove preconceived notions.

Rather than ensure that the negative impacts associated with other hazardous waste facilities did not occur at the proposed Martinez incinerator and rather than create an agreement that would address local concerns, the LAC in the best "NIMBY tradition" worked to make sure that this project did not go through. The LAC certainly did not reflect any broader countywide concerns over achieving adequate incineration capacity to meet Contra Costa County's substantial needs.

Instead, the LAC gave citizens another outlet to say no. The LAC did not compare the proposed Martinez incinerator with Contra Costa County's own CHWMP or with the Bay Area's regional plan to see if this proposed project was consistent with county or regional needs. The Contra Costa plan projected that the county would generate 33,748 tons of incineratable wastes annually. While the RPBC project would have substantially exceeded the county's own requirements for incineratable hazardous waste treatment, on a regional level RPBC's incinerator would have contributed substantially to meeting the larger Bay Area's hazardous waste management needs.

According to the market needs analysis, 189,551 tons of incineratable waste were being generated per year in the Bay Area. ABAG estimated that this figure would increase to 210,000 tons by 2000, even after factoring in maximum source reduction projections. These amounts exceed even the large capacity of RPBC's proposed incinerator. The RPBC facility thus would have substantially met the Bay Area's fair share of California's incineratable hazardous wastes. As such, it would have helped meet many, if not most, of Contra Costa County's facility-siting obligations under the AB 2948 process. Other facilities able to handle Contra Costa's nonincineratable wastes might have gone to Alameda, Santa Clara, and other counties with large waste generation rates.

While regional compacts among the Bay Area counties do not currently exist, the Contra Costa CHWMP does refer to possibly inter-county agreements:

> The county shall enter into negotiations with other jurisdictions for the purpose of negotiating one or more inter-jurisdictional agreements for the siting of hazardous waste management facilities adequate and necessary to meet the needs of the signatory jurisdictions. . . . This county and its cities, in turn, agree to actively consider and, if appropriate, to commit as part of an inter-jurisdictional agreement to approve the siting of an environmentally appropriate facility or facilities within its own borders designed and sized to serve the hazardous waste management needs of other jurisdictions as well as of this county.

By accepting a large facility like RPBC's incinerator, Contra Costa County could have demonstrated to other counties in the Bay Area that it had met its regional fair share, and thus it would not have been responsible to accept any further large-scale hazardous waste management facilities (such as a landfill, perhaps). Contra Costa generates more hazardous wastes per capita than any other California county. Under the regional approach, other needed facilities would have gone to other counties, required to site facilities locally in order to uphold their part of the regional interjurisdictional compact. With RPBC's incinerator approved, Contra Costa County would have been off the hook for most further facility siting. Yet this possibility never seems to have even entered the RPBC siting debate.

The proposed incinerator was formally consistent with Contra Costa's DHS-approved CHWMP. In the face of an appeal, and with RPBC's regulatory permits in hand, the state appeals board might well have decided to override local rejection of RPBC's incineration project and compel siting of the facility. RPBC's decision to withdraw its permit applications rendered such speculation moot, of course.

With the sides so polarized, the Martinez project was bound to fail. Obstacles and disagreements kept emerging. By the time the company withdrew its plans for the incinerator, RPBC had several factors working against it. The charges in Los Angeles intensified many participants' skepticism of this company's ability to operate an incinerator safely. The 1992 fire confirmed the doubts held by those who were against the project and convinced others to join in the opposition.

Finally RPBC gave up. It was not worth the company's time and energy to continue the process. The process took too long, and adequate economic incentives were no longer there. The county and the region had much potentially to gain from successful siting of a large incinerator, but neither was able to bring its powerful interests to bear on the case.

DHS' undercutting of the statewide planning and siting process contributed to the failure of the RPBC incineration-siting effort. Local residents in this case felt that they were being asked to accept far more than their fair share. Had the fair-share principle remained as designed under AB 2948, before DHS added its guidelines, this conflict might have been reduced. The county's needs would have been clear and politically defensible. Under a negotiated regional agreement, this incineration plant could perhaps have been structured to accept wastes produced primarily in the Bay Area. In this way, once the large new Martinez incinerator had been approved, the Bay Area counties would have been taking their fair share of California's hazardous waste and Contra Costa County would have been assured that it would not be forced to accept additional hazardous waste treatment facilities of any kind.

Some amount of NIMBY reaction is to be expected whenever a hazardous waste facility is proposed, whether a fair-share principle exists or not. AB 2948 had attempted to quell NIMBY and add some perspective to the controversy over siting facilities in reluctant counties by getting each county to agree to manage its own fair share of California's overall hazardous wastes. Larger waste-generating counties such as Contra Costa would have had a strong incentive to support development of some facilities as the basis for being able to resist others.

The opposite approach set by DHS required each county to plan to accommodate any and all kinds of facilities regardless of its local needs for hazardous waste management. As a result, the fight in Contra Costa over RPBC's incinerator escalated. None of the agencies would approve this project if it felt the process would force Contra Costa County to accept more wastes for treatment than was equitable on either the county or the regional level.

This case demonstrates that even in counties like Contra Costa with approved CHWMPs, NIMBY continues to be a powerful force against approval of proposed new facilities. When there is enough public opposition to a project, facility proponents such as RPBC face insurmountable problems in trying to obtain approval for their project. Reliance on fair-share siting as initially designed in AB 2948 has the potential—still untested, to be sure—to balance this situation by providing positive as well as negative siting incentives.

The winners in this case were all those parties who had reservations about the incineration project and did not want the facility to go through. The loser obviously was RPBC. In the larger sense, however, the San Francisco Bay Area and the state of California lost as well. Even though source reduction is the preferred method to decrease the amounts of hazardous wastes that must be dealt with, not all companies and industries can alter their production processes to reduce or eliminate toxic wastes at the source of their creation. As ABAG's numbers show, sizable quantities of hazardous waste (over 200,000 tons/year of incineratable wastes in the Bay Area alone) remain to be treated even after source reduction is considered. As many landfills that accept hazardous wastes are being phased out in accordance with the federal Hazardous and Solid Waste Amendments passed in 1984, failure to implement alternatives to the treatment of hazardous wastes forces greater volumes to be shipped to facilities located out of state.

CHEMICAL WASTE MANAGEMENT CASE STUDY (KINGS COUNTY)

The third incinerator-siting case study involves Chemical Waste Management's (CWM) Kettleman Hills facility in Kings County in California's Central Valley, about midway between Los Angeles and San Francisco. CWM is a large nationwide hazardous-waste management firm that also does remedial-action work and household waste collection.

CWM purchased the 1,200-acre Kettleman Hills hazardous waste landfill in the 1960s. In 1987, CWM applied to Kings County for an additional permit to build at its landfill a new off-site commercial hazardous waste rotary-kiln incinerator. This would allow the company to expand its waste treatment capabilities.

Kings County required that an EIR be prepared as part of its permitting process. The county hired a consultant to prepare the EIR, using funds provided by CWM. The environmental-review committee of the Kings County Planning Commission approved the EIR; and in January 1990, CWM applied for, and was granted, a conditional-use permit by the planning commission.

A clear indication of community opposition arose, however, when the permit approval was challenged at a planning commission public hearing. This challenge resulted in an appeal to the Kings County

Board of Supervisors. After investigating the EIR and CWM, the board voted three to one to allow CWM to proceed with the permitting process.

In 1993, however, six years after CWM initially applied for the permits to build and operate an incinerator at Kettleman Hills, the company was still far from receiving full permission to construct its facility. The reasons for this delay can be traced to community activism, including explicit accusations of "environmental racism." One article in the growing literature on environmental racism focuses specifically on this case (Cole 1994; also see Atkinson 1994; Been 1993; Been 1994b; Lazarus 1994; Mohai and Bryant 1992; Wright 1991).

When CWM applied for its incinerator permit, an LAC was formed as dictated by the Tanner Act. The county board of supervisors selected the seven LAC members, who included a community activist named Joe Maya. After many meetings with CWM representatives, and relying explicitly on the assumption that CWM would meet 100 technical, environmental, and economic conditions, the LAC voted to approve the EIR and to recommend to the planning commission that the permit be issued.

In the meantime, however, Greenpeace had become aware of the CWM incinerator proposal and had begun to organize the Latino community whose residences surround the facility. Greenpeace's primary organizing theme was "environmental racism." The group charged that CWM managers and county officials all lived in white areas unaffected by the proposed incinerator, while the Hispanic community living near the facility would bear the brunt of the increased health risks and lower property values posed by the new incinerator. As further evidence of its environmental racism charge, Greenpeace pointed to the fact that some of the permit hearings had been held 30 miles away in Hanford, the county seat, allegedly to prevent local attendance. The county and CWM responded by holding other meetings at locations more accessible to the community.

Two incidents involving the need for translation services for Spanish-speaking community members added fuel to the charges of environmental racism. CWM had produced bilingual newsletters and public notices for nearby residents throughout the process and had made translators available at meetings and public hearings. An outside judge hired by Kings County to facilitate the public hearings, however, ignited a storm of outrage when he told Spanish speakers to gather in the back of the room, where a translator would be available. The judge apparently was attempting to group those who needed the service so the translator could be heard by all Spanish speakers, but the attempt was interpreted as an effort to relegate the Hispanic community to the "back of the bus" for racist reasons.

The second issue fueling racism charges involved the EIR. Neither the county nor CWM were required by law to translate the EIR. The county decided not to do so because of the document's length and technical complexity. By the time the county reached this decision, the draft document was already in circulation in English. CWM did translate the draft EIR's twelve-page Executive Summary into Spanish, which it distributed to community members. But CWM chose not to translate the full EIR.

The California Rural Legal Assistance Foundation, on behalf of the local group, sued Kings County for failing to translate the EIR into Spanish. A superior-court judge ruled in 1991 that the EIR was inadequate (because of low estimates of air emissions) and that it needed to be translated into Spanish. The judge noted that the county had failed to involve local residents in the permitting decision because the 40% of the community who spoke only Spanish could not read the full text of the EIR and the related permit documents. CWM appealed this decision; the case is still under review in the judicial process.

Analysis of the Chemical Waste Management Case Study

The CWM Kettleman Hills case offers important issues for consideration. The AB 2948 siting process was followed, with the LAC fulfilling its role of examining the proposal and recommending to the county either permit approval or denial. According to a former Kings County supervisor, "reasonable balanced people were involved in the LAC, the AB 2948 time lines were met, and the committee functioned effectively." The LAC's community activist member originally approved the committee's recommendation, then later reversed his position and withdrew support (after pressure began to mount from Greenpeace's organizing efforts).

Greenpeace's Bradley Angel sees the process very differently. In his view, the state appeals board's power to overrule the decisions of local agencies amounts to a process created to "ram incinerators down the throats of communities and overrule democracy." Many local citizens adopted Angel's opinion that the process "excludes the community and gives a rubber stamp to racism."

Though many community members were originally impressed by the jobs associated with the proposed facility expansion, they became convinced by outside activists that the incinerator represented an impending health threat. As the community became more inflamed, the public meetings became dominated by an emotional and disruptive core group dedicated to avoidance of compromise and to total defeat of the proposed incinerator.

Attempts by CWM to satisfy activist demands were used as further ammunition to support the charges of racism. When the community expressed concern that the county had not translated the EIR, CWM translated its Executive Summary. This action was used as evidence that CWM did not want the public to have access to technical information but instead only wanted to provide a simplistic summary. When CWM hired a liaison to improve relations and communication with the Hispanic community, Greenpeace charged that it had hired a Hispanic "wolf in sheep's clothing" to lie for it. When CWM stressed the extra jobs the incinerator would provide, it was accused of "economic blackmail."

Greenpeace's Angel views the Kettleman Hills incinerator proposal as a "test case on environmental racism." Greenpeace publicized this case nationally and internationally to garner increased attention and financial support. Greenpeace sent protesters across the country to lead an anti-incinerator campaign and published an encyclopedia of CWM's alleged misdeeds at other facilities to expand national interest in the Kettleman Hills case.

Greenpeace and the California Rural Legal Assistance Fund have focused legal and political pressure on Kings County and other governmental organizations involved in the AB 2948 process, as well as on CWM. Bradley Angel feels that Cal-EPA, DTSC, and Kings County all have a vested interest in siting the incinerator and that because of racist motivations, they all ignore the risks the proposed incinerator poses to the local community. He alleges corruption and conflict of interest, since "several top state and federal permitting officials have gone to work for CWM after the Kettleman Hills incinerator permit was approved."

Whether the siting process would have resulted in successful siting of the Kettleman Hills incinerator if a national/international group from outside the community had not become involved is unknown. With the involvement of Greenpeace, however, the balanced components of the AB 2948 process were rendered ineffective by the unwillingness of the activists to engage in good-faith discussion and compromise. When asked how effective he felt the process had been from a dispute resolution perspective, Angel responds, "It's a war; there is nothing to negotiate or mediate." When a powerful national group takes this stance, the process alone cannot readily resolve the controversy.

Greenpeace has launched a massive campaign against the proposed incinerator, using protests and litigation to delay resolution of the situation. Its goal is to defeat the incinerator by making it too expensive for CWM to continue the fight. According to Angel, "Time works to our advantage; if it costs too much, the company will back down." This may indeed come to pass. The adequacy of the EIR and its translation are still being litigated, and the state appeals board will not rule on the county permit until the legal case is decided.

In spite of the adherence to the AB 2948 siting process and CWM's use of an aggressive community information program that exceeded formal requirements (consisting of many community meetings and voluntary provision of translation services), this incinerator permitting process became immobilized by litigation. In the absence of full permits from all the regulatory agencies, the state appeals board remains unwilling to resolve the challenge to the county's permit approval.

According to CWM spokesperson Sylvia Vickers, AB 2948's mechanisms succeeded in facilitating the process locally at the beginning, but the process fell short of its intent to site facilities in an acceptable way. Vickers feels that this may be an impossible goal, given NIMBY attitudes. In her view, state or federal intervention may ultimately be necessary to ensure that enough incinerators are available to meet state and national needs.

CONCLUSION

Most localities have traditionally managed to evade responsibility for hazardous waste management through a combination of relying on the private sector to "manage" these wastes—somewhere—and accepting an uncooperative NIMBY attitude that precludes local siting of proposed new hazardous waste management facilities. Because of perceived economic, environmental, and health risks associated with hazardous wastes, very few communities have wanted to site a treatment or disposal facility in their vicinity. More important, no community has wanted to become the "dumping ground" for a significant portion of California's wastes.

After years of gridlock regarding hazardous waste facility siting, in the early 1980s concerned representatives from local and state governments, environmental coalitions, industries, and citizen groups joined forces to reassess (and, it was hoped, to revolutionize) California's system of hazardous waste management and facility siting. Although the parties involved had different conceptions of exactly how much waste was being generated, how these wastes were being managed, and how the wastes should be managed, everyone agreed on the basic fact that California had insufficient capacity to deal with its increasing waste streams. Federal legislation in 1986 made it mandatory for each state to assess and plan for effective management of its hazardous waste streams.

AB 2948 emerged as a cooperative effort by industry, local governments, state officials, and environmental groups to affirm California's dedication to current and future hazardous waste management. The act was a courageous first step prompting localities across the state to confront at least a decade of denial. This legislation's fair-share approach provided incentives for localities to acknowledge, and take responsibility for, effective management of the hazardous wastes being generated in their own areas.

Although AB 2948's many positive features brought California a few steps closer to effective hazardous waste management, it fell badly short of breaking through the intense political, economic, and psychological gridlock. Most important, the fact remains that no facilities have been sited since 1986, when AB 2948 was passed.

On the positive side, the act succeeded in engendering an unprecedented cooperative effort by state and local authorities, private businesses, and concerned citizens. All fifty-eight of the state's counties prepared draft plans, most quite comprehensive. These plans established a blueprint for communities to determine the feasibility of hazardous waste facilities, based at least in part on the merits of the proposal instead of on NIMBY sentiments alone. In many counties, AB 2948 successfully promoted public participation in governmental decision making. The act established a process designed to involve the community as much as possible, from the county advisory committees, which helped facilitate development of CHWMPs, to the LACs, which negotiated with the facility applicant the conditions for approval of the proposed project.

Moreover, although new facilities have not been sited, county efforts under the AB 2948 framework have accelerated actions by industry to rely more on source reduction. The data provided and goals established in the CHWMPs made generators of hazardous waste more aware of their waste streams as well as more conscious of standards for source reduction. As one observer noted:

> Waste minimization is being accelerated. The impetus of the AB 2948 process has resulted in a heightened awareness and momentum to reduce hazardous waste through the waste minimization hierarchy of source reduction, recycling, treatment, and least desirably disposal. . . . It appears that while regulating hazardous waste, a number of jurisdictions are providing or intending to provide waste minimization information as a positive benefit of the regulatory process. (Penn 1988)

According to Peter Chamberlain of the Association of Monterey Bay Area Governments (AMBAG), AB 2948 forced counties to assess the wastes generated within their boundaries, forced local governments to deal with the prospects of siting facilities, and documented statewide data on hazardous waste generation.

Unfortunately, the parties lost the incentive to cooperate. Disagreements and controversy became too large and too convoluted. The impetus to implement a comprehensive enforced plan disappeared.

In furthering the actual siting of environmentally sound and safe new facilities, the process has not been very useful. California remains basically where it was before AB 2948: no new facilities have been sited to treat hazardous wastes, whose volume increases year by year despite intensified source reduction efforts. Currently only two off-site facilities accept the full range of residues of treated hazardous wastes: one located in the southern end of the Central Valley at Kettleman Hills, Kings County, and operated by CWM; the other, owned by Laidlaw, in Imperial County in southern California. No new treatment facilities or incinerators exist. Until industry adopts more of the alternatives for treatment of hazardous waste, including on-site treatment and additional source reduction, California clearly needs more facilities to treat these wastes.

The AB 2948 siting process was designed to encourage counties to prepare plans to allow California as a whole to get a better sense of the state's overall waste generation patterns and the resultant need for new facilities. The process was successful to the extent that the committees formed were able to design and recommend plans for approval.

When the heart of the process, the fair-share principle, was removed by the state hazardous waste regulatory agency, however, counties that had opened themselves to the idea of managing their own hazardous wastes turned negative. Positive local incentives for both siting and source reduction disappeared. The goal of the legislation—to smooth the way for siting needed new facilities in California—was never reached.

The process allowed a diverse mix of people to help create county plans to meet county and, in the aggregate, state needs. While the process was an important exercise in bringing all parties to the table, it did not achieve any facility siting primarily because of state subversion of the fair-share concept.The act set up a way for counties to prepare CHWMPs, collecting data and defining criteria in the process. But it failed to create a model for effective implementation of these plans after their approval. Both economics and NIMBY can be blamed for the impossibility of CHWMP realization.

It is frustrating for the people who envisioned the AB 2948 siting process to have had their efforts cut short by the fair-share disagreement. They will never know whether the process, fairly tried, would have been successful in getting facilities sited. The record does suggest hope for intergovernmental cooperation. Local governments worked together to create plans that would protect them from the threat of having a bad facility sited in their area. Industry and environmental groups sat at the same table, negotiating and understanding each other well enough to complete good plans. The process was successful to the extent that people with differing viewpoints and interests were able to resolve their conflicts, put aside their differences, and seek a mutually beneficial outcome.

Would the process really have worked had it been implemented as designed? No one really knows. Many signs did indicate that extensive early progress was being made toward true local assumption of responsibility for adequate hazardous waste management—including acceptance of the need for local siting of facilities to meet local needs. The sense of "shared pain" as fair and the awareness of an ability to manage one's own destiny (as a county)—rather than be totally vulnerable to the actions of the state or private hazardous waste management companies—were powerful dynamics clearly at work in the early months of the AB 2948 process. Local elected officials acted as if it were at least feasible to advocate siting of some locally needed facilities in their own localities, within the balanced structure of the statewide planning and siting process, which left their area protected from forced assumption of larger, unneeded facilities. This was a far cry from officials' traditional abdication of power to the strident voices of the NIMBYs.

Yet the question whether county-level responsibility would be sufficient to force NIMBYs (and cities like Martinez) to come to terms with the internal contradictions of their anti-siting movements remains unanswered. The experience documented in Contra Costa and Kings Counties is not very encouraging in this regard—though in all three cases the state's rejection of the fair-share principle rendered impossible a true test of AB 2948's underlying philosophy.

Public participation in the AB 2948 process was substantial but certainly falls far short of the voluntary-siting processes seen elsewhere (in several Canadian siting cases, for example). Perhaps a local referendum might have enhanced the legitimacy of this siting innovation, adding further balance between local interests and the state Siting Appeals Board. Fair-share and equity-based siting seemed equally vital to long-term success, however.

Even in counties, such as Contra Costa, with approved CHWMPs, NIMBY remains a powerful force opposed to approving proposed new hazardous waste management facilities. When public opposition against a project emerges, facility proponents find the permit approval process excessively time-consuming and costly.

In the future, when hazardous waste facilities are proposed in California (and elsewhere), one can hope that alternative dispute resolution will be utilized in all aspects of the approval process and in the facility proponent's interactions with the community. The act's mediation provisions provide a good start in assisting the LAC, the project applicant, and other disputants through their points of impasse. Conflict resolution techniques can be invoked in many other areas to facilitate approval and acceptance of a needed hazardous waste treatment project.

A forcing function seems necessary, however, to ensure that all the concerned parties are sufficiently motivated to employ the balanced mechanisms in the siting process and to abide by the results of negotiated solutions. Project opponents can too easily refuse to negotiate and stonewall the process through total opposition to a facility. In the final analysis, California's AB 2948 process is an innovative step forward in seeking solutions to complex hazardous waste treatment controversies, but one that in its current form is still not powerful enough to overcome the enormous forces of the NIMBY syndrome.

NOTE

1. Much of this chapter is based on research completed by Ilene Gregorian, Julie Knutson, Cynthia Leslie-Bole, and Jessica Tuteur, whose efforts are gratefully acknowledged.

REFERENCES

Abreu, A. 1990. Rhone Poulenc: City Not Accepting Share of Waste Problem. *Martinez News-Gazette,* October 20.

Amy, D. J. 1990. Environmental Dispute Resolution: The Promise and the Pitfalls. In *Environmental Policy in the 1990s,* edited by Norman J. Vig and Michael E. Kraft. Washington, D.C.: Congressional Quarterly.

Angel, Bradley. 1993. Interview with author.

Atkinson, Amanda. 1994. Environmental Inequity: An Emerging Concern for Government. *Maryland Journal of Contemporary Legal Issues* 5:1.

Been, Vicki. 1993. What's Fairness Got to Do with It? Environmental Justice and the Siting of Locally Undesirable Land Uses. *Cornell Law Review* 78:1001.

———. 1994a. Conceptions of Fairness in Proposals for Facility Siting. *Maryland Journal of Contemporary Legal Issues* 5:13.

———. 1994b. Siting of Locally Undesirable Land Uses: Directions for Further Research. *Maryland Journal of Contemporary Legal Issues* 5:105.

Bingham, Gail. 1986. *Resolving Environmental Disputes.* Washington, D.C.: Conservation Foundation.

Boyle, Edward P. 1993. It's Not Easy Bein' Green: The Psychology of Racism, Environmental Discrimination, and the Argument for Modernizing Equal Protection Analysis. *Vanderbilt Law Review* 46:937.

Brown, Donald A. 1988. *Report from the Front: The Failure to Integrate Environmental Ethics into Daily Decision-Making*. Pennsylvania Department of Natural Resources.

Bullard, Robert. 1990. *Dumping in Dixie: Race, Class, and Environmental Quality.*

California Department of Health Services. Toxic Substances Control Program. 1989. *Draft Status Report on Hazardous Waste Management in California.*

California Environmental Protection Agency (Cal-EPA). Department of Toxic Substances Control. 1989. *Capacity Assurance Plan.*

———. 1992a. *Dow Withdraws Incinerator Proposal.* 1, no 4.

———. 1992b. *Fact Sheet No. 3: Regarding the Hazardous Waste Incinerator Proposed by the Rhone-Poulenc Basic Chemicals Co.*

California Health and Safety Code. Section 25135.2. 1986. *Hazardous Waste Management Plans* (AB 2948).

Caroompas, Jim. 1989a. New Statewide Anti-toxics Group Formed. *Martinez News-Gazette,* July 25.

———. 1989b. State Will Have Final Say, *Martinez News-Gazette,* April 13.

Cole, Luke W. 1994. The Struggle of Kettleman City for Environmental Justice: Lessons for the Movement. *Maryland Journal of Contemporary Legal Issues* 5:67.

Contra Costa County. Community Development Department. 1989. *Contra Costa County Final County Hazardous Waste Management Plan.* Martinez, Calif.: Community Development Department.

Cuff, Denis. 1988a. Dow Toxic Plant Called Safe, *Contra Costa Times,* April 13.

———. 1988b. Environmentalists Attack County's HWMP. *Contra Costa Times,* June 23.

Dow Chemical Co. Representative (anonymous). 1993. Interview with author.

Elliot, Michael. 1989. Improving Community Acceptance of Hazardous Waste Facilities. *Hazardous Waste* 1, no. 3:397–410.

Fischer, R., and W. Ury. 1981. *Getting to Yes: Negotiating without Giving In.* Boston: Houghton Mifflin.

Johnson, Grantland. 1988. Hazardous Wastes. *Sacramento Bee,* December 18.

Keene, Karen. 1993. *Hazardous Waste.* Legislative Bulletin no. 10, March 15.

Kings County Supervisor (anonymous). 1993. Interview with author.

Lapin, Lisa. 1988. Little-Known Committee Lays Toxic Waste Plans. *San Jose Mercury News,* August 30.

Lazarus, Richard. 1994. The Meaning and Promotion of Environmental Justice. *Maryland Journal of Contemporary Legal Issues* 5:1.

Lennard, C., and C. Labadie. 1988. Coping with Tanner. *Western City,* November, 23, 25, 27.

Maclay, K. 1992a. Rhone-Poulenc Monitor Panel Is Completed. *Contra Costa Times,* July 22.

———. 1992b. Rhone-Poulenc Withdraws Plan for Incinerator. *Contra Costa Times,* September 1.

———. 1992c. State Might Mediate Dispute over Martinez Waste Burner. *Contra Costa Times,* May 14.

Marin County. 1988. *CHWMP Implementation.*

Martinovich, S. 1991. Martinez Residents Still Smoking over Rhone-Poulenc Incinerator. *Contra Costa Times,* December 4.

Mazmanian, Daniel, and David Morell. 1990. The NIMBY Syndrome: Facility Siting and the Failure of Democratic Discourse. In *Environmental Policy in the 1990s*, edited by by Norman J. Vig and Michael E. Kraft. Washington, D.C.: Congressional Quarterly.

———. 1992. *Beyond Superfailure: America's Toxics Policy for the 1990s*. Boulder, Colo.: Westview Press.

Mohai, Paul, and Bunyan Bryant. 1992. Environmental Injustice: Weighing Race and Class as Factors in the Distribution of Environmental Hazards. *University of Colorado Law Review* 63:921.

Penn, Paul J. February 4, 1988. *Tiptoe through the Tanner Process; or, What Are the Objective Ramifications of Hazardous Waste Management Planning in California?*

Rabe, Barry. 1989. *The Politics of Environmental Dispute Resolution*. Ann Arbor: University of Michigan, Center for Research on Social Organization.

Richards, Sam. 1992a. Dow Won't Build Waste Incinerator. *Contra Costa Times*, August 13.

———. 1992b. Foes of Burner Revel. *Contra Costa Times*, August 14.

Susskind, Larry, and J. Cruikshank. 1987. *Breaking the Impasse: Consensual Approaches to Resolving Public Disputes*. New York: Basic Books.

Toxics Coordinating Project. 1988. *A Community Perspective on Hazardous Waste Planning*, May.

Valcke, N. L. 1991. LAC Questions Value of Talks. *Contra Costa Times*, February 6.

Wright, R. George. 1991. Hazardous Waste Disposal and the Problems of Stigmatic and Racial Inquiry. *Arizona State Law Journal* 23:777.

Zeiss, Chris, and Bob Paddon. 1992. Management Principles for Negotiating Waste Facility Siting Agreements. *Journal of Air Waste Management Association* 42, no. 10: 1296–304.

Siting Hazardous Waste Facilities, Japanese Style

Don Munton

Japanese pollution control policies have been much applauded in recent years.[1] The country has "moved rapidly from being the most polluted nation on earth to being the nation most committed to environmental cleanup and the most advanced in pollution control technology." It "enacted some of the world's strictest environmental protection measures," possessed by the mid-1970s "one of the most complete statutory frameworks for environmental policy in the world," and developed what "must be judged the world's most successful pollution-abatement program." As a result, according to its own officials, "the quality of Japan's environment is steadily improving."[2]

These plaudits are not undeserved—if the focus is, as it tends to be, on traditional air and water pollution control measures.[3] With hazardous waste, on the other hand, the picture is less clear. It is widely accepted that European countries, having started earlier, are generally the world leaders in this area. Japan, on the other hand, rarely figures in comparisons. There is very little analysis of Japanese hazardous waste management in Western or English-language sources. There is similarly little information about whether the siting of hazardous waste treatment and disposal facilities is as much a matter of controversy in Japan as it is elsewhere. Does Japan, like other industrial countries, have NIMBY (Not in My Backyard) movements—strong local resistance to the siting of new treatment facilities?[4]

It might easily be assumed that NIMBY is not a Japanese phenomenon. According to a well-known analysis of American hazardous waste policies, Japan "has successfully sited numerous hazardous waste treatment facilities tailored to specific industries." The explanation, Daniel Mazmanian and David Morell suggest in passing, may lie in the high level of cooperation between public authorities and private corporations (Mazmanian and Morell 1992, 137–39). A major comparative study of environmental policies seems to support the prediction that opposition to facilities should be less evident in Japan, though it offers a somewhat different explanation. The politics of pollution, Cynthia Enloe (1975) argues, depends "not only on the public's level of concern but also on its willingness to press demands." She contrasts Japan with the United States in both regards. Implicit support for this view can be found in the most extensive analysis of NIMBY attitudes in the United States. Kent Portney concludes that "the inability to site facilities is fundamentally tied to contemporary American political and social culture" (Portney 1991, 136, 137, 150). Such an argument could be taken to mean that opposition to hazardous waste facilities is particularly, maybe peculiarly, American, and thus to imply that other cultures may not manifest the same rebelliousness. It might then well be assumed that NIMBY, like subway graffiti, is foreign to Japan.

This chapter will attempt to examine the Japanese approach to hazardous waste facility siting and, in so doing, to assess both the image of Japan as a pollution control leader and the common assumption that cultural differences are of great importance in this area. It asks a series of questions. First, is hazardous waste production a problem in Japan? And if it is, how much of a problem is it? Second, how does Japan go about siting and operating treatment and disposal facilities? Is there citizen opposition in Japan to facility siting comparable to that experienced recently in other industrial countries? While certain truths and myths about Japan would suggest NIMBY is not or should not be a common phenomenon, local opposition to facility siting exists and, in fact, thrives. Third, what, then, has been the impact of NIMBY in Japan on its management of hazardous wastes? And fourth, what strategies has Japan adopted to deal with what is often described as "the NIMBY problem." Some conclusions are drawn both about the NIMBY phenomenon and about lessons to be learned from the Japanese case.

The present focus on the siting of treatment and disposal facilities ought not to be taken as implying either that this is the key to solving the hazardous waste problem or that facility siting is more important than waste minimization—reducing the production of wastes and recycling them. Industrial recycling and source reduction need to go hand in hand with safe treatment; if anything, the emphasis ought to be on the former.[5] Waste minimization is widely practiced in Japan. Some relatively new, modern plants, such as the NKK Keihin steelworks near Yokohama, have achieved extensive recycling, up to 99.5% in some aspects of their operations.[6] This aspect of hazardous waste management in Japan, however, is not the main focus here.

INDUSTRIAL AND HAZARDOUS WASTE IN JAPAN

It should come as no surprise that Japan has a hazardous waste problem to deal with. The country's recent economic history has created an industrial structure where by-products and wastes, hazardous and otherwise, are inevitable. Japan experienced extraordinary economic growth over the 1950s, 1960s, and 1970s, averaging 8–9% per year. Although GNP growth rates eventually slowed down somewhat in the wake of the 1973 oil crisis, they remained relatively high. Industrial production overall increased 8% per year between the mid-1950s and the mid-1970s. Beginning in the 1960s, Japanese economic planning shifted from the development of light industry to heavy industry, especially steel and petrochemicals. These developments vastly increased the demand for energy and spurred a construction boom of thermal power stations (Imura 1989, 54). Production of steel and electricity increased nearly fourfold during the 1960s alone. Automobile manufacturing rose almost twentyfold (1800%), and plastics by a factor of fifty. Production of ethylene, a useful indicator of petrochemical output, increased tenfold just in the period 1959 to 1965 and another seven times by 1977.[7]

This phenomenal postwar growth can be seen as the contemporary manifestation of a drive to develop Japan that began with the Meiji restoration in the late 1800s. The high rates of growth and industrialization of what is now, if not the dominant, then certainly the most dynamic economy in the world were the necessary conditions for the emergence of an industrial and hazardous waste problem. Some would argue they were also sufficient conditions, given the lack of environmental consciousness. "Japanese industrial policy after World War II," one observer notes, "was pursued with . . . disregard for human health and environmental needs."[8] The extent and pace of industrialization, however, is only part of the waste story.

"The elementary fact of Japan's national life," Chitoshi Yanaga suggests, is "too many people on too little land" (1956, 20). Japan has thirty-five times as many people as the United States on every square kilometer of habitable area (about 1400/km^2 as compared to 40/km^2 when mountainous regions, etc., are excluded). It has about three times as many as densely populated Germany and the Netherlands. Japan also has, by far, the industrial world's highest ratio of GNP per unit area (OECD 1977, 10). Even more striking is the density of the Tokyo, Osaka, and Nagoya areas. They amount to only about 1 percent of the area of the country but held one-third of its population and fully three-quarters of its manufacturing in the 1970s. Industry thus tends to be in close proximity to population centers in Japan. The classic "solution to pollution through dilution" formula is therefore even more problematic here than in other industrial countries. Specifically with respect to siting hazardous waste facilities, the extraordinary density also means that space is, even more than usual, at a premium. Land costs are correspondingly very high. "Geographically, Japan is an environmentalist's nightmare," one student of the country has written (Pempel 1982, 219). Geographically, it is equally, if not more so, a facility-siting nightmare.

According to the accepted official estimate, Japan was in 1983 discharging about 1.4 million tons per year of material containing hazardous waste (or at least what its statutes and regulations define as such).[9] While this quantity is not inconsiderable, the total is very low by comparison with other industrial states. The United States, for example, generates an estimated 240–280 million tons per year of hazardous waste (as defined by U.S. regulations).[10] Even taking into account the fact that the relevant American laws define hazardous waste more broadly than did those in effect in Japan in the 1980s does not make up the substantial difference. Some simple calculations suggest that the Japanese figures ought to be much higher. If we consider total Japanese production in the major industries contributing to the hazardous waste stream (e.g., petrochemicals, iron and steel, pulp and paper, and electrical machinery and electronics)[11] and compare these output levels to those of the United States, Japan is probably producing, at a minimum, 40 to 60 million tons of hazardous wastes each year, not the 1.4 million tons the official statistics record.[12]

The scale of these quantities points to a substantial Japanese hazardous waste problem. Though the amount of wastes being produced is apparently underestimated in Japan itself, it is likely greater than that for any industrial country other than the United States. And the problem is worsening in both magnitude and complexity.[13] While there are no official figures available on the rate of growth of hazardous wastes in Japan, a reasonable ballpark estimate suggests their production is probably increasing 3 to 5% annually.[14] The waste problem in Japan is also becoming more complex and more difficult to manage. As one Japanese official, Masaru Tanaka, has noted, "new compound materials developed by new technology and diversified life patterns will change the quality [and] increase the quantity of such wastes that are difficult to [treat], and produce intermediate wastes in the course of treatment" (Tanaka 1985, 92). "These," he adds, with some understatement, "may cause new environmental problems."[15]

TREATMENT AND DISPOSAL OF HAZARDOUS WASTES

The Japanese system for the treatment and disposal of hazardous wastes is similar to that of some other industrial countries, in the sense that it is more a private-sector than a public-sector matter. Governments establish standards, but the onus is on private companies to treat and dispose of their own wastes. To the extent government is involved, the management of hazardous wastes in Japan is less the responsibility of the national government and much more that of the prefecture governments. Unlike many countries, particularly the United States and Canada, which rely heavily on land filling, there is much more emphasis in Japan on treatment and reduction of the industrial and hazardous waste stream. Most of the large Japanese companies thus utilize on-site treatment. The Tokyo-area plant of Nissan, for example, has its own industrial waste treatment facility, including two incinerators. Small- to medium-sized waste producers, on the other hand, tend to hire private disposal companies. Japan's Waste Disposal Law standards specify the type of treatment to be given to different types of industrial wastes. The law also specifies criteria for landfill operations and what sort of wastes can be deposited into what type of landfills. There are three types: simple uncontrolled landfills (for nonhazardous, stable materials), leachate-controlled landfills (for nonhazardous but nonstable wastes), and contained sites (for hazardous wastes). While the first two types are fairly common, there are only about forty of the third type in all of Japan.[16] About half of these (twenty or so) are used and managed by private companies for their own wastes. Most of the remaining are managed by private waste disposal companies. Two are publicly operated.

The hazardous waste problem cannot be described solely in terms of figures on production and treatment (or lack thereof). A major part of the problem in most industrial countries, and arguably the largest part, arises from already contaminated sites.[17] These sites include poorly managed present or former landfills and dumps,[18] old or abandoned industrial sites, and storage areas for such toxins as PCBs.[19] Most industrial countries have at least begun to deal with this problem, and most carried out surveys of its extent during the 1980s. Japan has not. There is no legislation in Japan focused on site remediation, and there is no ongoing national program to identify sites, let alone do the necessary cleanup (as of the mid-1990s).[20] There have been a few isolated and specific cleanup operations. Indeed, the contaminated-sites problem in Japan has not yet really surfaced (if that is the correct term). There has been no Japanese Love Canal story. As a consequence, public concern about the problem remains low.

There can be little doubt, however, that Japan has contaminated sites, and probably large numbers of them. According to one Japanese official, a program is "absolutely necessary" because there are "many areas where groundwater and soils have been contaminated by chemicals."[21] One was discovered a few years ago in Kokura city on Kyushu, for example. A property on which there had been an explosives plant since before World War II became surrounded by residential areas. The company decided to sell the site. When government officials investigated, they found toxic contamination from improper landfill. The area was quietly cleaned up by private contractors. In another case, highly toxic hexavalent chromium was found buried near a factory site in Tokyo (Huddle, Reich, and Stiskin 1987, 6). Private companies capable of dealing with such problems claim to know of many such affected areas that will require treatment, but they do not find their services in much demand.

National agencies have begun planning a comprehensive survey of possible contaminated sites.[22] If the surveys of other countries provide an indication, the number of sites in Japan could range from a few thousand (as in Canada, Denmark, and the Netherlands) to, as is more likely, tens of thousands (as in the United States and Germany). Experience elsewhere also suggests that the first surveys done will underestimate the extent of the problem; that of the sites discovered, one in ten or more will require urgent attention; and that cleaning up contaminated sites will be very expensive, roughly twenty times the cost per unit of treating waste.[23]

CONSTRAINTS AGAINST PUBLIC OPPOSITION

There is no question that Japan, like any industrial country, has a hazardous waste problem. But is the siting of facilities faced by local citizen opposition? Are management efforts being influenced and constrained, as in these other countries, by the NIMBY phenomenon? It might easily be assumed that NIMBY would not be a pervasive response in Japan. The rationale for such an assumption would point to three well-known features of the country—its centralized political system, passive political culture, and lack of strong environmental organizations.

In her pioneering comparative analysis of environmental politics, Cynthia Enloe sketches two common models of the Japanese system—"factional politics" and "Japan, Inc."[24] These two somewhat different models, she argues, share some common features. Among these are various elements of the political system relevant to public opposition to facilities: a centralization of authority at the national level, which renders local governments correspondingly weak; a formidable alliance between the traditionally dominant national party and major corporations; a strong bureaucracy; the hierarchical nature of Japanese society; and a tendency for political careers to depend more on personal ties and allegiances to senior party officials than on positions on issues. While acknowledging certain changes in the system, Enloe concludes that the Japan, Inc., model still has much validity (1975, 262). The effect of this centralized but factional politics would seem to discourage, rather than encourage, effective local citizen action, especially against declared governmental decisions.

Similarly emphasizing the "strong recent tradition of political centralization" as crucial to understanding Japan, Margaret McKean points to a cultural pattern, what she describes as a "historical tradition," for "governments to lead and ordinary people to practice the art of acquiescent followership." The Japanese, like citizens in most other countries, did not historically get much encouragement from ruling elites to participate in governance. But more than most other cultures, "Japan's heritage . . . discouraged the open expression of conflict and resistance to authority" (McKean 1981, 252, 259, 15). Other observers agree. Japan's ruling class, according to Chitoshi Yanaga, "made it clear that the people had no right to even show an interest, much less meddle, in the affairs of the government" (1964, 18). From Meiji times, T. J. Pempel asserts, the system rested on "the principles of political centralization and the citizen's duties to the state" (1982, 13). The people both understood and generally accepted the limitations of their role.

In part, this pattern can be traced to the Confucian idea that "the ruler was, by definition, virtuous" and "knew better than anyone else what was good for the people" (Yanaga 1964, 18). In part as well, the pattern is an outgrowth of the hierarchical nature of Japanese society and a pervasiveness of social obligations and duties, reflected in the concepts of *on* and *giri*. The former is a feeling of indebtedness and obligation to one's superiors; the latter, a broad, encompassing sense of what has been described as "duty, justice, honor, face, decency, respectability, courtesy, charity, humanity, love, circumstance, gratitude, claim" (Yanaga 1964, 59). The resulting custom of obedience to superiors, though never an absolute one, has demonstrated remarkable persistence in contemporary Japan.[25] A traditional Japanese proverb, "Nagai mono ni wa makerero"—literally, "Yield to the powerful"—illustrates this perceived obligation.

Such cultural traditions are not naturally fertile ground for citizen protest. Despite the changes of the twentieth century and especially of the postwar period, Japan's political culture still emphasizes responsibility more than rights; collective approaches more than individual ones; passive involvement in politics more than active involvement; respect for, if not always total obedience to, authority more than protest; accommodation more than open, direct competition or expression of conflict; and action through formal processes, such as elections, more than informal action, such as interest group mobilization. Many of these traditional tendencies were evident in initial responses by victims of some of Japan's worst pollution cases. Urged to press for an answer to whether her child in fact had Minamata disease (or mercury poisoning), one mother demurred. "Somehow," she said, "it seems a bit forward to me" (Upham 1976, 590).

It is often noted that Japanese culture stresses the importance of living in "harmony with nature." It is sometimes argued that this value ought to incline the Japanese generally toward environmental protection, if not environmental activism. That it has not done so is then explained by the dominance of the driving concern for economic growth in modern Japan. The prevailing view in Japan, as it was in most industrial countries at least until the early 1970s, was that industrial pollution was the price of economic growth and the symbol of progress. More than in other countries, however, in Japan "that silent sacrifice was [deemed] part of the citizen's obligation to cooperate with national objectives" (McKean 1981, 34). As another victim of one of the country's most serious pollution cases observed: "For the sake of the country, for the sake of industrial expansion . . . there is nothing we can do. No one said anything" (Upham 1976, 592). The contradiction between a respect for nature and the lack of environmental concern is more apparent than real, however. The traditional notion of living in harmony with nature arguably involved "an acceptance of the cycles of nature—their malevolent as well as beneficent manifestations"—and thus reinforced a social ethic of "resignation and endurance" (Huddle 1987, 269). Japan's traditional political culture thus neither fostered a participant orientation generally nor an activist environmental one specifically.

Given the constraints of its centralized political system, Japan has not fostered the emergence and growth of strong national interest groups. The most obvious important exceptions to this rule are the powerful business associations that act together through the Keidanren, the Federation of Economic Organizations. On noneconomic issues, and particularly in the environmental area, there is a striking absence of active and influential pressure groups *(atsuryoku dantai)*. The few national environmental organizations that do exist tend to be financially and politically weak and devoted more to communication and networking than to direct lobbying.[26]

The reasons for this weakness are at least partly cultural. Traditionally, Japanese have viewed direct lobbying as "a contravention of the rules of society" and "a deviation from the accepted norm of behavior." Pressure groups have thus tended to be "undeveloped" and "virtually innocuous" (Yanaga 1964, 108). Prior to the 1970s, "ordinary citizens did not become involved in politics on their own initiative." "Truly voluntary associations were simply not part of the Japanese tradition and are very rarely seen" (McKean 1981, 6). The postwar period brought a new constraint. Japanese industry was "stronger, more secure, more defiant, and perhaps less socially responsible than it had been earlier." As a result, "postwar antipollution protesters faced very difficult struggles" (McKean 1981, 42).

The Japanese, moreover, are reluctant to express themselves publicly on political issues (Yanaga 1964, 38). The differences between Japan and other countries in this respect are evident in the results of survey research. Various studies show that the Japanese political culture was as participatory as that of most other democracies in terms of voting but much more passive in other respects (Richardson 1974, 93–94). The option of forming an interest group to make demands and lobby is not a favored one. In one survey, only 4% of the Japanese respondents thought it a good idea to "form a group and try to realize our wants through its activities." Other surveys have found slightly higher but still modest proportions favoring group action as compared with, for example, voting in elections. There was a strongly declared preference to "think carefully at election time . . . rather than upset the peace of the community." Particularly in rural areas, notes Bradley Richardson, "community harmony and consensus is preferred over satisfaction of the wants of any of its component units." As a consequence, there is "no tradition of popular participation in politics" (Richardson 1974, 99, 232). Compared with those in other countries, Japanese organizations are less inclined either to encourage political discussion or to inject themselves into community debates (Flanagan, 1978, 154).

This is not to say that interest groups, environmental or otherwise, do not form and operate. They do. But the national political system in Japan does not promote them, and the Japanese are not natural joiners or supporters of pressure groups. That such groups do emerge and can succeed is illustrated, for example, by the movement seeking action on the problem of Minamata disease. This and similar movements, however, struggled unsuccessfully for years not only against an unsympathetic governmental structure but also against "tremendous community pressure to endure in silence." Victims "encountered informal prejudice and social ostracism"; they were regarded "as hypochondriacs and freaks, [or] as the possible carriers of some dread contagion or hereditary defect" (McKean 1981, 28).[27]

Another much more specific factor that, if present in Japan, would arguably promote environmental activism is media attention to environmental problems and, in the present context, to the dangers and dilemmas of hazardous waste. It seems clear that media coverage devoted to Japan's severe air pollution in the late 1960s was a major reason for the legislation and control actions of the 1970s.[28] In contrast, however, there is a perhaps surprising lack of media—as well as public and governmental—attention to hazardous waste issues and especially to the problem of contaminated sites in Japan. Few of the environmentally related complaints made to the Environmental Disputes Coordination Commission concern soil contamination.[29] Although there have been some major cases in Japan involving damage to the environment and human health from hazardous wastes, such as the internationally famous Minamata case, the settlement of these cases and the specific remedial measures seem to have created the impression that existing policies have dealt with the problem.

NIMBY IN JAPAN

While aspects of the politics and culture of Japan militate against environmental activism, one cannot assume that it does not exist. To some extent, the image of a passive and obedient people is an outdated view of Japan. Japan has changed in recent decades. To some extent, as well, the preceding analysis simply misses some fundamental facts, in part because it is focused on the national level. To conclude that NIMBY is not a problem in Japan might be tempting but would be wrong. It is, in fact, very much in evidence and has been for some time.

One of the most prominent manifestations is the large number of citizens' groups that have arisen in Japan since the 1960s. Most observers seem to agree that by the mid-1970s there were more than 3,000 of these citizens' movements *(shimin undö)*. Most, but not all, of them are environmentally oriented.[30] By one "conservative" estimate, they boasted from 60,000 to 90,000 core members and up to 1,500,000 rank-and-file participants (McKean 1976, 91). The groups strongly tend to be locally based and to confine their activities to the local level, rather than act at the national level.[31] Another manifestation of the rise of local activism is that by the early 1970s local governments throughout Japan were receiving about 75,000 complaints about pollution annually from citizens, four times the number they had received in 1966. As of the mid-1970s, there were an estimated 10,000 local environmentally related disputes under way (Krauss and Simcock 1980, 187).

Many of the groups originally emerged to seek reduction of, or compensation for, specific, existing environmental threats. (Not all of the groups, therefore, could be described strictly as NIMBY groups.) Their emergence was also, more generally, a response to the country's severe pollution problems in the 1960s. More people in Japan suffered more serious pollution-related health effects than in any other industrial country, a fact that also helps to explain the "victim-consciousness" of these groups (McKean 1981, 256; Weidner 1989b, 481). Both the protests and the groups themselves were fostered as well by dramatic changes in mass attitudes at the same time. Japanese public opinion shifted, around the beginning of the 1970s, from perceiving pollution as something inevitable to regarding it as unacceptable.[32]

The citizens' movements had a strong and profound impact. They were, in the words of two sympathetic observers, "vehemently active," demonstrably effective in blocking projects if appropriate pollution prevention measures were not being taken (Yamanouchi and Otsubo 1989, 231). Thus, "nearly all development projects[,] large and small, face local resistance," according to a respected activist (Ui 1989a, 111). Beyond their local communities, these grassroots organizations influenced public opinion, which, in turn, influenced national as well as local governmental agencies. Collectively, the movements significantly disrupted the traditionally tranquil state of Japanese politics. They represented "the replacement of polite consensus politics by open confrontation" and, indeed, nothing less than "Japan's cultural revolution."[33] Some examples may be useful to illustrate both the roots and the impacts.

In 1964, local residents in Mashima, Numazu, and Shimizu in Shizuoka Prefecture, near Mount Fuji, opposed the construction of a refinery, a petrochemical complex, and a thermal power plant. They were eventually successful. This case has often been hailed as "the first major domestic challenge to the nation's postwar economic policies" and as Japan's first environmental protest.[34] About the same time, the Chuba Electric Company proposed to build a nuclear power plant on the border of Nanto and Kisei towns in Mie Prefecture. Citizens in both towns, especially the local fishers, protested against the project, and eventually triumphed. The company abandoned the plan in 1967. It was the first nuclear plant in Japan to be stopped by local opposition (Lesbirel 1992, chap. 4). A proposal to make the city of Mutsu in Aomori Prefecture the home port for a new nuclear ship met with strong local protests after it was discovered to have leaked radiation on its maiden voyage in September 1974. The national government agreed to find a new home for the ship (McKean 1981, 249; Huddle, Reich, and Stiskin 1987, 237–55). A similar movement in the late 1960s in the industrial town of Fuji led not only to the cancellation of additional industry and greater control of existing pulp mill pollution but to compensation of affected fishers (McKean 1981, 30–32). Yet another movement arose to press for relief of severe air pollution in Yokkaichi. Once a seaport city renowned for its beach area, it had become after 1955 the site of Japan's largest petrochemical complex, and thus more renowned for its water pollution and respiratory diseases ("Yokkaichi asthma").

Other cases might be cited, but the point should be clear. The impact of postwar industrialization in Japan, coupled with the proximity of pollution sources to pollution centers, agriculture, and fishing, led to some of the worst environmental conditions to be found anywhere in the industrial world. The often intolerable conditions, in turn, fostered protests and direct action.[35] As one participant, an elderly fisherman, put it: "We used to just put up with it and cry ourselves to sleep *[nakineiri suru]*. But this time it affected our economic circumstances and our health. So we just had to rise up and protest against it!" (McKean 1981, 268).

The organization of local, as opposed to national, citizens' groups is reinforced by two tendencies. Environmental activists tend to perceive the causes of pollution less in the high consumption levels of an industrial society than in "neglect or malicious intent, in the attitudes and action (or inaction) of industry and government." They thus regard pollution as "something evil that polluters *(kagaisha)* did to victims *(higaisha)*" (McKean 1981, 132, 140). Accordingly, issues are defined in health-related, geographically limited, usually local, terms. Second, they also tend to restrict membership in these groups to persons directly affected by the pollution and to avoid forming active alliances with "outside" groups.[36] What are weaknesses with respect to building strong national coalitions are strengths with respect to building a strong local organization and gaining influence in local decision making.

It would be easy to exaggerate the success Japanese groups have had in blocking projects. Some protests have been failures, even, or perhaps especially, when protesters have gone beyond peaceful means and provoked violent clashes, as, for example, in the well-known Narita airport case. The opposition to projects has not always proven intractable; in some cases, it has ultimately been worn down and defeated. The Narita airport, as present-day travelers to Japan know, eventually opened. And in some cases, such as that of a chemical plant in Oiso, local officials allied with companies outmaneuvered opponents, occasionally with skulduggery, or won them over (McKean 1976, 70). Many, but not all, proposed electric power plants, both fossil fuel and nuclear, have eventually gone ahead. The time required to gain siting approval, however, has gotten longer and longer. Fossil fuel plants in Japan, which took an average of forty months to site in the 1960s, averaged seventy months by the mid-1970s. Nuclear plants, which took ninety months on average to site in the 1970s, required almost twice as long in the early 1980s (Lesbirel 1987–88; 1992).

Citizens' groups have opposed not only potential environmental damages from planned projects but also the projects themselves and the jobs that come with them. In the process, these groups have come to wield considerable influence on municipal and prefectural policies in general and on industry. Their muscle has been shown not just through pressure and petitions but via the ballot box. They have forced referenda, as in Kagoshima, where a proposed petroleum complex in Shibushi was defeated in 1972. They have brought about the recall of local politicians who took positions opposed by the groups, as in Usuki, Tokaidö, Fuchü, and Fuji. And they have driven pro-industry candidates away or run their own candidates. The "anti-kombinat" movement in Mishima, for example, overturned the traditional Liberal Democratic Party–dominated local government and elected a member of the citizens' movement mayor (McKean 1981, 240, 250, 154; Huddle, Reich, and Stiskin 1987, 286).

According to one study, virtually all of the nearly 200 communities surveyed had made some response to citizens' movements. A Kyoto survey found city administrators agreed that citizens' movements were the most important source of innovation after the mayor. A major factor underlying this influence is what Bradley Richardson terms "localism." Japanese tend to be more, not less, interested in local politics than in national politics and more inclined toward "collective participation" in "the social world of the small community" than in national politics. "Japanese politics," he adds, "is above all a politics of local 'connections' " (Richardson 1974, 95, 58, 234 n. 5).

The impact and importance of the citizens' movements has been bolstered by the important role local governments play in environmental issues. Although prefectural and municipal governments have only limited autonomy in Japan, especially in terms of governmental finance, they can have substantial clout on environmental policy. For one thing, they, not national government agencies, have responsibility for negotiating and implementing environmental regulations. And unlike many other controversies in postwar Japan, pollution became a political issue in specific localities long before it was viewed as a national problem (McKean 1977, 211). Local authorities can also, if they so decide, establish higher or more stringent standards than those set by the national government, and many prefectures have done so. Heavily industrialized Yokohama, Japan's second largest city and, during the 1960s, its most polluted, has adopted some of the most stringest air pollution standards in the country. Local activism has thus been a source of increased environmental regulation at the local level. To be sure, such regulations are not new. The first ordinance was put into effect in Osaka Prefecture as long ago as 1877, to regulate emissions from manufacturing plants. A number of other city and prefectural actions against factory smoke followed in the 1950s. But the numbers of all types of environmental ordinances by municipalities and prefectures increased dramatically during the 1970s.[37]

Another manifestation of the potency of local environmental politics, and thus of the role of local citizens' movements, is the growing use of pollution prevention agreements *(kōgai bōshi kyōtei)*. These are formal, written accords (or contracts) negotiated and signed usually by corporations and municipal or prefectural governments.[38] The first such agreement seems to have arisen in 1952 and involved Shimane

Prefecture, the Shanyo Pulp Company, and the Daiwa Spinning Company.[39] The pattern thus spawned has become a unique feature of the Japanese regulatory landscape. There were about 900 such agreements by 1970, about 3,200 by 1972, almost 9,000 by 1975, over 15,000 by 1980, more than 20,000 by 1982, and around 25,000 by 1984.[40] Many of these agreements directly involve nongovernmental groups as signatories: no less than 37 by 1969, 1,000 by 1974, and 8,200 by 1984 (Weidner 1989b, 491). While these agreements are entirely voluntary, local governments often can pressure the companies to accept tough provisions—and are increasingly able to do so because of the force of citizen action.[41] "Government pressure is effective," Yamanouchi and Otsubo observe, "because of public support against pollution" (1989, 227).

If the early 1970s were the heyday of these citizens' movements, the 1980s have been less kind. According to many observers, the movements were in decline by then. The dramatic court decisions that had once given them strength had been followed by less favorable outcomes. The "cold spell," as some termed it, left the citizens' movements "stagnant." One analyst went so far as to say that "environmentalism in Japan had . . . largely vanished from the political scene" (Yokoyama 1982).[42] The environment has certainly declined as a priority on the national agenda. It did not disappear at the local level, however. There were fewer spectacular successes, perhaps, but many movements were still active. NIMBY has not died out in Japan.

Tokyo, for example, currently finds itself caught by different NIMBY movements, one at home, the others elsewhere. Only about half of the city's industrial waste is treated and disposed of within the metropolitan area. The rest is transported into neighboring and more distant prefectures for disposal in landfills. (A small amount is even exported to the southern island of Kyushu.) Tokyo's waste-exporting habits are leading to increased complaints from the surrounding areas. Why, they ask, should we be responsible for Tokyo's waste? Some local and prefectural governments in Japan, like subnational governments elsewhere, are attempting to limit or prohibit transportation and dumping of wastes originating outside their own jurisdictions. Plans for a Tokyo-area hazardous waste center were, in the mid-1990s, in the initial stages. But they were already under attack. A site has been chosen for a treatment facility on Jonan Island, in the Bay of Tokyo. It is largely an industrial zone, but there are some residential areas nearby. Local citizens' groups have come out in opposition to the project, as has the mayor of the ward in which the island is located.

Other manifestations of NIMBY-type opposition are not hard to find. The twenty-four municipalities in the Tama district, a suburban area of Tokyo, are facing a solid waste crisis as the only landfill serving 3.6 million people nears capacity. Opposition from residents has deadlocked plans for a new site (Azuma 1993). Even new-technology demonstration projects promoted by the Clean Japan Center, a national recycling-promotion agency, have not been immune from the NIMBY syndrome. Some have been opposed by citizens' groups and local politicians simply because they deal with "wastes," whether or not these wastes are hazardous in nature. Siting facilities that deal with hazardous waste remains doubly difficult, and the root cause is clear. In the words of one expert, hazardous waste has an "extra, extra bad image."[43]

EFFECTS OF NIMBY IN JAPAN

The consequences citizens' movement activism bring to hazardous waste management have been predictable, given the experience of other jurisdictions. The opposition has led to a growing shortage of available sites for treatment and disposal of hazardous wastes, to rising opposition and conflict over "exports" to other jurisdictions, to escalating prices for treatment and disposal, and to increased improper and illegal dumping of wastes.

By the late 1980s, public opposition had made it almost impossible to open new hazardous waste sites in Japan, as it had in other industrial countries (Yakowitz 1988). Expansion of disposal sites has not kept pace with industrial output, even with substantial gains in the recycling of industrial materials. Although land is at a premium and, until recently, the country lacked a national policy framework, the real problems have been almost entirely local and political in nature. Faced with stiff public opposition, local governments in Japan have become very reluctant either to plan new facilities or to accept expansion of existing facilities. Local politicians do not need to see protestors in the streets on such questions; they know full well that protests will arise if any such plans are contemplated. Treatment facilities have not been established, notes a Kobe City report, because "the understanding of the citizens is hard to obtain" (Kobe City Environmental Bureau 1994). Private waste disposal companies as well as public agencies are thus severely constrained. For example, one such company, Sanyu Plant Service, operates a treatment plant (including a small incinerator) on the same crowded 1,000m^2 site in an industrial park south of the center of Yokohama where it was established fifteen years ago. Despite high and growing demand from nearby industry and numerous proposals to expand, approval to do so has not been forthcoming.[44]

The Environment Agency has been concerned about the shortage of sites since the late 1970s. "Securing proper disposal," it noted then, "is *still* a problem" (quoted by Huddle, Reich, and Stiskin 1987, 6; italics added). The Ministry of Health and Welfare acknowledges frankly the problem of the shortage of sites. The ministry's annual report, in its English translation, notes that "it has become increasingly difficult to build new facilities due to oppositions [*sic*] from nearby residents who worry about their effects on the environment." The report goes on to note that the insufficient number of final disposal sites is an acute problem that "must be tackled."[45]

A second consequence of local opposition is that Japan has been experiencing growing conflict among different jurisdictions, especially among its prefectural governments, over industrial and hazardous waste management, especially transportation and disposal issues. With the lack of landfill facilities in some prefectures and with differing standards being required in different areas of the country, there has been an increasing tendency to transport to other prefectures wastes that cannot be treated in their areas of origin. (It is possible, of course, that some jurisdictions deliberately impose more rigorous standards than their neighbors in order to encourage this export of wastes.) One not-so-recent example of this problem was the long-standing protest in the 1970s against Tokyo sending its garbage through the streets of Kötö ward and especially the Edagawa-chö neighborhood. The protests included sit-ins to block garbage trucks (McKean 1981, 102–6).

Third, hazardous and industrial waste producers in Japan are facing higher bills for treatment and disposal. The responsible ministry accurately describes the trend as one of "soaring" costs (Japan, Health Ministry 1991–92, 129). Informed experts estimate that charges for hazardous waste disposal have increased as much as 200–300% in recent years (Tanaka interview). Municipal waste disposal expenditures, by way of comparison, rose from around 8,000 yen per person in the mid-1980s to close to 11,000 yen per person by the early 1990s (Japan, Health Ministry 1991–92, 132). This cost trend, paralleling similar trends in other countries, has been due to a number of factors, not just the shortage of sites and NIMBY-type political resistance, though that has played a part. Other factors include the increasing amounts and complexity of wastes and more stringent environmental policies (Tanaka 1985, 91).[46]

A fourth problem is that of unsafe and illegal dumping of hazardous waste. While much waste in Japan is properly managed, it is clear that much also is not being handled and disposed of safely. Materials deposited in nonsecure landfills sometimes contain hazardous chemicals and organic materials. Barrels of toxic wastes have been dumped illegally in old mines. On-site dumping of hazardous waste by companies has also led to soil, groundwater, and stream pollution.[47]

Although such sensitive matters as this might not have been publicly discussed in Japan in the past, the problem is widely acknowledged today. The Environment Agency, the most outspoken of the national agencies, in its annual reports has been issuing warnings since 1978 about illegal dumping. A 1983 Council on the Living Environment report to the Ministry of Health and Welfare pointed out the need for "the correction of the unproper, or illegal, disposal of industrial wastes." Masahiro Fujiwara, the general director of the Water Supply and Environmental Sanitation Department of the health ministry, has acknowledged that "improper treatment of waste such as illegal dumping . . . posed a serious problem (Fujiwara 1992)" The ministry's 1991–92 report notes there have been "many incidents" of "illegal dumping." The problem is increasingly being recognized by prefectural and local governments as well. Tokyo officials report that there has been "a lot of poor environmental . . . disposal" (Azuma 1993). Fukushima prefectural officials report they, too, have "a big illegal dumping problem."[48] It seems likely that recent amendments to the Waste Disposal Law were in part motivated by a desire to deal with this problem.

And there is a recognition that the problem is worsening. An authoritative analysis of waste management research in Japan concluded that "illegal dumping has increased" (Tanaka 1985, 91). The official 1987 annual report of the health ministry has similarly acknowledged that "illegal dumping is on the rise" (Japan, Health Ministry 1987, 129). More precise statistics on the extent of the problem do not seem to exist—for perhaps understandable reasons, given the difficulties of researching this sort of activity. Official figures show a substantial increase in arrests by the national police for illegal dumping; there were 420 arrests recorded in 1972 but, by one estimate, as many as 9,000 in the mid-1980s (Weidner 1989b, 494). By other figures, the number of arrests declined from the 1980s to the 1990s but was still in the thousands. In Tokyo there have reportedly been large numbers of violations and numerous disposal contractors arrested for illegal dumping, but precise figures are difficult to obtain.[49] Most of these violations, as many as 95%, are for industrial waste practices. Another estimate, now dated, was that "illegal dumping . . . accounts for more than half the pollution offenses each year" (Japan, Environment Agency 1978). Where there is illegal dumping, of course, there are contaminated sites.

The problem of unsafe and illegal disposal of hazardous waste is sometimes ascribed to a lack of knowledge and training on the part of disposal contractors. That undoubtedly plays some part. There is, however, a very strong emphasis in Japan on training and certification of persons with environmental responsibilities, and there are extensive programs designed to provide this training. It is likely that, generally speaking, capabilities and knowledge have increased in recent years. Thus, it follows that the level of education cannot explain the fact that the problem is mounting. The worsening trend, in fact, would seem largely due to the shortage of sites and to high and increasing waste treatment charges. The economics of waste disposal are clear; few waste producing or waste transport companies could be oblivious to the relative costs of using proper disposal sites compared with other options. In this situation, there are substantial profits to be made by the unscrupulous. It is not surprising, then, that organized crime in Japan, the *Yakuza,* is widely thought to be involved. These speculations are confirmed by some government officials. In this respect, Japan's situation is not much different from that of the United States, where there is also evidence of organized-crime involvement in illegal hazardous waste disposal (Epstein, Brown, and Pope 1982; Block and Scarpitti 1985; Egan 1990; Wolf 1983).

STRATEGIES FOR SITING HAZARDOUS WASTE FACILITIES

What has Japan done to deal with these problems? The Waste Disposal and Public Cleansing Law, the basic statute in Japan dealing with industrial waste, was amended in October 1991 and came into effect in 1992. The amendments had three basic objectives: (i) "securing proper waste disposal," (ii) "providing waste disposal facilities," and (iii) "promoting waste reduction and recycling."[50] The new law represents both a novel set of approaches toward siting hazardous waste and other facilities in Japan and an important step away from the traditional spirit of ecological laissez-faire, in which industrial wastes generally were handled by the private sector operating relatively free of government regulation.

Most important for the focus of the present chapter, the new law is designed to encourage the development and siting of new treatment and disposal facilities.[51] (Such facilities, however, are not seen as alternatives to improved waste reduction and recycling. Even with an expansion of these capabilities, the expectation was that there would be an increasing need for intermediate and final disposal facilities.) The new law also broadens and strengthens the role of governments in ensuring proper industrial waste disposal, but leaves primary responsibility with private-sector companies to dispose properly of their wastes; assigns greater responsibility to municipal and prefectural governments to ensure adequate facilities are available and to ensure compliance with the laws and disposal standards (a devolution process begun in 1970); assigns greater responsibility to waste producers to ensure adequate arrangements for disposal of their wastes; and encourages them to be more environmentally responsible with their products. The "polluter pays" principle, long the basic tenet of Japanese industrial waste management, is thus essentially maintained.

While the changes are something less than a radical intrusion into the private sector, the new law nevertheless represents an important evolution from long-standing Japanese practice.

It is well recognized, in Japan and elsewhere, that the days when facilities could be imposed on host communities are long gone. The preemption of local approval processes or the overriding of local opposition belongs in the dark ages of democratic society. If any further strike against these two first-generation approaches is needed, it is also the case that they simply do not work (Mazmanian and Morell 1992; Piasecki and Davis 1987).

A number of alternatives, often couched as strategies to "overcome" the NIMBY problem, have been proposed since the early 1980s, most of them by North American analysts.[52] (There is little or no corresponding literature in Japan.) These strategies include the following:

- *increasing the public role and trust* through educational programs, governmental public relations, and other measures (Matheny and Williams 1988);
- *creating public utilities or encouraging private-public-sector cooperative efforts* to manage wastes (Piasecki and Davis 1987);
- *broadening the burden of costs* through the sharing of responsibility for treatment facilities, what has been dubbed the YIMBY (Yes, in Many Backyards) strategy (Mazmanian and Morell 1992);
- *offering "risk substitution"* to already affected communities, that is, the reduction of existing risks through the establishment and operation of treatment facilities (Portney 1991);
- *seeking prior public approval,* rendering the siting process voluntary on the part of the local citizenry (see Castle and Munton, Rabe, Gunderson and Harbage; Richards, and Zeiss, in this volume);
- *offering financial support and compensation,* direct or indirect payments or incentives to the communities or individuals affected (O'Hare 1977; O'Hare, Bacow, and Sanderson 1983; Kunreuther 1986);
- *attempting risk mitigation,* the reduction of both the real and perceived risks of such facilities, through pollution control technology, facility monitoring, and other measures (Mazmanian and Morell, 1992, 1994; Portney 1991; Davis 1986).[53]

Perhaps the most striking aspect of the emerging Japanese approach to siting hazardous waste treatment facilities being pursued under the revamped Waste Disposal Law is that it incorporates not any single one of these ideas but all of them. The basic approach seems to be to foster a broad and integrated set of strategies. In so doing, it implicitly rejects arguments that some of these strategies are ineffective. While by no means all are explicit in the new Waste Disposal Law, these strategies would seem to be key to its implementation. Clear evidence of success of this approach will only emerge in the coming years. Some facilities were being planned when the new legislation was introduced, and are now proceeding under its provisions. In Iwate, in northern Japan, for example, construction of a facility has already commenced. In Ōita, on Kyushu, design of a major facility has begun. It will have a large rotary kiln, largely for the incineration of plastics, and a leachate-controlled landfill. Overall, though, most prefectures did not begin as quickly. While there is planning activity under way in various parts of Japan, then, there has been no rush to site.[54] It would be useful to review these various elements of the emerging Japanese policy.

Increasing the Public Role and Trust

The effort to site future facilities in Japan may well be accompanied by efforts to educate local citizens about the existing hazardous waste problem around them and to assure them of the safety of treatment operations and thus generate support. But the new Waste Disposal Law takes a more indirect, less manipulative approach. In part to build public trust and support, it assigns greater responsibility to municipal and prefectural governments in their existing regulatory role of ensuring compliance with the laws and disposal standards, and it enlarges their role in ensuring that adequate treatment and disposal facilities are available. In both ways, it extends the jurisdiction of subnational governments in waste management.[55]

An important (but implicit) assumption underlying these modifications to the Waste Disposal Law is that attempts by local governments to site facilities are more likely to succeed than are attempts by industry, at least by private waste disposal companies. That is, the bureaucrats who framed the law assumed that a greater involvement of local and prefectural governments might help to overcome public opposition to industrial waste treatment and disposal facilities. The public has generally a bad impression of private facilities, in part because industry is commonly assumed to be mainly interested in minimizing costs. Local governments in Japan, on the other hand, tend to be, and tend to be perceived as, responsive to citizen concerns. While this devolution of authority might be viewed cynically as an effort by the national government to off-load an unwelcome task onto local governments, it could also be viewed, more neutrally, as an attempt to find a siting process in which there will be some public trust. In this sense, the strategy seeks to involve more directly, in the planning and operation of hazardous waste facilities, those governments closest to local concerns about these facilities, those most influenced by citizens' groups in the past and most credible in the public mind.[56]

The new law also provides for stricter control of waste collection and disposal companies and of waste treatment facilities. It introduced a licensing system for companies in the disposal business, provided for a tighter manifest system covering the transportation of hazardous waste or waste requiring "special control," imposed new conditions on treatment operations, and brought in fines for illegal disposal. The goal here, in the words of the Ministry of Health and Welfare, was "to remove vicious entrepreneurs" (Japan, Health Ministry 1991–92, 133). While these changes are mainly targeted at reducing unsafe and illegal dumping of wastes, they may also, less directly, have been aimed at increasing public confidence and trust in the system.

Institutionalizing Private-Public-Sector Cooperation

The involvement of subnational governments will likely go beyond the siting and regulation of new facilities and extend to their operation. New facilities to be established may be "quasi-third-sector," that is, they will involve a partnership of government and the private sector, or something very close to it. This, at least, is the expectation and the objective of the national bureaucrats working to implement the new law. (There are at present about forty prefecture-owned or third-sector waste facilities of all kinds in the country.) The intention is to encourage cooperation of industry and governments not only in the financing but also in the management of hazardous waste facilities. Similar arrangements are found in management structures for facilities in some German states.[57] There are various advantages to such public-private-sector cooperation; with respect to siting, it may have the effect of increasing public confidence that a facility will be tightly regulated and its management responsive to public concerns.[58]

Broadening the Burden

Another implicit approach in the new law is a variant of the "Yes, in many backyards" (YIMBY) concept pioneered in California (Mazmanian and Morell 1992; Morell, chapter in this volume). The previous Waste Disposal Law required prefectural governors to draw up a "regional disposal plan" for industrial waste. In subsequent years, forty-seven of Japan's prefectures did so. Each prefecture was required to estimate the amounts of waste generated and treated, to make proposals for waste reduction measures, and to identify options for the construction of treatment facilities (Ishino 1989, 322–23). The plans tended, however, to be both advisory and very general.

The recent amendments and other national programs provide for and encourage, but do not require, the establishment of comprehensive waste treatment "centers" in all prefectures. (Such centers could have been developed by private companies or by prefectural governments under previous statutes, but there was nothing to encourage this. Some prefectures, nonetheless, had set up such centers.) Creation of a "center" would not necessarily mean construction of a treatment facility, although national government officials regard such a consequence as highly likely in most prefectures. Nor would it necessarily incorporate a disposal capability. Prefectural authorities who are responsible for planning and siting any such facilities will make these decisions on the basis of local needs. There is, moreover, no reason under the new laws why two or more prefectures could not cooperate to establish a joint center, and such cooperation will be encouraged.

Risk Substitution

The idea of risk substitution, proposed by Kent Portney, is an approach to siting based on the calculation that local communities that already have an environmental problem, such as a polluting industry, may be willing to trade this for a new hazardous waste facility if the latter is perceived to lessen their existing risk. This approach, while certainly not explicit in the new law, has already been tried in Japan, with some, albeit short-term, success.

The Kanegafuchi Chemical Company produced 96 percent of the PCBs manufactured in Japan up to 1972, when a national ban came into effect. In subsequent years, a significant quantity (5,500 tonnes) of PCBs was collected and stockpiled at the Kanegafuchi plant site; this represented somewhat less than 10 percent of the total production of nearly 59,000 tonnes. In the mid-1980s, the government proposed that a special high-temperature incinerator be designed and built (at a cost of 20 billion yen) to burn these wastes. The local community approved the incineration, presumably as preferable to leaving the material in storage at the plant site. Local groups were also involved in the close monitoring of the process, which was shown to be achieving a decomposition rate of 99.9999998 percent. When the PCBs at this plant were fully destroyed, however, the community opposed a new plan for the incinerator to burn PCBs imported from other parts of Japan. Although the citizens had been satisfied about the safety of the process, they were unwilling to continue accepting their new risk. The incinerator was shut down. PCBs thus remain stored in other locations in Japan.

Public Approval

It is inconceivable, given the politics of these issues in Japan, that any major facility will go forward without public approval. In some areas, it is explicitly mandated. An industrial waste policy statement adopted by Kobe City in March 1994, for example, sets out various requirements for siting and operating facilities, including environmental impact assessments. Also required, before siting, is "a written consensus or agreement of the neighboring citizens" (Kobe City Environmental Bureau 1994). It is not clear, however, whether the relevant authorities are considering a truly voluntary process, that is, one that allows and invites communities to come forward rather than simply be selected.

Public approval almost necessarily implies the use of negotiation to identify and address local concerns. This, too, is not a new approach in Japan, having been used in one form or another in most of the country's major pollution cases. During the Tokyo "garbage wars" of the 1970s, for example, authorities negotiated with the residents of Edagawa, through which ran most of the Tokyo metropolitan area's garbage trucks. The residents demanded a system of traffic lights to slow the trucks down, financial compensation for damages—and a community hall. They got all of these in the negotiations, as well as a pollution prevention agreement regarding the transporting of garbage. The trucks still run through their area, however. In the case of new hazardous waste facilities, it is clear that more will be required than merely after-the-fact negotiations over details.

Financial Support

Two of the most important features of the emerging Japanese approach to siting hazardous waste treatment facilities are financial support for facilities and compensation for communities in which they are sited. The two can and ought to be distinguished. Financial support is applied directly to reduce the capital costs of the facility. Compensation, on the other hand, is provided to individuals, or to a community in general, to offset externality costs or damages inflicted because of the siting of a facility. Compensation is not necessarily financial in form and would include the provision of desirable community services and facilities. These two will be discussed separately.

Optimum waste disposal will require the best available technology and thus will be expensive, in both capital and operating costs. The new law requires that the prefectures contribute financially to these facilities and that some of the funding for each center also come from the private sector. Financial contributions from industry have already been sought and secured, for example, in the case of a new hazardous waste center in Iwate. Financial support may also come from the national government in the form of grants. As noted above, prior to the passage of the 1991 amendments, there was little or no direct national involvement in hazardous waste facilities (although the national government has traditionally provided partial funding for municipal waste treatment facilities). Under the new law, the national government will provide financial assistance if a facility treats general as well as industrial waste, but the assistance will only apply to the municipal waste disposal component of such facilities.[59] While the primary responsibility for financing as well as building these new facilities will rest with local governments, the new grant program is an obvious inducement. A new national Waste Foundation will also encourage the establishment of new facilities by, for example, guaranteeing low-interest loans to local governments and firms. The foundation will be funded both by the national government and by contributions from private industry.[60]

Sophisticated waste disposal facilities charging high disposal fees cannot compete in a free-market situation with facilities with lower standards charging lower fees. The former thus will likely not receive all the wastes they should if forced to compete with other options. The large number of small private waste disposal companies in Japan and their competitive spirit would almost ensure that someone would undercut the price of a state-of-the-art prefectural facility.[61] There is a danger, therefore, that these new "centers" might prove uneconomic and might even exacerbate one of the problems they were designed to solve, namely, improper disposal and illegal dumping. To avoid this development, the strategy also calls for agreements to be negotiated with major industries to dispose of their wastes in the new facilities. A long-term contract with a large waste producer or set of producers could guarantee a sufficient volume of waste to ensure economies of scale. Efficient operation will in turn allow a reasonable fee structure for other, smaller waste producers who, being very price sensitive, might otherwise be tempted to use low-cost disposal alternatives.

Compensation

While compensation in environmental management is not a common approach in industrial countries, it has its proponents.[62] The theoretical argument for it arises from a basic asymmetry in benefits and costs—the fact that persons close to hazardous waste facilities gain only a small proportion of the benefits (which they share with all those, wherever they live, who are served by the facility) but bear a disproportionate share of the social costs of such a facility. The argument for compensation assumes, first, that people are rational economic actors and second, even more questionable, that the burden of costs is a paramount concern, perhaps the paramount concern, to affected individuals. Compensation attempts to close the gap between costs and benefits by providing some additional tangible benefits to those bearing the greatest costs.

As a strategy, compensation also has its strong critics. Kent Portney, in particular, argues against compensation on the grounds that NIMBY opposition is based on perceptions of risks to health as much as, if not more than, perceptions of an unfair distribution of costs. "There is abundant evidence," he states flatly, that compensation "does not and will not work in practice" (Portney 1991, 38). There is some, though perhaps not abundant, circumstantial evidence to support this claim. The case is less than clear-cut, however.[63] In fact, there are examples where various forms of compensation have been part of successful siting efforts, even in the United States (which is the focus of Portney's analysis). Other experts suggest that compensation can be a part of a successful siting strategy.[64] Despite the lack of compensation provisions in American environmental legislation, there have also been U.S. cases where monetary compensation was provided to persons affected by hazardous wastes.[65]

In contrast to the practices of other industrial countries, compensation is an inherent part of Japanese environmental policy, and has been for a long time. Compensation is provided both for damages done and for anticipated costs from facility siting. One of the oldest recorded pollution cases in Japan, going back to the late 1800s, involved the Ashio copper mine and smelter, which polluted a wide area.[66] After repeated and increasingly well organized protests, the company eventually undertook substantial control measures and offered annual monetary compensation to the affected villages. The bulk of the original payments went to village leaders and large landowners (McKean 1981, 36; Tsuru 1989, 17–24), but further compensation was later paid to almost 1,000 farmers for their losses. Other cases in the early twentieth century, those of the Sumitomo company's Besshi mine and Niihama refining complex in Aichi Prefecture and of the Hitachi mine of Nippon Kogyo, involved compensation.

Compensation was also at the heart of the settlements of the Big Four environmental court cases in the 1960s and early 1970s. The most famous of these outside Japan concerned what came in Japan to be called *itai-itai* disease (literally, though inadequately, translated as "ouch, ouch," from the consuming pain it produces), the result of cadmium poisoning (McKean 1981, 45–50; Huddle, Reich, and Stiskin 1987, 186–89), and the (methyl) mercury poisoning at Minamata. Both resulted in negotiated and court-imposed compensation (McKean 1981, 50–59; Weidner 1989b, 539; Ui 1992; Huddle, Reich, and Stiskin 1987, chap. 4). Another of the Big Four cases, in Yokkaichi city in Mie Prefecture, concerned victims of air pollution, not water pollution, and the cause was a complex of industrial and petrochemical plants, not a single source. The companies were found guilty specifically of "negligence in siting" (Pempel 1982, 244). Thousands of "Yokkaichi asthma" victims eventually received substantial monetary compensation (McKean 1981, 59–61; Huddle, Reich, and Stiskin 1987, 51–77, 95–101).[67]

As important as these cases were in setting legal precedents, the use of compensation was not itself new to the Japanese system. Indeed, compensation was a feature of the old Japanese civil code.[68] In the late 1960s, new laws provided allowances for the medical expenses of certified victims of air and water pollution. These were followed in 1973 by a Pollution-related Health Damage Compensation Act, which included "indemnity liability" and provided for automatic compensation to certified victims of recognized pollution problems.[69] By 1985, nearly 100 billion yen were provided to victims under this legislation. Some jurisdictions, such as Yokohama, supplemented the new law with special funds contributed by the major industrial polluters.[70] Many of the pollution prevention agreements concluded between local governments and particular firms since the 1960s, moreover, in addition to specifying emissions standards and pollution reduction or prevention obligations, also set out the firm's liabilities for pollution-related damage (Yamanouchi and Otsubo 1989, 221).

A less than explicit but vitally important feature of the new policy enshrined in the Waste Disposal Law is the provision of indirect compensation for the siting of facilities, or what might be termed "side payments." This compensation is intended for communities as a whole rather than for individuals. The national government will link the siting of waste treatment facilities in particular locations with the construction of other public works deemed more desirable by the public. One example of this strategy now under way is in Ōita, where the energy provided by the incineration of wastes will heat a swimming pool in a new aquatic facility and supply hot water (Tanaka and Hashizume interviews).

This sort of arrangement, linking the siting of an incinerator to the provision of other public facilities, is not a new approach in Japan.[71] It has been utilized, for example, in development of a substantial number of municipal waste incinerators. In what amounts to a combination of compensation and energy saving, 168 such incinerators currently are supplying heat to more than 200 external facilities, including swimming pools, botanical gardens, welfare facilities, and private firms, most often for hot water, heating, and air-conditioning. One example is the Komaki-Iwakura Environmental Works Center in Aichi Prefecture, which provides hot water and air-conditioning for the Komaki Welfare Center for the Aged and heat for the Komaki indoor aquatic center (Japan, Health Ministry 1991–92, 130). Such compensation packages have also become widely used in the siting of Japanese electric power plants.[72] It largely remains to be seen, however, whether this aspect of the new Japanese siting strategy will increase acceptance of hazardous waste treatment and disposal facilities.

Risk Mitigation

Another approach advanced to facilitate siting is the attempt to reduce the risks, perceived and actual, involved in the operation of hazardous waste facilities. The assumption here is that the most important underlying concern of local citizens is not an unfair distribution of costs or financial loss but, rather, the health and environmental risks associated with such facilities. The prescription here is often twofold: to take all possible technical and engineering measures to reduce environmental impacts and to monitor operations and establish management and oversight committees that include local resident representatives and have meaningful powers. Again, these practices seem well known in Japan.[73]

Citizens' movement participation in facility decision making and implementation is often provided for in pollution prevention agreements concluded with private industry and local governments. In the 1970s, the residents of the Tokaidö area of Suginami ward were waging, and ultimately losing, a fight against an incinerator in their ward. The final agreement negotiated gave them the right to influence decisions on plant design and operation and the right to be involved in monitoring the plant. It also included guarantees that the plant would be shut down if it was exceeding environmental standards (McKean 1981, 105). In the case of the pulp-processing plant in Kochi City, the eventual agreement provided for a management committee, one-half of which was to be local citizens. Another example is the close citizen involvement in the monitoring of the incineration of PCBs at the Kanegafuchi plant. This practice of citizen representation on management and committees in Japan seems certain to be a key feature of the siting arrangements for new hazardous waste treatment and disposal facilities, especially those operated by local governments.

CONCLUSION

This chapter has attempted to answer a number of basic questions about hazardous waste policy and facility siting in Japan. First, hazardous waste production is a problem, as might be expected. Indeed, it is almost certainly a much greater problem than official Japanese statistics suggest. Second, while Japan can boast of an admirable environmental record, for example, in reducing air pollution, its policies for dealing with hazardous wastes have been inadequate to the considerable challenge posed. Third, despite substantial cultural differences and some differences in environmental management between Japan and other industrial countries, there are prominent similarities with other industrial countries in the constraints posed by NIMBY movements and the way in which Japan is going about rethinking its hazardous waste management and facility siting policies. Contrary to common supposition, there is, in fact, much citizen opposition in Japan to facility siting. And citizen opposition has led to a number of reasonably predictable consequences: a growing shortage of available sites for treatment and disposal of hazardous wastes, rising opposition and conflict over "exports" to other areas, escalating prices for treatment and disposal, and increased unsafe and illegal dumping of wastes. Japan has adopted a set of strategies to deal with these problems, and especially to facilitate siting efforts. These strategies include increasing the public role and trust, creating public utilities, broadening the burden of costs, offering risk substitution, seeking prior public approval, offering financial support and compensation, and attempting risk mitigation. This package is consistent with, and in some respects seems to have anticipated, various prescriptions offered in recent years to "overcome" what is often described as "the NIMBY problem."

NIMBY is not an American or North American phenomenon. It is a reality in Japan—and in Europe (as Seeliger's chapter in this volume shows). The level of NIMBY-type opposition to treatment and disposal facilities in Japan is at least comparable to that experienced recently in other industrial countries, and perhaps greater in some respects. In the matter of hazardous waste facility siting, Japan is not the exception and the anomaly that many Western predispositions about Japan, Inc., assume it to be. The national government and bureaucracy do not control all, the local governments are not irrelevant, and citizen opposition is not absent. Indeed, municipal and prefectural governments have considerably more regulatory clout in hazardous waste management issues than does the national government, citizens' groups are active and effective, and thus NIMBY is a well-known and widely accepted reality. While "the older style of private compromise and surface harmony is widespread" in Japan in some respects (Langdon 1967, 94), there has been marked conflict and lack of compromise in this issue area. Siting in Japan invokes, as some other issues, "conflictual outbursts in midst of consensus" (Pempel 1982, 3).

Americans' opposition to hazardous waste facilities may well be traceable in part to the American political culture, as Kent Portney suggests. Japanese opposition might similarly be seen to emerge from elements of the Japanese political culture—its traditionally strong local organizations, for example, or its concerns about health and its sense of social justice. And perhaps the NIMBY-type opposition in Germany or elsewhere in Europe could be traced to German or other social characteristics. But the similar national experiences here suggest that some commonalities in human nature are at root. The question of what those sociopsychological roots are must be left to further and genuinely comparative studies. But it seems likely that neighbors of facilities in industrial countries commonly anticipate costs and commonly fear risks. To the extent that siting hazardous waste facilities is a classic "public goods" problem, it is one that seems non-culture specific.

To argue that unique aspects of political culture are not at the root of the NIMBY phenomenon does not mean that political culture is of no importance in hazardous waste management in general. The combination of siting strategies being adopted in Japan might well run into implementation difficulties if adopted, say, in the United States. In Japan, though, it may well work, in part because of the unique relationship that prevails between government and business and the fact that business tends to try to meet environmental norms even when they are not legally enforceable. Culture, or some aspects thereof, may thus matter more to the success of policy implementation than it does to the emergence of policy issues.

The siting of major facilities in Japan went from what might be called the "declare and build" approach of the 1950s and early 1960s to that of the familiar "Decide, Announce, Defend" (DAD) approach. Currently the Japanese process is becoming something that might be termed a "cooperative management" approach (e.g., Berkes and Feeny 1990; Berkes et al. 1991; Feeney et al. 1990; McKean 1992).[74] That is to say, siting and management of hazardous waste facilities in Japan seems to be heading toward a system of negotiated cooperation among the major actors—public and private (as well as quasi-public) organizations, national and subnational governments, officials and publics. The system features the involvement of local citizens in both the establishment and the operation of facilities, the compensation of communities for costs incurred, and the mitigation of perceived risks through contractual commitments. This *negotiated order* thus encompasses most, if not all, of the current individual strategies for siting hazardous waste facilities. Unlike the DAD approach, which has underlying central planning assumptions, or the compensation approach, which is based on economic or rational utilities assumptions, it is fundamentally political in its assumptions. The Japanese seem to have learned, from their own experience, that "the name of the game in facility siting" has become "gaining public acceptance" (Armour 1991, 383).

Cooperative management approaches have proven useful in solving public goods problems in various resource and environmental management areas (e.g., Berkes and Feeny 1990; Feeny et al. 1985). These approaches counter the common notion that public goods problems are inevitable and nearly unsolvable, that communities cannot manage renewable resources or take actions for collective, long-term benefit. It is not yet clear whether the new and developing Japanese approach will be more effective than past efforts, although it is doubtful it could be less so.

What else might be done? What lessons might Japan learn from others' experiences? One element that may be missing from the various strategies being adopted is the use of voluntarism of the kind being tried elsewhere in facility siting. (See Castle, Munton and Richards, and Rabe, Gunderson, and Harbage, this volume.) It is not clear in the emerging Japanese policy whether potential host communities will be allowed self-selection through a voluntary process or sites will simply be selected by governments and operators and *then* local residents consulted. The potential advantages of the former as part of a broader approach might well be considered by governments and waste operators in Japan.

The risk-substitution approach in general seems to be limited by the lack of potential appropriate substitutions and by local opposition to the importing of wastes, as seen in the Kanegafuchi PCB case. A variant might prove more workable in more situations. This would involve the use of mobile treatment facilities to deal with wastes stored in diverse locations. The facilities could be conventional mobile incinerators or one of the more high-technology treatment options currently being tested.

Another response to the shortage of treatment facilities in general (and not just hazardous waste treatment ones) might be to develop and make greater use of on-site equipment operated by existing industrial companies. The on-site treatment facilities of large enterprises in Japan that handle their own wastes have not aroused much NIMBY-type opposition. Communities have clearly been more willing to accept hazardous waste treatment facilities that are part of industrial plants than to accept dedicated, stand-alone treatment plants.[75] Some communities have also tolerated (though their publics may not have widely and knowingly accepted) the treatment of off-site but community-generated wastes at these locations. The sprawling NKK Keihin steelworks near Yokohama, for example, now incinerates some municipal waste, especially sewage sludge. Public acceptance thus seems higher when waste treatment is carried out at an existing manufacturing plant than when it involves the siting and construction of a dedicated waste management plant. This option might be combined with mobile treatment technologies. Among the merits is that it heightens the degree of responsibility of companies to handle their own wastes and thus possibly increases incentives for waste reduction.

On-site facility utilization is a somewhat different approach from what Portney calls risk substitution. Existing risks are not being replaced or substituted for. The risks associated with treatment and disposal are being offset for the community by various benefits of economic activity. As a strategy, this might be termed a "cost-benefit balance" approach, one that closely relates, and balances, the benefits of industrial activity and the costs of managing the hazardous waste produced. It is an approach that might well find broader application not only in Japan but elsewhere.

In one respect, Japanese hazardous waste management clearly needs to be strengthened—that of emission standards from waste treatment facilities. At present there are national guidelines but few legally enforceable standards, unlike the often strict regimes for standard air pollution emissions. Collecting and controlling leachate at hazardous waste disposal sites would also be highly desirable. Such standards should apply both to dedicated and to on-site facilities. It makes no environmental or public-health sense to apply rigid standards to hazardous waste facility operators and not to the treatment and disposal operations of manufacturing plants. Beyond the environmental need, there is a case for such standards in a risk-mitigation strategy. If the public is to be satisfied that the risks of hazardous waste treatment and disposal have been minimized, then Japan needs clear, stringent, effective standards for emissions into the environment.

And what might others learn from Japan? Are there some lessons, suggested by developments in Japan, that might be applied to other jurisdictions? The emerging Japanese approach analyzed here has yet to prove itself thoroughly in practice. As noted earlier, the amendments to the Waste Disposal Law have not led to any flood of facility siting. While it is therefore too early to judge whether this approach may begin to solve the country's hazardous waste management problems, the adoption of a complementary set of strategies has much to recommend it over the magic-bullet approach of looking for one simple solution. A specific lesson might be that in some circumstances, and perhaps in some cultures, compensating the communities bearing most of the social costs of treatment and disposal facilities may be a useful, fair, and necessary part of such a package. Another highly desirable and likely essential element is to pursue increased treatment and disposal capacity along with increased industrial waste recycling and waste reduction. However much of a link the public may perceive between these two, there is no question that the key to long-term success in managing hazardous wastes lies not in siting more facilities but in reducing very substantially the amounts of wastes requiring treatment and disposal.

NOTES

1. Given the paucity of information on Japanese hazardous waste policies in the North American literature, in English or otherwise, much of this chapter is based on interviews conducted in Japan in May and June 1993 and on materials obtained there. Interviews were conducted at the Ministry of Health and Welfare, National Institute of Public Health, Environment Agency, Ministry of International Trade and Industry (MITI), Tokyo Metropolitan Government, Clean Japan Center, National Institute for Resources and Environment, National Institute for Environmental Studies, Industrial Pollution Control Association of Japan, Keidanren, NKK Corporation, Sanyu Corporation, and IRM, Inc. Assistance in arranging and conducting these interviews was kindly provided by Gregory Rust and Dr. Yukiya Tobe of the Canadian embassy; Jim Anholt, Office of the Agent General, Government of British Columbia; Peggy Tsang, Department of External Affairs; Yasuo Ogawa, director, Planning Division, Tokyo Metropolitan Government; and Geoffrey Castle, Frank Langdon and Masaru Kohno of the University of British Columbia (UBC). I am also most grateful to the many Japanese officials and representatives who in various ways assisted the research. This chapter is a much revised version of part of a paper by the author for the conference on hazardous waste siting on which this volume is based. Margaret McKean, Miranda Schreurs, Masaru Kohno, Frank Langdon, Hideo Azuma, Masaru Tanaka, Lonnie Carlile, and Audrey Armour provided helpful comments on this or earlier drafts.

This field trip and the broader project of which it is part were funded by the British Columbia Ministry of Environment in a grant to the University of British Columbia (UBC) through the Science Council of British Columbia. I am grateful to Bernadette Mah of the Science Council and John Grace of UBC for their support of this project. Invaluable research assistance, interpretation, and translation were provided by Sayuri Hiraga, Yuko Kawamatsu, Anson Koo, Tomoko Okagaki, Roger Smith, and Mie Takahashi. Here and elsewhere in this chapter Japanese names are written in English-language style, that is, with the family name last.

2. These praises may be found, respectively, in McKean 1981, 141; Pempel 1982, 231–32; Reich 1984, 384; Mills 1978, 264; Shigeru 1983.

3. The successes detailed by the Organization for Economic Cooperation and Development (OECD), for example, are all in conventional areas of air and water pollution—control of SO_2, NO_x, BOD, etc. Among the specific areas where less progress has been made is the low percentage of Japanese

homes served by sewer systems (OECD 1977). The OECD also notes that "other types of pollution appear to have been comparatively neglected" (p. 21). Hazardous waste appears to be one of these "other types."

4. Some of the basic references on the NIMBY phenomenon are Morell and Magorian 1982; Armour 1984; Mazmanian and Morell 1994, 1992; U.S. GAO 1978; U.S. EPA 1979. On local opposition to facilities in Europe, not always successful, see Piasecki and Davis 1987, chap. 8.

5. On industrial waste reduction, see, for example, Davis, Huisingh, and Piasecki 1987; Freeman 1990; U.S. EPA 1986; U.S. OTA 1986. Warren (1990) lists various estimates of the effects of potential waste reduction measures, ranging from 11 to 75% of total wastes. For reports that waste reduction efforts are beginning to succeed, see the references cited in the introductory chapter to this volume.

6. Interviews at NKK Steel by author, May 31, 1993.

7. Sources for these figures include Hoshino 1992, 64–76; Barrett and Therivel 1991, 31–32; OECD 1977.

8. Weidner 1989b, 481. See also Pempel 1982, 220. Article one of the 1967 Basic Law for Environmental Pollution Control stated that "environmental pollution control policy should be conducted in harmony with a sound development of the economy." "What this actually meant," says Tokue Shibata, "was that environmental pollution control should be kept within such limits so as not to hinder economic development" (1989, 246–47). The "harmony" clause was removed during the famous 1970 "pollution Diet."

9. Japan, International Cooperation Agency 1922 [*sic*], 6. (This document is dated "1922," but the correct date seems more likely to be 1992.) Weidner (1989, 494), among others, also uses the figure of 1.4 million tons. This figure does not take into account the broader definition of "hazardous waste" now incorporated into the revised Waste Disposal Law. (The new definition of what are now termed "special control wastes" encompasses not only toxic substances but also those that are explosive, corrosive, infectious, or easily combustible and can have an impact on human health or the environment.) It might be noted that these figures do not include radioactive material, which is covered by different statutes in Japan, as elsewhere. As of 1987, Japan had more than 600,000 drums of low-level nuclear reactor waste stored at nuclear power plants. The government was developing a permanent storage site in Aomori, the northernmost prefecture on Honshu (Weidner 1989b, 495). On siting of low-level and high-level radioactive wastes in the United States and Canada, see the chapters by Rabe, Gunderson, and Harbage and by Kraft in this volume.

10. This appears to be the current consensus range. See Barnett 1994, 17; U.S. EPA 1985, 1991; Reiter 1992; Warren 1990.

11. On these industries as the major sources, see Liptak 1991, 468; Barnett 1994; Harris, Want, and Ward 1987, 14. The estimates of the proportions contributed by these various industrial sectors changed considerably in less than a decade; see, by way of contrast, those cited by Morell and Magorian 1982, 4.

12. These calculations are spelled out in more detail elsewhere (Munton 1994) but can be summarized here. In some of these key sectors, total Japanese production exceeds American (e.g., plastics, iron and steel, electric machinery), as does consumption in many areas (e.g., iron ore and coke, pesticides). On most other comparisons, Japanese output is about one-half or more that of the United States (e.g., crude-oil refining, paper manufacturing, and total industrial chemical production, including propylene, butadiene, and benzene), according to various sources (OECD 1977, 1985, 1991a, 1991b; Hart 1992). The calculations assume, in short, that Japanese production in these key industries is overall about half that in the United States (a very conservative assumption) and that the industrial processes used in the two countries are roughly equivalent. Using the lower limit of the accepted figures for total U.S. hazardous waste production (i.e., 240 million tons) would thus yield an estimate for Japan of 120 million tons. Using an estimate for the United States that excludes wastes that are ignitable, corrosive, and reactive (159 million tons, from Reiter 1992), on the grounds that the Japanese definition of hazardous waste operative in the 1980s did not encompass wastes in these categories, would thus yield an estimate for Japan of about 80 million tons. If Japanese industrial recycling is assumed to have reduced waste output by half (likely an optimistic assumption), the range would then be 60 to 40 million tons.

An alternative ballpark calculation procedure would be to assume, even more simply, that Japan produces about the same amount of hazardous waste per capita as other industrial countries. This would not seem unreasonable. While Japanese production of municipal waste per capita is about one-half that of the United States, it is on par with most European countries. Its production of industrial waste per unit GDP is substantially higher than that for the United States or any other industrial country (OECD 1991, 45). American hazardous waste production is roughly 1 ton/capita/year. Germany's is estimated at about one-third of that (Piasecki 1984, 103). Taking these two estimates as the range, the Japanese total by this measure is likely somewhere from 40 to 120 million tons.

13. The magnitude of this industrial-pollution problem has not been reduced significantly by the growing tendency of many leading companies, over the 1970s and 1980s, to move dirty industrial plants offshore, especially into Southeast Asia. Ui (1989b) cites examples of industries that fled. These offshore plants have tended not to have very extensive pollution control equipment. See also the Japan Environment Agency–Japan Overseas Enterprises Association study cited by Weidner 1989, 522. (The same pattern by American companies has been noted by Shuman 1983).

14. The generation of general industrial waste in Japan, for which figures are available, has risen at a rate of 4 to 5 percent since the mid-1980s, notwithstanding the recession and recycling efforts (my calculations, based on data supplied by Takahiko Kimura, Bureau of Public Cleansing, Tokyo Metropolitan Government, personal communication to the author, October 1993, and on Hashizume 1994). The total volume of all industrial waste in Japan increased by a factor of eight during the period 1955–70 (Barrett and Therivel 1991, 37). Official estimates suggest the production of hazardous wastes in the United States increased at an average annual growth rate of 10 percent per year after 1945 (Block and Scarpitti 1985, 46). Walter and Getz (1986, 224) cite a 1981 Environmental Protection Agency (EPA) report that projected an increase of 3 percent per year in U.S. toxic wastes through the 1980s. A detailed county-by-county California study projected, however, that most areas in the state would increase waste generation by around 50 percent between 1986 and 2000 (Mazmanian and Morell 1992, 199).

15. The Environment Agency has warned publicly about "high-tech pollution," such as the toxic trichloroethylene used in the production of integrated circuit boards (Japan, Environment Agency 1985). See also Weidner 1989b, 497–98.

16. Tanaka 1990, 4. Other sources cite figures of thirty-five and thereabouts. On figures for the number of U.S. disposal facilities for hazardous waste, see Mazmanian and Morell 1992.

17. LaGrega, Buckingham, and Evans state that "remediation of contaminated sites dwarfs the management of currently generated hazardous waste on virtually whatever scale one would use" (1994, 23).

18. Chemical contamination was found in about 80% of the American landfills studied by the EPA (Block and Scarpitti 1985, 53). Illegal dumping aside, the reasons are not hard to find. Hazardous wastes have been estimated to compose about 3% of the municipal waste stream in Canada, not 1% as previously thought (Zeiss 1991). Kelleher 1991 estimates that there are 35 kg of stored hazardous wastes in each North American household.

19. In all, 34,000 tonnes of PCBs were shipped from the major Japanese producer to more than 1,200 factories in ten prefectures in just the seven years prior to being banned in 1972. The heaviest users of PCBs were manufacturers of electrical appliances (especially transformers and condensers), heat transfer agents, and noncarbon copy paper. An official survey discovered seventy-two locations in Japan where PCB contamination in soils was high enough to require remediation (Japan, Environment Agency 1986, 128). The Japanese government's Pollution Prevention Corporation has a mandate to provide pollution control assistance to industries building new facilities and relocating to new sites factories that are causing environmental problems, but it does not seem to have any mandate to deal with those old sites (Shigeru 1983).

20. This conclusion is developed more fully elsewhere (Munton 1993). Japan and the United Kingdom are the only two industrial countries that have neither site inventories nor a cleanup program (LaGrega, Buckingham, and Evans 1994, 172).

21. Dr. Osami Nakasugi, National Environmental Research Center, interview by author, Tokyo, June 2, 1993. Prefectural governments are worried both about how to deal with the problem and how to finance the cleanup efforts (Teruhiko Murayama, Waste Management Office, Fukushima Prefecture, personal communication to the author, March 23, 1994).

22. Researchers at one of the national research institutes are currently developing an assessment approach to determine and identify sites. The Environment Agency is planning to conduct a survey to determine the number and extent of possible problems, and a fact-finding team has been organized (Nakasugi interview). The basic problem is a familiar one: "The history of the lands where final waste disposal facilities were located," one official has noted, "is not clear" (Tanaka 1985, 92).

23. For a summary of the scope of the contaminated sites problem in Germany, Denmark, Belgium, France, the Netherlands, and the United States, see OECD 1991b, 152–52; Gourlay 1992, chap. 4. A 1988 study showed that an earlier 1985 survey had missed fully half the sites requiring attention in Denmark (Yakowitz 1988, 26). In the United States, the number of sites targeted for priority cleanup increased from 115 in 1981 to 160 in July 1982, to 418 in December 1982, to 551 in 1983, to 786 in 1984, to 964 in 1987, and to 1,187 by 1990 (various sources).

24. The term "Japan, Inc.," appears to have been popularized by Herman Kahn 1970.

25. Langdon 1967, 73–74. Yanaga notes that the norm of obedience to authority was not always followed and did not prevent, for example, peasant uprisings during the Tokugawa shogunate period (1964, 47).

26. See McKean, 1981; Vogel 1994, 265; Linden 1989, 50; Yamaguchi 1991. Keiji Yokoyama (1982) suggests that while true national organizations are not strong, some groups (such as anti-nuclear and anti-thermal power plant opponents) do exchange information on a national scale and could become strong national movements.

27. As Chitoshi Yanaga (1964, 43) has said, "Conformity to the collective wish of a group or community becomes a powerful conditioning factor." Even after publicity about the plight of the victims, community support was limited; indeed, the publicity "crystallized" the "hostility of other local residents," as they blamed the victims that markets for local products were lost because of the publicity. It was only after the victims took their case to court and after the sources of the problems were better understood that the hostility gave way to sympathy (McKean 1981, 81; see also 146). While those who act counter to the community's sense of propriety can be (and sometimes were) threatened with ostracism (*murahachibu*), at the same time group action in Japan can gain community support under certain circumstances, for example, "if the motivation is laudable" or "if justice and honor were at stake," and if the group members persevere (Yanaga 1964, 87, 84).

28. Statistics provided by Barrett and Therivel suggest media attention was a factor in the policy attention the environment received in Japan in 1970: "the media sometimes blew environmental problems up to disturbing proportions, causing increasing public concern" (1991, 22). See also OECD 1977, 17.

29. The most common were complaints about noise (32%) and odors (19%); those concerning soil contamination and ground subsidence amounted to less than 1% of all complaints (Environmental Disputes Coordination Commission, *Report on Complaints about Environmental Pollution*, 1984, cited in Japan, Environment Agency 1986).

30. The figure of 3,000 or so citizens' movements originated with a May 21, 1973, report in the newspaper *Asahi Shimbun*, which also suggested that there had been a tenfold increase in their number since 1970. This 3,000 figure is accepted by many observers, including McKean (1981), Krauss and Simcock (1980, 187–227), and Yokoyama (1982). The distinction between citizens' movements (*shimin undö*) and residents' movements (*jümin undö*), which is of some importance conceptually, is of less importance for present purposes (see McKean 1980). One estimate puts the number of active NIMBY groups in the United States, which has roughly twice Japan's population, at 6,300 (data from Citizens' Clearinghouse on Hazardous Waste, cited by Jessup 1992, 251).

31. It has been suggested that Japan is unique in the sense that environmental groups in most countries are strong at the national level, while in Japan they are strong at the local level (Pierce et al. 1986). This is not to say that local groups are unimportant in other countries; they clearly are important on many issues, including facility siting. See, e.g., Kraft and Kraut (1988, 73), who note the influence of local groups in Ohio.

32. A comparison of two public opinion surveys shows that pollution was regarded by more people as "inevitable" than "unacceptable" in 1966. By 1973, it was perceived to be "unacceptable" by more than four times as many respondents as thought it "inevitable." These results are summarized in Japan, Environment Agency 1973, cited by Tsurutani 1981, 47–48. Although similar shifts in attitudes were occurring in other countries, the so-called Big Four Japanese pollution cases in the late 1960s and early 1970s (to be discussed) likely played a role in this shift.

33. McKean 1981, 267; Michio Hashimoto, Japan, Environment Agency, quoted in Huddle, Reich, and Stiskin 1987, 264. Other analysts agree that the movements, beyond their significance as protests, manifested and led to fundamental changes in the political culture and political system of Japan (see, e.g., Krauss and Simcock 1980, 188–89; McKean 1981). Ui argues that the 1967 Basic Law of Environmental Pollution Control was intended "to avoid any such further action on the part of local populations" (1989a, 109).

34. Weidner 1989b, 484, citing Huddle, Reich, and Stiskin 1987, 256–57. Others agree that the Mishima-Numazu-Shimizu case was significant because it was the first successful movement (Ui 1989a, 109; McKean 1981, 29–30). Krauss and Simcock (1980, 195) characterize the Mishima-Numazu-Shimizu movement as a "watershed," marking a shift from the pursuit of compensation for damages done to the pursuit of pollution control and pollution prevention.

35. A different explanation is provided by Herman Kahn, who, putting his usual positive spin on events, attributes the rise of environmental interest in Japan not to greater pollution caused by industrialization but, rather, to the increased wealth it brought. The Japanese, he says, now "are rich enough to do something about all this" (1970, 131).

36. These groups tended to place great importance on "local residence" and "intimate involvement," believing "that only the victims or potential victims in a particular dispute belonged in a citizen's movement" (McKean 1981, 137). One reason for this was to counter any charges that they were outside agitators (McKean 1981, 148, 162).

37. See McKean 1981, 244. Prefectural and municipal governments also employ fully 90% of the officials who deal with environmental matters in Japan (OECD 1977, 73). Moreover, while there is no national environmental impact assessment (EIA) law, some prefectures have their own laws. On prefectural government EIA procedures and ordinances, see Barrett and Therivel 1991, table 7.5.

38. On these agreements, see Imura 1989, 63–64, 68ff; Shibata 1989, 246–51; Hashimoto 1989, 252–60.

39. Some sources date the first such agreement to one between Yokohama and the Electric Source Development Corporation concerning a coal-fired power plant in the early 1960s. Yamanouchi and Otsubo describe Yokohama as "the birthplace of a new method of concluding agreements on pollution prevention" (1989, 234); McKean agrees, saying this agreement was the "most important precedent" (1981, 149).

40. The sources for these figures are, variously, Weidner 1989b, 490; Imura 1989, 64; McKean 1981, 150–51; Pempel 1982, 232; Yamanouchi and Otsubo 1989, 222. To be sure, not all of these agreements impose stringent pollution control measures, nor are they necessarily of equal significance.

41. There is some debate about whether such agreements have legal effect, whether they are constitutional, and whether national laws preclude additional controls by local governments (Yamanouchi and Otsubo 1989, 222, 228). The question of legal effect seems moot; the fact is that the companies involved generally feel very much bound by the agreements.

42. Huddle, Reich, and Stiskin argue that the weaknesses of the citizens' movements is shown by their decline since the heyday of the 1970s (1987, chap. 9). Movements and victims have also been less successful in lawsuits (Barrett and Therivel 1991, 22). Weidner finds there has been an "erosion of purpose" (1989, 498).

43. Dr. Masaru Tanaka, director, Department of Waste Management Engineering, Institute of Public Health, Tokyo, interview by author, June 2, 1993.

44. Kyoji Kinbara, Sanyu Plant Service, interview by author, May 29, 1993. This site is about 1/30 that of the site of the average small incinerator in the United States, which typically occupies about 8 acres, which is 3.2 hectares, or 32,000 m^2 (Mazmanian and Morell 1992, 124).

45. Japan, Health Ministry 1991–92, 128, 135. While these national agencies are undoubtedly concerned about the problem of ensuring adequate sites, expressed concerns should not be confused with effective action.

46. The volume of hazardous wastes is also mounting, ironically, because of the application of more stringent air and water pollution policies and standards. In short, what industries once emitted into the air or released into waterways is now collected by pollution control equipment and must be dealt with in other ways. By one estimate, about 25% of American hazardous wastes, as defined by the U.S. RCRA (U.S. Resource Conservation and Recovery Act) regulations, are sludges from wastewater treatment and dust and sludges from air pollution control devices. See DiNal 1981.

47. Japan is, of course, not alone in facing this problem. See, for example, Yakowitz 1988.

48. See, respectively, Ishino 1989, 330; Fujiwara 1992, 1; Azuma 1993, 2; Teruhiko Murayama, Waste Management Office, Fukushima Prefecture, personal communication to the author, March 23, 1994.

49. National figures provided by Hideo Azuma, Industrial Waste Management Section, Bureau of Public Cleansing, Tokyo Metropolitan Government, interview by author, May 21, 1993. See also Azuma 1993, 6. Yamaguchi 1991 reports a figure of 870,000 tons per year for industrial wastes that were illegally dumped.

50. The explicit rationale was "to improve previous disposal measures, to clarify the responsibilities of the citizens, entrepreneurs, the Government and local public organizations toward waste measures, and for the whole community to develop proper disposal measures including volume decrease, recycling and reuse" (Japan, Health Ministry 1991–92, 130). The new law was designed to be complementary to the 1991 recycling law (Law for Promoting the Utilization of Renewable Resources); on the linkage of the two laws, see Japan, Clean Japan Center, n.d.

51. The solutions sought are not merely domestic. Along with promoting the construction of industrial-waste disposal facilities, the Ministry of Health foresees greater "transportation of wastes abroad" and has sponsored research on air transport of waste (Japan, Health Ministry 1991–92, 136; see also Fujiwara 1992). Japan, however, is a signatory to the Basel Convention and has passed implementing legislation.

52. One commonly proposed solution, negotiation between facility proponents and a local community, is not distinguished here as a separate strategy. As a generic process, it underlies all feasible alternatives to decision by fiat. On negotiation and mediation, see, for example, O'Hare et al. 1973; Bingham 1986; Bacow and Wheeler 1984; Rovet 1987.

53. Portney and Davis find that residents preferred measures such as safety inspections and measures to prevent groundwater contamination. Elliot's simulations suggested the same conclusion (1984). Some long-operating treatment facilities in Europe, such as the Danish Kommunekemi plant, have adopted a full public-disclosure policy, which seems to have helped maintain public acceptance (Piasecki 1984, 103).

54. The conclusion that few prefectures are actively working on new siting proposals was substantiated by the responses to letters sent to most prefectural governments in spring 1994 by the author.

55. This extension of the role of local and prefectural governments can be seen as part of a broader pattern of devolution of power in the environmental area that goes back to the 1970 amendments to the Basic Law for Environmental Pollution Control, which gave subnational authorities the right to adopt their own environmental standards; see McKean 1981, 243. This trend is, of course, opposite to that of increased centralization in certain other industrial countries.

56. Another policy advantage of involving local governments is that their authority can be more effective in the implementation of waste policies. When reliable waste facilities are available, these authorities can more easily require that particular actions be taken by waste producers (Shigemoto Kajihara, Marine Pollution and Waste Management Office, Water Quality Bureau, Environment Agency, Tokyo, interview by author, June 1, 1993).

57. The emerging Japanese system will resemble that in use in some parts of Germany, for example, Bavaria and Hesse. It will differ from the arrangements adopted in the provinces of Alberta and Manitoba in Canada, discussed by Castle and Munton and by Richards in this volume, under which a provincial government agency designs and owns a facility but contracts out its management to a private waste disposal firm. Some areas in Japan have experimented with industrial waste treatment cooperatives; one in Gyoda City manages wastes from 150 member companies (Pempel 1982, 233).

58. Among the other advantages is expertise sharing. Given that most local government officials are not well informed about the more technical aspects of wastes and the companies that produce them usually have a much better knowledge of their chemical nature, the latter's expertise will be of much value in decisions about handling and treatment.

59. Given these funding requirements, the new facility in Ōita, for example, will treat both industrial and municipal waste (Hiroki Hashizume, Ministry of Health, Tokyo, interview by author, May 28, 1993). It is something approaching the norm for industrial-country governments to provide grants at least to defray capital costs of hazardous waste facilities. This is the practice in, e.g., Canada, Belgium, Germany, Denmark, Finland, Sweden, and the Netherlands, though not in the United States (Yakowitz 1988, 26).

60. This support may not be enough. Financing remains both an economic and a political problem for prefectures. The financial resources of local governments, as always, are limited, and some officials report this to be a major stumbling block in the establishment of such facilities (Teruhiko Marayama, Waste Management Office, Fukushima Prefecture, personal communication to the author, March 23, 1994). The political problems stem from the reluctance of local taxpayers to provide the funding for industrial waste facilities, something that has traditionally been the responsibility of the private sector itself.

61. One example of such a problem developed in Aichi Prefecture, where the government built an intermediate facility for waste oil treatment, but its relatively high prices meant it was poorly utilized. Tanaka interview.

62. In the United States, for example, it has proven impossible to include compensation in legislation, except in certain states, e.g., New Jersey and California (Mazmanian and Morell 1992, 55, n. 47). Bills introduced into Congress to provide for victim compensation have failed, and compensation provisions included in the Superfund legislation were dropped in a series of compromises deemed necessary to allow its passage.

63. The bulk of Portney's evidence that offers of compensation do not induce people to be more positive toward hazardous waste facilities derives from public opinion polls, yet he himself acknowledges that people may well behave differently when faced with a real-world situation than is suggested in their responses to a pollster. Moreover, the data Portney presents (1991) suggest clearly that significant numbers of people do, in fact, say their views would be more positive if compensation were offered. To find that compensation does not move all or most people is not the same thing as finding that "it does not work."

64. Kasperson 1986, Connor 1990, and Armour 1988 all advocate compensation as part of a package of measures in siting cases. Armour, who makes the general case for compensation as part of an "impact management" strategy, notes that Swartzman, Croke, and Swibel 1985 find that compensation (e.g., property tax reductions, user fees to the community) as well as risk reduction measures (e.g., increased monitoring, greater control over who monitors) reduced opposition to hazardous waste landfills. Shuff 1988 argues that compensation helped siting efforts in Wisconsin. Other studies suggest compensation has played a role in facility siting in Europe (Visocki 1988; Paige and Owens 1984). See also Morell and Magorian 1982, 59–60, 167–70.

65. For example, about 1,300 former residents of Love Canal received $20 million from Hooker Chemical (as well as a $30 million buyout from the U.S. federal government). Residents near the West Covina dump site of BKK, a major California hazardous waste disposal firm, received $43 million, the "largest ever" such settlement (Mazmanian and Morell 1992, 45–46).

66. See McKean (1981, 35–38), who cites numerous Japanese-language sources; Ui 1992; Huddle, Reich, and Stiskin 1987, 25–33. As the latter make clear, the Ashio conflict was only one of several protests against environmental damage, especially to fishing and agriculture, from mining operations in eighteenth- and nineteenth-century Japan (p. 264).

67. Actual compensation of the victims has not been an easy or automatic process, however. Implementation of the court decision was much delayed. Many of those claiming compensation had to wait years for the administrative process to unfold, and some had to go to court to press their claims. The compensation awards in the Minamata case and others tended to restrict the number of potential victims either by limiting the geographic scope of the affected area or by making it difficult to qualify as a certified victim. See Huddle, Reich, and Stiskin 1987, 20–23.

68. Tsuru 1989, 34. The fact that Japanese courts tended, until the Big Four cases, to define negligence rather narrowly does not undermine the main point here—that is, that the concept of compensation was an inherent part of Japanese statute and thinking.

69. On the 1967 law, see Kanazawa 1973; on the 1973 law, see Gresser 1975.

70. Imura 1989, 75–76; Hashimoto 1989, 257–58. The amount of compensation provided in various major environmental court cases (termed judiciary compensation) during the 1970s is summarized by McKean (1981, chap. 2, esp. table 4). The amounts of compensation provided under the 1973 law (termed administrative compensation) up to 1976 are summarized in OECD 1977, 43. By 1987, compensation had been paid to 99,000 individuals, most of whom suffered from respiratory problems (Weidner 1989b, 489–90, 496). See also Barrett and Therivel (1991, chap. 2, esp. table 2.1).

71. It seems, however, that this long-standing practice is not well known outside Japan, even by experts on compensation. O'Hare, Bacow, and Sanderson (1983, ix) suggest that "the first explicit mention" of the concept of compensation in facility siting was, to their knowledge, in a 1972 paper by one of their MIT colleagues.

72. These arrangements have been formalized as "Compensation Standards," which apply to property holders and go back to 1963, and as the Regional Development Law (part of the Three Electric Development Laws of 1974), which covers community compensation for social risks, paid to prefectural governments, which channel them to local governments. MITI has an important role in establishing compensation arrangements. There is some evidence that this scheme has reduced the time for fossil fuel plant siting during the 1980s, compared with 1975, and reduced the rate of increase in the time for siting nuclear plants, again compared with 1975 (Lesbirel 1992, chap. 2).

73. These measures have also been adopted in Canada, and in Europe (Piasecki and Davis 1987, chap. 8).

74. For use of this term in a somewhat different, more practical manner, see the chapter by Richards in this volume. The co-management approach as conceived here has some of the elements of what Audrey Armour calls an

Impact Management Program, namely, mitigation measures such as pollution control, contingency measures such as responses to problems, compensation measures, and "community relations" measures. See Armour 1988.

75. A similar pattern has been observed in the United States. Walter and Getz note, "on-site facilities stimulate less intense local opposition than off-site operations" (1986, 241).

REFERENCES

Armour, Audrey. 1988. The Social Aspects of Devising an Impact Management Program. In *10th Canadian Waste Management Conference*, 379–85. Ottawa: Environment Canada.

———. 1991. Opting for Co-operation: A Case Study in Siting a Low-level Radioactive Waste Management Facility. *Progress in Planning* 35:49–67.

———, ed. 1984. The Not-In-My-Backyard Syndrome. Symposium, February, Faculty of Environmental Studies, York University.

Aronson, Bruce E. 1983. Environmental Law in Japan. *Harvard Environmental Law Review* 7, no. 1:135–71.

Azuma, Hideo. 1993. Administrative Control and Guidance for Business Enterprisers Discharging Industrial/Business Waste in Tokyo. Paper presented at the Second Seminar on Industrial and Hazardous Waste Management, March, Jakarta.

———. 1993. Establishment of Recycling Ordinances in Tokyo Municipalities. *Newsletter, Japan Society of Waste Management Experts,* no. 6 (March).

Bacow, Lawrence S., and James R. Milkey. 1982. Overcoming Local Opposition to Hazardous Waste Facilities: The Massachusetts Approach. *Harvard Environmental Law Review* 6, no. 2:265–305.

Bacow, Lawrence S., and Michael Wheeler. 1984. *Environmental Dispute Resolution.* New York: Plenum.

Barnett, Harold C. 1994. *Toxic Debts and the Superfund Dilemma.* Chapel Hill: University of North Carolina Press.

Barrett, Brendan F. D., and Riki Therivel. 1991. *Environmental Policy and Impact Assessment in Japan.* London: Routledge.

Berkes, Fikret, and David Feeny. 1990. Paradigms Lost: Changing Views on the Use of Common Property Resources. *Alternatives* 17, no. 2:48–55.

Berkes, Fikret, Peter George, and Richard J. Preston. 1991. Co-Management: The Evolution in Theory and Practice of the Joint Administration of Living Resources. *Alternatives* 18, no. 2:12–18.

Bingham, Gail. 1986. *Resolving Environmental Disputes: A Decade of Experience.* Washington, D.C.: Conservation Foundation.

Block, Alan R., and Frank R. Scarpitti. 1985. *Poisoning for Profit: The Mafia and Toxic Waste in America.* William Morrow and Company.

Connor, Desmond. 1990. Managing the "NIMBY Syndrome" in the 1990s: Principles and Cases for Waste Managers. In *12th Canadian Waste Management Conference,* 1–8. Ottawa: Environment Canada.

Davis, Charles E. 1986. Public Involvement in Hazardous Waste Facility Siting Decisions. *Polity,* Winter.

Davis, Charles E., and James P. Lester. 1988. Hazardous Waste Politics and the Policy Process. In *Dimensions of Hazardous Waste Politics and Policy,* 1–34. New York: Greenwood Press.

———, eds. 1988. *Dimensions of Hazardous Waste Politics and Policy.* New York: Greenwood Press.

Davis, Gary, Donald Huisingh, and Bruce Piasecki. 1987. Waste Reduction Strategies: European Practice and American Prospects. In *America's Future in Toxic Waste Management: Lessons from Europe,* edited by Bruce Piasecki and Gary Davis, 17–42. Westport, Conn.: Quorum Books.

Davis, Gary, and Bruce Piasecki. 1987. The Next Steps after Land Disposal. In *America's Future in Toxic Waste Management: Lessons from Europe,* edited by Bruce Piasecki and Gary Davis, 223–32. Westport, Conn.: Quorum Books.

DiNal, Joan G. 1981. Hazardous Waste: Environmental Issue of the Eighties. Paper presented at the Environmental Law Symposium, American Bar Association.

Egan, T. 1990. Mob Looks at Recycling and Sees Green. *New York Times,* November 28.

Elliot, Michael. 1984. Improving Community Acceptance of Hazardous Waste Facilities through Alternative Systems for Mitigating and Managing Risk. *Hazardous Waste* 1 (Fall): 398–410.

Enloe, Cynthia H. 1975. *The Politics of Pollution in a Comparative Perspective.* New York: McKay

Epstein, Samuel, Lester O. Brown, and Carl Pope. 1982. *Hazardous Waste in America.* San Francisco: Sierra Club Books.

Feeney, David, Fikret Berkes, Bonnie J. McCay, and James M. Acheson. 1990. The Tragedy of the Commons 22 Years Later. *Human Ecology* 18, no. 1:1–19.

Flanagan, Scott. 1978. The Genesis of Variant Political Cultures: Contemporary Citizen Orientations in Japan, America, Britain, and Italy. In *The Citizen and Politics,* edited by S. Verba and L. W. Pye, 129–63. Stamford, Conn.: Greylock.

Freeman, Harry, ed. 1990. *Hazardous Waste Minimization.* New York: McGraw-Hill.

Fujiwara, Masahiro. 1992. Current Status of Waste Treatment in Japan. Unpublished paper. Ministry of Health and Welfare.

Gourlay, K. A. 1992. *World of Waste: Dilemmas of Industrial Development.* London: Zed Books.

Gresser, Julian. 1975. The 1973 Japanese Law for the Compensation of Pollution-Related Health Damage. *Law in Japan* 8:91–135.

Harris, Christopher, William Want, and Morris Ward. 1987. *Hazardous Waste: Confronting the Challenge.* New York: Quorum Books.

Hart, Jeffrey. 1992. *Rival Capitalists.* Ithaca: Cornell University Press.

Hart, Steve. 1990. Federal Waste Management Program. In *12th Canadian Waste Management Conference,* 191–96. Ottawa: Environment Canada.

Hashimoto, Michio. 1989. Adminstrative Guidance in Environmental Policy: Some Important Cases. In *Environmental Policy in Japan,* edited by Shigeto Tsuru and Helmut Weidner, 252–60. Berlin: Sigma Bohn.

Hashizume, Hiroki. 1994. A Brief Note on Solid Waste Management in Japan: Industrial Waste. *Newsletter, Japan Society of Waste Management Experts*, no. 10 (July).

Hirakawa, Naomi. 1995. Waste not, want not. *Japan Times Weekly International Edition*, 35, no. 18 (May 8–14): 7.

Hoshino, Yoshiro. 1992. Japan's Post–Second World War Environmental Problems. In *Industrial Pollution in Japan*, edited by Jun Ui, 64–76. Tokyo: United Nations University.

Huddle, Norie, M. Reich, and N. Stiskin. 1987. *Island of Dreams: Environmental Crisis in Japan*. New York: Autumn Press.

Iijima, Nobuko. 1992. Social Structures of Pollution Victims in Japan. In *Industrial Pollution in Japan*, edited by Jun Ui, 154–72. Tokyo: United Nations University.

Imura, Hidefumi. 1989. Administration of Pollution Control at Local level. In *Environmental Policy in Japan*, edited by Shigeto Tsuru and Helmut Weidner, 54–96. Berlin: Sigma Bohn.

Ishino, Koya. 1989. Waste Management. In *Environmental Policy in Japan*, edited by Shigeto Tsuru and Helmut Weidner, 320–31. Berlin: Sigma Bohn.

Japan. Clean Japan Center. N.d. How the Lecycling [*sic*] Law and the Waste Management Law Work Together. Unpublished chart. Tokyo.

Japan. Environment Agency. 1973. *Quality of the Environment in Japan*. Tokyo.

———. 1978. *Quality of the Environment in Japan*. Tokyo.

———. 1985. *1985 Report*. Tokyo.

———. 1986. *Quality of the Environment in Japan*. Tokyo.

Japan. International Cooperation Agency. 1922 [*sic*]. *Solid Waste Management and Night Soil Treatment*. Tokyo: Industrial Waste Management Administration.

Japan. Ministry of Health and Welfare (Health Ministry). 1987. *Annual Report, 1987*. Tokyo.

Japan. Ministry of Health and Welfare (Health Ministry). 1991–92. *Annual Report, 1991–92*. Tokyo.

Jessup, Deborah H. 1992. *Waste Management Guide: Laws, Issues, and Solutions*. Washington, D.C.: Bureau of National Affairs.

Kahn, Herman. 1970. *The Emerging Japanese Superstate*. Englewood Cliffs, N.J.: Prentice-Hall.

Kanazawa, Yoshio. 1973. A System of Relief for Pollution-Related Injury. *Law in Japan* 6:65–72.

Kasperson, Roger. 1986. Hazardous Waste Facility Siting: Community, Firm, and Governmental Perspectives. In *Hazards: Technology and Fairness* by National Academy of Engineering, 118–44. Washington, D.C.: National Academy Press.

Kelleher, Maria. 1991. Household Hazardous Waste—How Much Is Really Out There? In *Wastetech '91: Canadian Waste Management Conference*, 457–69. Ottawa: Environment Canada.

Kobe City Environmental Bureau. 1994. *Industrial Waste Measures in Kobe: Summary of the Fundamental Plan for Guidance on Industrial Waste Treatment in Kobe*. March 16.

Kraft, Michael, and Ruth Kraut. 1988. Citizen Participation and Hazardous Waste Policy Implementation. In *Dimensions of Hazardous Waste Politics and Policy*, edited by Charles Davis and James Lester, 63–80. New York: Greenwood Press.

Krauss, Ellis, and Bradford Simcock. 1980. Citizens' Movements: The Growth and Impact of Environmental Protest in Japan. In *Political Opposition and Local Politics in Japan*, edited by K. Steiner, E. Krauss, and S. Flanagan, 187–227. Princeton: Princeton University Press.

Kunreuther, Howard C. 1986. Hazard Compensation and Incentive Systems: An Economic Perspective. In *Hazards: Technology and Fairness* by National Academy of Engineering, 145–63. Washington, D.C.: National Academy Press.

LaGrega, Michael, Phillip Buckingham, and Jeffrey Evans. 1994. *Hazardous Waste Management*. New York: McGraw-Hill.

Langdon, Frank. 1967. *Politics in Japan*. Boston: Little, Brown and Company.

Lesbirel, S. Hayden. 1987–88. Project Implementation in Japan: Origins of Delay in Developing Energy Facilities. *Journal of Business Administration* 17, nos. 1–2:249–86.

———. 1992. Externalities, Bargaining, and Compensation: Managing Environmental Conflict in Japan. Manuscript, National University of Singapore.

Lewis, Jack G. 1980. Civic Protest in Mishima: Citizens' Movements and the Politics of the Environment. In *Political Opposition and Local Politics in Japan*, edited by K. Steiner, E. Krauss, and S. Flanagan, 274–313. Princeton: Princeton University Press.

Linden, Eugene. 1989. Putting the Heat on Japan. *Time*, July 10, 48–50.

Liptak, Bela G. 1991. *Municipal Waste Disposal in the 1990s*. Radnor, Penn.: Chilton.

Matheny, Albert R., and Bruce A. Williams. 1988. Rethinking Participation: Assessing Florida's Strategy for Siting Hazardous Waste Disposal Facilities. In *Dimensions of Hazardous Waste Politics and Policy*, edited by Charles E. Davis and James P. Lester, 37–52. New York: Greenwood Press.

Mazmanian, Daniel, and David Morell. 1992. *Beyond Superfailure: America's Toxics Policy for the 1990s*. Boulder, Colo.: Westview Press.

———. 1994. The "NIMBY" Syndrome: Facility Siting and the Failure of Democratic Discourse. In *Environmental Policy in the 1990s*, edited by Norman J. Vig and Michael E. Kraft, 233–49. 2d ed. Washington, D.C.: CQ Press.

McKean, Margaret A. 1976. Citizens' Movements in Urban and Rural Japan. In *Social Change and Community Politics in Urban Japan*, edited by James White and Frank Munger, 61–99. Monograph no. 4. Institute for Research in Social Science, University of North Carolina at Chapel Hill.

McKean, Margaret A. 1977. Pollution and Policymaking. In *Policymaking in Contemporary Japan*, edited by T. J. Pempel, 201–38. Ithaca: Cornell University Press.

———. 1980. Political Socialization through Citizens' Movements. In *Political Opposition and Local Politics in Japan*, edited by K. Steiner, E. Krauss, and S. Flanagan, 228–73. Princeton: Princeton University Press.

———. 1981. *Environmental Protest and Citizen Politics in Japan*. Berkeley: University of California Press.

———. 1992. Success on the Commons: A Comparative Examination of Institutions for Common Property Resource Management. *Journal of Theoretical Politics* 4, no. 3:247–81.

Mills, Edwin S. 1978. *The Economics of Environmental Quality*. New York: Norton.

Morell, David, and Christopher Magorian. 1982. *Siting Hazardous Waste Facilities: Local Opposition and the Myth of Preemption*. Cambridge, Mass.: Ballinger.

Munton, Don. 1994. Go (Not) Placidly amidst the Waste Statistics: Estimating Japanese Hazardous Waste. Unpublished paper. York University.

———. 1995. Changing Policies for Hazardous Waste in Japan. Unpublished paper. University of Northern British Columbia.

OECD. 1977. *Environmental Policies in Japan*. Paris.

———. 1985. *The State of the Environment, 1985*. Paris.

———. 1991a. *Environmental Indicators*. Paris.

———. 1991b. *The State of the Environment*. Paris.

O'Hare, Michael. 1977. Not on My Block You Don't—Facility Siting and the Importance of Compensation. *Public Policy* 25: 407–58.

O'Hare, Michael, Lawrence Bacow, and Debra Sanderson. 1983. *Facility Siting and Public Opposition*. New York: Van Nostrand-Reinhold.

Paige, H. W., and J. E. Owens. 1984. Assessment of National Systems for Obtaining Local Acceptance of Nuclear Waste Management Siting Activities. In *Radioactive Waste Management: Proceedings of an International Conference*, 319–31. Vienna: IAEA.

Passin, Herbert. 1962. The Sources of Protest in Japan. *American Political Science Review* 56, no. 2 (June): 391–403.

Pempel, T. J. 1982. *Policy and Politics in Japan*. Philadelphia: Temple.

Piasecki, Bruce. 1984. Europe's Detoxification Arsenals: Lessons in Waste Recovery and Exchange. In *Beyond Dumping: New Strategies for Controlling Toxic Contamination*, 101–11. Westport, Conn.: Quorum Books.

Piasecki, Bruce, and Gary Davis, eds. 1987. *America's Future in Toxic Waste Management: Lessons from Europe*. Westport, Conn.: Quorum Books.

Pierce, J., T. Tsurutani, N. P. Lovrich, and T. Abe. 1986. Vanguards and Rearguards in Environmental Politics: A Comparison of Activists in Japan and the United States. *Comparative Political Studies* 18, no. 4:419–47.

Portney, Kent. 1991. *Siting Hazardous Waste Treatment Facilities: The Nimby Syndrome*. New York: Auburn House.

Reich, Michael R. 1984. Mobilizing for Environmental Policy in Italy and Japan. *Comparative Politics* 16, no. 4 (July): 379–402.

Reiter, Anne J. 1992. Hazardous Waste Generation in the US and Japan: A Comparative Analysis. Paper presented at a conference, Waste Management: U.S.A. and Japan, September.

Repa, Edward W., and Susan Sheets. 1992. 1991 Landfill Capacity Report. In *14th Canadian Waste Management Conference*, 37–50. Ottawa: Environment Canada.

Richardson, Bradley. 1974. *The Political Culture of Japan*. Berkeley: University of California.

Rovet, Ernest. 1987. Legitimized Structured Public Involvement in the Environmental Assessment Process. In *9th Canadian Waste Management Conference*, 301–12. Ottawa: Environment Canada.

Shibata, Tokue. 1989. Pollution Control Agreements: The Case of Tokyo and Other Local Authorities. In *Environmental Policy in Japan*, edited by Shigeto Tsuru and Helmut Weidner. Berlin: Sigma Bohn, 246–51.

Shigeru, Matsumoto. 1983. Efforts of the Pollution Prevention Corporation. *Japan Quarterly* 30, no. 4 (October–December): 409–12.

Shuff, Richard G. 1988. "Bribes" Work in Wisconsin. *Waste Age*, March, 51–55.

Shuman, Eric. 1983. Potentially Hazardous Merchandise: Domestic and International Mechanisms for Consumer Protection. *Vanderbilt Journal of Transnational Law* 16, no. 1:179–229.

Swartzman, Daniel, Kevin Croke, and Sheri Swibel. 1985. Reducing Aversion to Living near Hazardous Waste Facilities through Compensation and Risk Reduction. *Journal of Environmental Management* 20:43–50.

Tanaka, Masaru. 1985. The State of the Art of Investigations and Researches in the Field of Waste Management in Japan. *Bulletin, Institute of Public Health* 34, nos. 3–4:83–94.

———. 1990. Risk Reduction Strategies for Hazardous Wastes. Paper presented at the Asia and Pacific Regional Workshop on Hazardous Waste Minimization and Reduction, December, Kyoto.

Taniguchi, Tomohei. 1976. A Commentary on the Legal Theory of the Four Major Pollution Cases. *Law in Japan* 9:35–62.

Tatsukawa, Ryo. 1976. PCB Pollution of the Japanese Environment. In *PCB Poisoning and Pollution*, edited by K. Higuchi, 147–79. Tokyo: Kodansha.

Trauberman, J. 1981. Compensating Victims of Toxic Substance Pollution: An Analysis of Existing Federal Statutes. *Harvard Environmental Law Review* 5:1–19.

Tsuru, Shigeto. 1989. History of Pollution Control Policy. In *Environmental Policy in Japan*, edited by Shigeto Tsuru and Helmut Weidner, 15–42. Berlin: Sigma Bohn.

Tsurutani, Taketsugu. 1981. *Political Change in Japan*. New York: McKay.

Uga, Katsuya. 1991. Japan's Environmental Agency: Its History and Contemporary Issues. *Law in Japan* 24:4–38.

Ui, Jun. 1989a. Anti-pollution Movements and Other Grass-Roots Organizations. In *Environment Policy in Japan*, edited by Shigeto Tsuru and Helmut Weidner, 109–19. Berlin: Sigma Bohn.

———. 1989b. Pollution Export. In *Environmental Policy in Japan*, edited by Shigeto Tsuru and Helmut Weidner, 395–412. Berlin: Sigma Bohn.

———. 1992. *Industrial Pollution in Japan*. Tokyo: United Nations University.

Upham, Frank K. 1976. Litigation and Moral Consciousness in Japan: An Interpretive Analysis of Four Japanese Pollution Suits. *Law and Society* 10:579–619.

U.S. Environmental Protection Agency (U.S. EPA). 1979. *Siting of Hazardous Waste Facilities and Public Opposition*. SW–809, November.

———. 1985. *National Biennial Report of Hazardous Waste Generators and Treatment, Storage, and Disposal Facilities Regulated under RCRA*. Washington, D.C.: U.S. Government Printing Office.

———. 1986. *Minimization of Hazardous Waste*. EPA 530-SW–86–0330. Washington, D.C.: U.S. Government Printing Office.

———. 1991. *National Survey of Hazardous Waste Generators and Treatment, Storage, and Disposal Facilities in 1986*. Washington, D.C.: U.S. Government Printing Office.

U.S. General Accounting Office (U.S. GAO). 1978. *How to Dispose of Hazardous Waste—a Serious Question That Needs to Be Answered*. CED–79–13, December 19.

U.S. Office of Technology Assessment (U.S. OTA). 1986. *Serious Reduction of Hazardous Waste: For Pollution Prevention and Industrial Efficiency*. OTA-ITE–317. Washington, D.C.

Visocki, Kathryn. 1988. *Incentives, Compensation, and Other Magic Tricks: Will They Help in Establishing New Waste Disposal Sites?* Raleigh, N.C.: Southeast Compact Commission for Low-Level Radioactive Waste Management.

Vogel, David. 1994. Environmental Policy in Europe and Japan. In *Environmental Policy in the 1990s*, edited by N. J. Vig and M. E. Kraft, 257–77. Washington, D.C.: Congressional Quarterly Press.

Walter, Benjamin, and Malcolm Getz. 1986. Social and Economic Effects of Toxic Waste Disposal. In *Controversies in Environmental Policy*, edited by S. Kamieniecki, R. O'Brien, and M. Clarke, 223–45. Albany: State University of New York Press.

Warren, John L. 1990. Potential for Waste Reduction. In *Hazardous Waste Minimization*, edited by Harry Freeman, 15–23. New York: McGraw-Hill.

Weidner, Helmut. 1989a. Environmental Monitoring and Reporting by Local Governments. In *Environmental Policy in Japan*, edited by Shigeto Tsuru and Helmut Weidner, 461–76. Berlin: Sigma Bohn.

———. 1989b. Japanese Environmental Policy in an International Perspective: Lessons for a Preventive Approach. In *Environmental Policy in Japan*, edited by Shigeto Tsuru and Helmut Weidner, 479–552. Berlin: Sigma Bohn.

Whymant, Robert. 1973. Sharing the Burden of Pollution. *Manchester Guardian* 108, no. 27 (June 30).

Wolf, Sidney M. 1983. Hazardous Waste Trials and Tribulations. *Environmental Law* 13:367–491.

Yakowitz, Harvey. 1988. Possible Future Trends for Management of Hazardous Wastes in OECD Member Countries. In *Tenth Canadian Waste Management Conference*, 25–32. Ottawa: Environment Canada.

Yamaguchi, Mitsutsune. 1991. Japanese Companies and the Environment. *Law in Japan* 24:62–88.

Yamanouchi, Kazuo, and Kiyoharu Otsubo. 1989. Agreements on Pollution Prevention: Overview and One Example. In *Environmental Policy in Japan*, edited by Shigeto Tsuru and Helmut Weidner, 221–45. Berlin: Sigma Bohn.

Yanaga, Chitoshi. 1964. *Japanese People and Politics*. New York: Wiley and Sons.

Yokoyama, Keiji. 1982. Review of *Environmental Protest and Citizen Politics in Japan*, by Margaret McKean. *Japan Quarterly* 29, no. 1 (January–March): 112–13.

Zeiss, Chris. 1991. Estimates and Sources of Hazardous Waste Loading in a Municipal Landfill in a Resource-Based Community. In *Wastetech '91:*

Canadian Waste Management Conference, 441–54. Ottawa: Environment Canada.

Zeiss, Chris, and James Atwater. 1988. The Significance of Property Value Guarantees as a Tool for Siting Waste Disposal Facilities. In *10th Canadian Waste Management Conference,* 355–71. Ottawa: Environment Canada.

Siting Hazardous Waste Incinerators in Germany: From Political Imposition to Public Involvement?

ROBERT SEELIGER

INTRODUCTION

Germany has long been a popular destination for North American engineers and planners studying concepts of integrated waste management and the design and operation of hazardous waste disposal facilities.[1] Some have endorsed hazardous waste disposal systems in Europe, among them those operated in Germany, as being examples for North American jurisdictions (Piasecki and Davis 1987). And in a few instances German hazardous waste management systems actually served as models for similar systems in North America. Alberta and Manitoba emulated the successful structure of the Bavarian and Hesse systems, which feature collection and transfer stations feeding prescreened waste streams to large and comprehensive central treatment facilities. The technologies of Ontario's proposed central treatment facility were modeled on the Biebesheim treatment center in Hesse.[2]

If Germany was a source of innovation and inspiration in the past, today German government officials, waste management companies, academics, and environmental groups could turn to the western shores of the Atlantic for insights and lessons. For almost a decade no new hazardous waste incinerators or landfills have been sited in Germany. Public opposition and political stalemate seem insurmountable. The challenge is to break the gridlock in siting waste management facilities and to settle associated conflicts in a politically acceptable way. The question is how to do that.

Pressed by the need to get additional treatment and disposal capacity on line as well as by political distress over failed siting attempts, some jurisdictions have been experimenting with weaving elements of negotiation and mediation into the legalistic German siting and permitting procedures. Interestingly, these initiatives draw heavily on the theoretical literature on conflict resolution techniques that developed in North America (Ziellеßen and Barbian 1992). At the same time, German social scientists have only begun to analyze social and political conflicts in waste management (e.g., Müller and Holst 1987; Herbold and Wienken 1993). The task remains to identify systematically the circumstances under which mediation and negotiation strategies (which originated in the social and political settings of the United States and Canada) could be used more widely to resolve political conflicts in waste management in Germany. Such an endeavor would have to build on a sound understanding of the underlying patterns of such conflicts in Germany. These, however, have not been the object of much scholarly attention.

This chapter contributes to the understanding of German waste management politics by exploring past and present successes and failures to site hazardous waste incinerators. In particular, three sets of questions are investigated: (1) In the 1970s and early 1980s several hazardous waste treatment and disposal facilities were successfully sited, e.g., in Bavaria and Hesse. How did governments go about identifying and securing sites? How did the public react? How did governments and facility proponents deal with public opposition? (2) Have there been more recent successful siting attempts based on a government-directed top-down approach? If so, how did governments and public interact? Which factors account for siting success or siting failure? (3) Which dynamics characterize siting attempts that added negotiation and/or mediation strategies to the legalistic procedures set out by law? How did this affect popular attitudes toward the siting of hazardous waste incinerators?

These questions are explored through four case studies of hazardous waste incinerator sitings. They include the siting of the hazardous waste incinerator in Ebenhausen (Bavaria) in 1972–73, the siting of the Hesse Industrial Waste Corporations incinerator in Biebesheim (Hesse) in the late 1970s, the recent siting of a hazardous waste incinerator in Brunsbüttel (Schleswig-Holstein), and the ongoing siting odyssey in Baden-Württemberg. The case studies are placed in the larger context of hazardous waste management in Germany. They are preceded by a discussion of the technological challenges and political constraints of hazardous waste management (particularly incineration), as well as of its legal framework. A concluding section offers observations on patterns in hazardous waste facility siting, on the discourse over obstacles to facility siting, and on the prospects for innovative siting strategies in Germany.

HAZARDOUS WASTE MANAGEMENT IN GERMANY: TECHNICAL CHALLENGES AND POLITICAL CONSTRAINTS

Germany is a heavily industrialized country, and its factories produce commensurate streams of hazardous wastes. Aggregate data on waste generation and waste management methods are scarce and habitually years behind. Returns of the 1990 waste disposal survey by the Federal Statistics Office show that about 15.7 million metric tons of hazardous waste are generated annually (Statistisches Bundesamt 1994).[3] The 1990 *Reststoffbestimmungsverordnung* (RestBestV) [Residue Determination Ordinance] and the *Abfallbestimmungsverordnung* (AbfBestV) [Waste Determination Ordinance] introduced national classification systems for hazardous residues and wastes and designated specified treatment and disposal technologies (e.g., chemical-physical treatment, hazardous waste landfill, single-substance landfill, hazardous waste incinerator) for specific types of waste. Extrapolations from earlier estimates (SRU 1990, 170) would indicate that based on the requirements of the aforementioned ordinances, roughly 2.0 to 2.5 million metric tons of hazardous waste should have been disposed of by incineration in 1990. Total annual capacity of Germany's twenty-seven hazardous waste incinerators in 1990, however, stood at only about 460,000 tons.[4] Existing hazardous waste incinerator capacity evidently falls short of what would have been required, if statistics and extrapolations are even only vaguely accurate. This situation has led to the burning of substantial quantities of hazardous wastes in municipal waste incinerators, cement kilns, and industrial boilers and furnaces,[5] improper disposal of wastes in landfills, and substantial exports to neighboring countries. The Umweltbundesamt [Federal Environment Administration] put the export of hazardous wastes in 1990 at 557,600 tons, and at 661,700 tons in 1991 (UBA 1993a, 1; 1993b, 24).[6]

Only seven of the twenty-nine incinerators in western Germany are accessible for small and medium-sized companies, the great majority of hazardous waste generators. The other facilities are operated by large corporations for their own use. Some of them accept limited quantities of wastes from smaller companies, or even household hazardous wastes through contracts with *Länder* governments. Access to thermal-treatment facilities is uneven across both industries and *Länder* (German states). Incineration and landfill capacity shortages have plagued most German *Länder* ever since they recognized in the early 1970s that hazardous wastes required special supervision and management.

While few doubt the reality of a disposal crisis, the extent of the crisis and the need for additional disposal capacity over the next two decades are a matter of speculation. Over the past few years hazardous waste deliveries to treatment and disposal facilities declined, leaving some of them running below capacity (*Frankfurter Allgemeine Zeitung*, March 22, 1994). Thus, the need for replacement or additional disposal facilities is in dispute. Several reasons for the decline in delivered and treated wastes have been identified. Illegal waste dumping appears to be on the rise (UBA 1994), although it is unclear how much of the waste dumped illegally is hazardous. Second, managers of waste disposal facilities have speculated that waste taxes levied by some *Länder* are an incentive to evade the control system (Hess 1993). Third, driven by soaring disposal costs and consumer demands for corporate environmental ethics, many companies have implemented waste reduction and clean-production schemes that undoubtedly are showing effects. Fourth, waste streams are becoming more homogeneous, which allows for recycling or specialized treatment. Fifth, increasing numbers of specialized waste-recycling and treatment facilities are coming on line. They are withdrawing wastes from incinerators and landfills that were designed to treat and dispose of different types and mixtures of wastes. While these trends all indicate falling waste volumes, other possible developments should not be disregarded. The cleaning up of thousands of contaminated sites and uncontrolled dump sites in all parts of Germany may require substantial additional incineration capacity. The implementation of the Basel Convention may reduce the volume of wastes approved for export and commensurately increase the need for domestic disposal capacity. Finally, stricter environmental controls in various industries may result in additional volumes of hazardous residues (e.g., organic sludges from industrial wastewater treatment) that have to be thermally treated.

In the light of poor data, dramatic technological developments, and changes in the composition of hazardous waste streams, planning for future hazardous waste disposal capacity has become little more than an exercise in tea leaf reading. Consequently, politicians, bureaucrats, and planners are in a weak position making the case for new "universal" disposal facilities such as incinerators and landfills.

Political conflicts over the siting of new facilities do not afflict all *Länder* equally. During the early 1970s, Bavaria and Hesse created state-led hazardous waste management corporations that provided comprehensive hazardous waste management services, including collection and transfer stations, incineration, and land disposal facilities (Davis, Linneroth, and Piasecki 1987). Hamburg also built a large rotary-kiln incinerator. Their problem today is not to site new incineration facilities but to keep existing facilities running at capacity.[7] Other *Länder* only took an active role in hazardous waste management in the 1980s, mainly by requiring companies to deliver their wastes to specialized administrative agencies that direct the wastes to authorized treatment and disposal facilities. Other states have left the management of hazardous wastes entirely to the private sector.

The historical pattern of different levels of state intervention into hazardous waste management at the *Länder* level has left Germany with an uneven distribution of disposal capacity across the country, and an overall capacity shortage. As long as neighboring European countries were willing to absorb large volumes of waste exports from those *Länder* with insufficient treatment and disposal capacity, politicians had no incentive to expose themselves to the hazards of facility siting. By the late 1980s, Germany exported more than a million tons of waste per year for disposal to other European countries, making it probably the number one waste exporter in the world (UBA 1990, 1991, 1993a, 1993b).[8] Substantial pressure from foreign governments, environmental groups, and intense media coverage of illegal hazardous waste exports to countries in Africa and Eastern Europe forced governments to discontinue this practice (EC 1993). In September 1992, the Joint Conference of Federal and State Ministers of the Environment declared that waste exports had to cease and Germany should work toward waste disposal autarchy (BMU 1992).

While political consensus across federal and state divides was an important step toward guaranteeing the safe disposal of hazardous wastes within Germany, it did not do much to address public opposition to new facilities. Concerns over the public-health and environmental impacts of existing pollution sources had generated substantial distrust of all levels of government, which was only fueled by suggestions of a hazardous waste treatment facility being sited. At the same time, the public decried waste exports as environmental crimes against poorer countries and evidence of politicians' collusion with profit-hungry industries. The connection between domestic disposal capacity shortages and huge waste exports has rarely been an accepted argument for facility siting at localities faced with a proposal. National and international environmental groups opposing waste exports, most prominently Greenpeace, however, joined governments in their lament of popular opposition to new hazardous waste disposal facilities, opposition that channels waste streams outside the country to disposal facilities with no or lax environmental protection standards (Baerens and von Arnswald 1993). Environmental groups are, however, more insistent than governments that all means to reduce and recycle hazardous wastes should be exhausted before new disposal capacities are created.

The late 1980s and early 1990s saw a number of attempts to site hazardous waste incinerators and landfills throughout Germany, including the new *Länder* in the former East Germany. Siting processes were initiated and frequently abandoned a few years later when popular protest and political pressures became unbearable in the face of elections—much to the dismay of waste disposal companies and planners (see, e.g., Marx 1990; H. H. Müller 1988). Obstacles to facility siting in Germany not only arise from public opposition but also from complex and time-intensive planning laws and permitting procedures (SRU 1990, 101). The average duration for the formal permitting process currently is four to six years—under the best of circumstances.

These estimates do not include the time required for actual site identification or delays due to objection procedures or litigation at various stages of the permitting process (Lahmeyer International 1992). In all, successful planning, siting, and permitting of a hazardous waste disposal facility can take as long as ten years. After a decade, however, waste composition and volumes and the business environment may have changed substantially, casting strong doubts on the viability of constructing and operating a facility as planned.

THE LEGAL FRAMEWORK OF WASTE AND CONFLICT MANAGEMENT

Germany is a federal state. Article 74 of Germany's constitution, the *Grundgesetz* (GG) [Basic Law], enumerates policy areas and specifically allocates jurisdiction to either the federal government or the *Länder*, or both. With few exceptions, all policy areas concerning the environment are a matter of shared jurisdiction. The federal government may provide framework legislation or may set technical standards if, for example, state legislation does not exist or does not effectively deal with problems or if nationally uniform regulation is desirable (Article 72(2) GG). States may augment federal legislation or standards but rarely do so. The responsibility for implementing environmental regulations rests with the *Länder*. There are no federal agencies that exercise an oversight function. Only in exceptional cases may the federal government assign a specialist to state agencies to aid in the implementation of a law or program (Article 85 GG).

Prior to 1972, waste management fell under state jurisdiction, and operational responsibility rested with cities and municipalities. Increasing concern over the unorderly disposal of municipal and industrial wastes[9] led to an amendment of the Basic Law that made waste management a policy field of shared jurisdiction (Article 72(24) GG).[10] The federal *Abfallbeseitigungsgesetz* (AbfG) [Waste Disposal Act] of 1972 was the first attempt to establish order and uniform regulation in waste management. The Waste Disposal Act was revised in 1986 to include waste reduction and waste recycling, and the planning for waste disposal by the *Länder*.[11] AbfG §6(1) requires the *Länder* to develop waste disposal plans for wastes, including domestic, industrial, and hazardous wastes. Such plans may include decisions or policy statements regarding the type of ownership and the identification of appropriate sites for treatment and disposal facilities. The procedures and thoroughness of drafting waste disposal plans vary greatly among the *Länder* (SRU 1990, 77–87).

In the German legal system the regulation of waste management activities and the regulation of environmental effects from waste management and disposal facilities—and consequently the permitting of waste disposal facilities—fall under multiple separate legal regimes. Waste management facilities in general, and hazardous waste management facilities in particular, require a permit under §7 of the *Abfallgesetz* (AbfG) [Waste Act].[12] They also require permits under other environmental-protection legislation, such as the *Wasserhaushaltsgesetz* (WHG) [Water Management Act] and the *Bundesimmissionsschutzgesetz* [Federal Emmission Protection Act].[13] In addition, environmental effects and mitigating measures have to be evaluated under the *Gesetz über die Umweltverträglichkeitsprüfung* [Environmental Assessment Act]. The *Verwaltungsverfahrensgesetz* (VwVfG) [Administrative Procedure Act] provides that all permits should be issued in a *Planfeststellungsverfahren* [Plan Declaratory Proceedings] conducted by one office, usually the Regierungspräsidium [Regional District Administration] (see §§72–78 VwVfG). The advantage of this system is that the applicant has to deal with only one authority, there is only one review procedure, and permits pertaining to facility construction and facility operation may be issued at the same time.

The 1993 *Investitionserleichterungs- und Wohnbaulandgesetz* [Investment Facilitation and Settlement Land Act] exempted municipal and hazardous waste incinerators from key provisions of the Plan Declaratory Proceedings (see Müllmann 1993). In the future they will only require a simpler *Genehmigungsverfahren* [Permitting Procedure] established by §10 of the Federal Emission Protection Act. The Permitting Procedure does not require submission of detailed regional and statewide waste management analyses and does not make obligatory consultation with recognized environmental organizations. The Federal Emission Protection Act (§6), furthermore, entitles the applicant to permits as long as the proposed facility meets all environmental regulations. Environmental organizations have decried the Investment Facilitation and Settlement Land Act as undermining public consultation in permitting procedures (Koch 1993). But it may be too early to say whether administrative practice will change dramatically in this respect. To analyze past siting processes and to delineate the departure points for the introduction of innovative siting strategies, it is necessary to dwell further on the structure of the Plan Declaratory Proceedings.[14]

Environmental assessments, publication of the facility plans, and public hearings are part of the Plan Declaratory Proceedings as well as the Permitting Procedure. In the initial phases the proponent is required to submit a detailed project proposal. The Environmental Assessment Act (§6) and the *Technische Anleitung Abfall* [Technical Instructions on Waste][15] establish minimum requirements for documents that have to be submitted with an application. These include a detailed site description, construction plans, a justification of the need for the facility, a justification of the facility site, extensive data on the current conditions of—and anticipated impacts on—the surrounding environment, emergency and safety plans, and much more.

The permitting agency reviews the application and plans and consults with all other government agencies possibly having jurisdiction over areas touched by the proposal. This includes the specialized departments of municipalities where facilities are to be sited. After completion of the review, the permitting agency has to make the plans accessible to the public for one month (§73(3) VwVfG). Time period and venue for public inspection of documents must be advertised. Objections to the plan must be filed within two weeks after the public-notice period ends (§73(4) VwVfG). The right to object is restricted to persons and legal entities directly affected by the proposal. Permitting authorities have to discuss the plan and objections to it during a public hearing (§73(6) VwVfG). The public hearing is open only to persons affected by the project (§73(6) VwVfG). Media access may be restricted, and tape or film recordings prohibited. The length and structure of a hearing are within the discretion of the permitting authority. Hearings on a controversial project such as a hazardous waste incinerator may last several days and include expert testimonies from proponents, environment ministry officials, lawyers, and consultants of objectors. Citizens have the right to speak, but only selected representatives of environmental groups usually do. Public hearings generally focus on specific legal or technical issues, which limits the opportunities for more general comments from the public.

After reviewing the application, pertinent documents, commentaries by other government agencies, and the arguments brought forward during the public hearing, the permitting agency decides on the plan (Plan Declaration) (§69(1), §74(1) VwVfG). The applicant and affected parties, including government agencies and objectors, have to be informed of the decision (§75(4),(5) VwVfG). In the case of an approval, the decision includes all permits, operational restrictions, etc., and an authorization that the project may proceed. Objectors may, within one month, challenge the Plan Declaration in court. A court challenge acts as a stay unless the permitting agency orders the immediate implementation of the plan (§80(1) *Verwaltungsgerichtsordnung* [VwGO] [Administrative Court Regulation]). Alternatively, a proponent may seek a court order for the immediate implementation. Administrative orders and lower court orders are subject to judicial review if so requested by an affected third party (§80(5)–(7) VwGO).

It is evident that application and public-review procedures are structured by law and detailed administrative regulations. Formal public consultation is limited to notification and presentations at a public hearing. The degree to which affected parties can impact the decision making of the permitting authority, then, crucially depends on the factual argument presented in their objections. If the permitting authority cannot show that these arguments are unjustified, it has to include appropriate provisions in the permit decision or risk a court challenge. Environmental assessments and risk analyses, however, are characterized by substantial uncertainty. German courts have generally been deferential to the final determinations of permitting authorities as long as the procedures set forth by law were adhered to. Thus, court challenges based on allegedly unjustifiable environmental impacts are rare compared to challenges based on alleged procedural deficiencies or infractions of other legal requirements.

This review of technical, political, and legal challenges to, and constraints of, hazardous waste management should be sufficient to enable us to see the siting conflicts described below through the lenses of projects proponents, politicians, and the affected population, respectively.

SITING OF HAZARDOUS WASTE INCINERATORS: CASE STUDIES

Ebenhausen, Bavaria

Bavaria could be called Germany's cradle of systematic and comprehensive hazardous waste management. As early as 1966, the regional councils in the industrial Nuremberg-Fürth corridor initiated the Mittelfranken Cooperative for Special Waste Management (ZVSMM) to prevent groundwater contamination by wastes and runoff from the metal-processing and electroplating industries in the region (Linnerooth and Kneese 1989). The ZVSMM established a comprehensive treatment and disposal facility, including a rotary-kiln incinerator, at Schwabach (Amsoneit 1994). Shortly thereafter, in 1970, the state government and several large corporations founded a second cooperative hazardous waste management corporation, the semipublic Gesellschaft zur Entsorgung von Sondermüll in Bayern (GSB) [Corporation for the Disposal of Special Wastes in Bavaria]. The GSB provides for collection and centralized treatment and disposal of some 500,000 tons of hazardous wastes annually outside the Nuremberg-Fürth region. The GSB operates five collection stations, two chemical-physical treatment plants, one special waste incinerator, and one special waste landfill.

Planning for the GSB special waste incinerator commenced in 1971. The Bavarian environment ministry and the GSB looked for a site in central Bavaria, close to a major traffic link and outside water supply areas (Frank 1993). As probably typical of 1970s Germany, the search process was characterized by informal discussions among industry officials, administrators in the federal and state bureaucracies, and selected municipal governments. Eventually, word of the potential for new industry and employment reached the municipality of Ebenhausen. Its mayor eagerly advised the environment ministry of a vacant lot on an abandoned munitions factory north of Ebenhausen, near a major highway. The site was deemed almost ideal for the purpose. The munitions factory had been demolished after World War II and had left behind tracts of contaminated land. A few years earlier, a car-shredding plant and a facility burning used oils had commenced operation on sections of the old munitions factory lot. The hazardous waste treatment facility seemed to be a natural addition to the disposal-oriented industries in place.

The facility had to go through the usual permitting procedures under the 1972 Waste Disposal Act, which included public notice and hearing. Plan justification, questions of zoning, and environmental impacts, however, were not considered with as much care as would be customary today (Frank 1993). Consultations with potentially affected third parties were also limited because the old munitions factory site was owned by the state-run Industrieverwaltungsgesellschaft (IVG) [Industrial Property Administration]. State ownership of the site would have restricted the standing of almost any potential objector, including the municipality of Ebenhausen if it had protested the siting (Dick 1993). Permitting and construction proceeded without objection from the community, and in 1975 the facility was operational. Dieter Frank, former GSB general manager, remembers that "Mayor Gröser had a firm handle on municipal affairs. *He* ran the town. If there were any dissenting voices at the time, this was limited to the local pub." The only restrictions imposed on the incineration facility came from the Federal Defense Ministry: to ensure unimpeded operations at a nearby NATO air base, the stack height was limited to twenty-eight meters.

The GSB incinerator in Ebenhausen was among the first of its kind in the world. Since there was little technical experience in running hazardous waste incinerators to build on, the first few years were marked by trials and errors, but also by continuous operational improvements. Repeated stoppages due to equipment failure or unacceptable emissions were part of "a learning process" (Frank 1993). Problems were taken seriously because the short stack did not allow good exhaust dispersion, and stack downwash, at times, inadvertently "hit the ground in neighboring fields" (Frank 1993). The burning of PCBs was another reason for the great attention devoted to emissions reductions. Initial emission controls consisted of electrostatic precipitators and a two-stage wet scrubber. Later, a flue gas condenser was added, which reduced emissions by another 70–90%.

It is common knowledge in Ebenhausen, and even GSB does not

deny it, that surrounding fields and pastures were subject in the past to high emissions originating from the incinerator (Frank 1993; Schweiger 1993). This fact, however, never gave rise to controversy. In the late 1980s, GSB bought several of the affected fields from local farmers and used some of them to expand the facility (Dick 1993; Schweiger 1993). Today, the incinerator operates well below the stringent emission limits set by the 1990 Seventeenth Federal Emission Control Ordinance, including the 0.1 ng/m^3 limit on dioxins and furans (Frank 1993). Even environmentalists acknowledge that the Ebenhausen incinerator is a state-of-the-art facility and that GSB invests heavily in continuous upgrading and improvements (Schweiger 1993).

Keeping with its commitment to maintain a safe and state-of-the-art facility and meet the demand for increased disposal capacity in Bavaria, in 1985 GSB started planning for a replacement of the incinerator by two modern rotary-kiln incineration lines with an annual capacity of 50,000 tons each.

The towns of Ebenhausen and Manching as well as the Bund für Umwelt- und Naturschutz Bayern [Federation for Environmental Protection Bavaria] filed objections against the expansion in 1990, although for different reasons. The environmental group submitted a statement regarding the potential harms of dioxin emissions. The issue was never discussed in great detail, however, since the federation did not enjoy standing at the subsequent public hearing and did not attempt to make a presentation. The town of Ebenhausen also voiced environmental concerns, but GSB suspected their objection to be a political strategy in disguise (Frank 1993). Ebenhausen cannot collect business taxes from the GSB because the GSB is a cooperative, although a very profitable one. The only direct compensation it receives for hosting the facility is through state transfer payments and GSB "donations" to community projects, such as an upgrading of the fire department. Otherwise, Ebenhausen and surrounding communities benefit indirectly from employment opportunities and contracts for local suppliers. Ebenhausen may have hoped to strengthen its bargaining position for increased state-municipal transfer payments vis-à-vis the state government by objecting to the expansion.

The objection by the neighboring town of Manching may have had similar motivations, and more justification. Although the GSB facility is located partially on Manching territory and is close to the centers of Ebenhausen and Manching, Manching has not received any compensation or acknowledgment for being "co-host" to GSB (Frank 1993). Subsequently, Manching challenged the expansion permits in court—without success.

There has been virtually no discontent over the incinerator among the Ebenhausen populace ever since the 1970s. Several reasons can be identified. First there is the facility location itself. The GSB facility is located not in the town of Ebenhausen itself but a subsection called Ebenhausen-Werk [Ebenhausen-Plant] (after the old munitions factory). Ebenhausen-Werk is about a kilometer from the town proper, and home to a working-class population of 400–500, of low social status. The proportion of foreign and migrant workers living in Ebenhausen-Werk is also said to be high (Schweiger 1993). It is possible that this social setting is not conducive to the emergence of a local environmental initiative. But even in the neighboring town of Manching, which is relatively more affluent, no initiative directed specifically against the GSB facility ever took off.[16] Apparently, the people have resigned themselves to the fact that the GSB incinerator will not go away, and the chances of success in fighting it are remote. People recognize that hazardous wastes have to be treated somewhere, although the Manching-Ebenhausen area is seen as shouldering much more than its fair share of the burden (Dick 1993). Another factor explaining the low level of opposition to the facility is that GSB employs 150–60 people, mostly from Ebenhausen and neighboring towns and villages. The economic benefits to numerous members of the wider community certainly foster a positive attitude. Former GSB general manager Dieter Frank emphasizes another aspect of drawing on the local labor pool: employees spread the news about anything happening inside the GSB facility, good or bad. According to Mr. Frank, employees must have generated an impression in the community that the incinerator is safe, that any problems are taken seriously, and that the facility is run at the highest standards of technology and corporate responsibility. Otherwise, the absence of distrust and opposition to the incinerator would be difficult to comprehend (Frank 1993).

It may be for this reason that GSB only embarked on a more systematic approach for community information and liaison when it planned for the incinerator replacement in the mid-1980s. At that time, GSB invited local politicians, environmentalists, and interested citizens to visit the facility and to participate in the "Ebenhausen Round Table" to share concerns and discuss questions regarding hazardous waste management at the facility. This initiative was well received even by environmentalists. They interpret it as "GSB's realization that they [GSB] have to be more open with the public" and as a hopeful sign that more will change in the future (Schweiger 1993).

Biebesheim, Hesse

Hesse is a small state in the heart of Germany. The areas around its major cities, Frankfurt and Darmstadt, are heavily industrialized, with the chemical industry and car manufacturing dominating the scene. After repeated scandals over the improper disposal of hazardous wastes in 1973 and 1974 and concerns that the landfilling of these wastes would endanger future drinking-water supplies (*Der Spiegel* 1973, 1974a, 1974b, 1974c), the Hesse state government formed a joint public-private waste disposal company, the Hessische Industriemüll GmbH (HIM) [Hesse Industrial Waste Corporation]. The state holds about 26% of the capital, and various industrial corporations in Hesse hold 74%. The HIM was conceptualized to provide the full range of treatment and disposal services, including chemical-physical treatment, incineration, and specialized landfilling. In turn, all hazardous waste producers were required to deliver their wastes to the HIM.

The siting of the HIM treatment and disposal facilities, including its incinerator, faced political pressures after scandalous hazardous waste disposal practices and the collusion of bureaucrats with industry in improper waste disposal deals were exposed (Mahl and Darnstädt 1974; *Der Spiegel* 1973, 1974c). In addition, the closing of unsafe disposal sites caused an acute shortage of disposal capacity, forcing Hesse to beg the neighboring state of Rhineland-Palatinate to take its wastes (Darnstädt 1975). In the absence of a legal framework that could have guided the site selection process, environment ministry officials chose the route of successive exclusion of inappropriate sites (Zubiller 1993). Water supply areas and nature reserves were excluded, as well as regions without proper traffic links or that were located far away from the waste-producing industries. The environment ministry stopped its iterative site selection process after the HIM bought a waste-oil incineration plant in Biebesheim (Zubiller 1993). Indeed, the ministry seems to have been supportive of this strategy, as it avoided lengthy procedures to meet zoning requirements for the hazardous waste incineration facility.[17] Biebesheim, a town of 6,500, was deemed suitable also in other respects. Located about 40 kilometers southwest of Frankfurt and 25 kilometers southwest of Darmstadt, it was most central to Hesse's industrial hubs with their chemical plants and car assemblies.

Environmentalists and critics of the incinerator add an interesting perspective. They point out that their region, the Biebesheim–Groß-Gerau corridor, was targeted by state governments as early as the late 1960s to absorb some of the more noxious industry and the accompanying pollution, then still concentrated in the densely populated centers of Darmstadt and Frankfurt (Unger 1993). Local officials supported this strategy as they tried to diversify the predominantly agricultural economy of their region. Subsequently, several chemical and small manufacturing plants located in the area in the 1960s and 1970s, followed by two nuclear power plants at Biblis, 15 kilometers south of

Biebesheim. The Biebesheim town council allegedly supported the siting of the HIM incinerator, primarily because it increased tax revenues.

The HIM filed permit applications for its Biebesheim incinerator in 1978. The application and facility plans were open to public comment as required by the federal Waste Law of 1972, but no formal objections were received (Zubiller 1993). Instead, groups of concerned citizens voiced their objections by organizing demonstrations. Also, the towns neighboring Biebesheim made presentations to the permitting authorities, and only consented to the construction after several restrictions were included in the permit (Stork 1991).

Neither the siting nor the application appear to have been newsworthy items in the major newspaper of Hesse, the *Frankfurter Rundschau*. This is surprising because concerns over toxic pollution and the dangers of hazardous waste had been dramatized by the 1976 Seveso catastrophe and Hesse had had its own share of hazardous waste scandals (Flottau 1976; Mahl and Darnstädt 1974; *Der Spiegel* 1973, 1974a, 1974b, 1974c, 1975). Nevertheless, there was some opposition from neighboring municipalities and from citizens' groups. Why this was not expressed more forcefully is difficult to establish fifteen years later. Interviewees speculated that there was widespread awareness that incineration was a necessary component in a waste disposal system geared toward the protection of water supply and environment; that the population was aware of potential risks but largely indifferent; and that the population was aware of potential risks but did not have the organizational capacity to oppose the facility and protest more actively.

In 1978 the HIM collected and treated some 234,800 tons of hazardous waste. That amount increased to about 600,000 tons in 1989. Only about 50,000 to 60,000 tons are incinerated; the remainder undergoes chemical-physical treatment and is landfilled. By the mid-1980s, the Hesse environment ministry and the HIM anticipated that changing federal regulations, particularly the designation of whole waste streams for incineration by the then anticipated Waste Determination Ordinance, would require additional incineration capacity of 25,000 to 35,000 tons per year (*Bonner Energie-Report* 1990a; HMUR 1990, 184). The HIM also planned to expand its chemical-physical treatment capacity and special waste landfills.

If opposition to the siting of the HIM incinerator in Biebesheim was low, it was all the more intense against the new incinerator applied for in 1988. More than 9,000 people filed objections. In Biebesheim two well-organized environmental groups opposed the incinerator, and the Regional Council of Groß-Gerau initiated an alliance of regional and municipal governments and environmental organizations against the HIM expansion, the Kommunale Arbeitsgemeinschaft "SAV Biebesheim" (KAG) [Municipal Working Group "Hazardous Waste Incinerator Biebesheim"] (Unger 1993).[18] These groups argued that Biebesheim already suffered from high levels of background contamination and that additional sources of pollution from an expanded incinerator and increased truck traffic could not be tolerated. A second battlefront opened over the issue of industry's ability to reduce waste generation. While the environment ministry and HIM maintained that additional incineration capacity of approximately 30,000 tons per year was necessary to maintain a system of safe hazardous waste disposal, critics charged that industry could reduce waste generation by 35,000 tons per year (Unger 1993).

During the public hearings April 23 through May 11, 1990, incinerator critics failed to convince the HIM and the environment ministry that the expansion was unnecessary, and thus they failed to shelve the project. The ministry agreed, however, that the establishment of a comprehensive biomonitoring program would be valuable to monitor systematically the potential of increased public-health risks in the area and their possible association with the HIM facility (Unger 1993). Under this program heavy metals and chlorinated organics are sampled and traced in the food chain and throughout the sample population close to the incinerator and a control population in a rural district of Hesse. The state government, furthermore, seemed to agree that stronger incentives and policies for waste reduction were necessary. Subsequently, it passed the *Hessisches Sonderabfallabgabengesetz* [Hesse Special Waste Tax Act], which imposed a hazardous waste tax ranging from DM 150 to DM 300 per ton (approximately U.S. $100–200).

In 1992, HIM shelved its plans for the incinerator expansion because of rapidly declining waste deliveries (HIM 1993). Wastes delivered for incineration declined from 66,026 tons in 1989 to 49,922 tons in 1992 (HIM 1992, 2). The official position of the HIM and the Hesse environment ministry is that declining waste volumes are an indication of success in campaigning for waste reduction and waste recycling (HIM 1992, 9). This view, however, is disputed, and incineration critics claim that waste generation did not decline appreciably. The KAG alleges that environment ministry efforts to implement a comprehensive hazardous waste reduction and recycling program failed because of industry's refusal to cooperate (KAG 1993). Waste generators are suspected of declaring their high-BTU wastes "recyclables" or "economic goods" and selling them to cement kilns in Belgium and France (Hess 1993). They thereby avoid the waste tax as well as otherwise obligatory and costly disposal at HIM facilities.

Over the two decades of operation, HIM slowly but continually tried to improve its relations with the Biebesheim populace. Bracing for the controversial incinerator expansion, HIM in 1989 started a biannual newsletter that is distributed free to all households. HIM also initiated a liaison committee to share information with municipal and regional governments (Hess 1993). Facility critics appreciate these initiatives but say they fall short of what would be required of "a good corporate citizen in the 1990s" and will be insufficient to stem the popular groundswell against the HIM facility and other polluting businesses (Unger 1993). That this groundswell is very real became evident in the 1993 Hesse municipal elections. In October 1993, the people of Biebesheim and other towns voted proindustry local governments out of office. The results, in the words of a Biebesheim environmentalist, "clearly reflect the popular quest for clean government, clean industry, and a clean environment": the Green Party took almost 25% of the vote in Biebesheim, and the town is now governed by a "green" mayor. The HIM can expect more local-government scrutiny in the future.

Brunsbüttel, Schleswig-Holstein

Schleswig-Holstein is Germany's northernmost state and covers an area of 15,675 square kilometers, about the size of Connecticut. A few industrial centers such as Kiel and Lübeck aside, the state's economy is predominantly agricultural and service-oriented. Some 180,000 tons of hazardous waste are generated each year, 10% of which has to be incinerated. The amount requiring thermal treatment could more than triple, however, if the government enforced waste regulations on car-shredding facilities and the state's 20,000-plus businesses with less than twenty employees (Meyer 1993).

In the early 1980s, the state government in Kiel set out to plan for an appropriate disposal infrastructure for hazardous wastes, including an incineration facility with two rotary kilns with an annual capacity of 35,000 tons each. Initially, this would have been excess capacity. The environment ministry reasoned that disposal security would be enhanced if one incineration line could run while the other was serviced (Meyer 1993). If both lines could run at the same time, the plant could accommodate waste imports from other states. The state of Bremen, for example, had already asked that 14,000 tons per year of the plant's total capacity be reserved for its use. The state government also prepared regulations to compel hazardous waste generators to deliver wastes needing thermal treatment to the incinerator (see §3(2) *Schleswig Holstein Sonderabfallverbrennungsverordnung* [Schleswig Holstein Special Waste Incineration Ordinance]).

While the state government wanted to maintain control over the organization and, to some extent, the operation of hazardous waste disposal facilities, it also wanted the private sector to take responsibility in hazardous waste management. In the mid-1980s, various industries and disposal companies were approached by government officials who encouraged them to propose and operate an incineration facility. Eventually, several medium-sized regional disposal and construction companies formed a consortium to carry out the plan.

The site selection process for the hazardous waste incinerator was guided by established principles of industrial planning in Schleswig-Holstein (Meyer 1993). These require, first, that new industrial facilities (incinerators are classified as such) can only be located on land already zoned for industrial use. Second, state authorities should spread emission sources around the state and not concentrate them in one locality. And third, the location of a hazardous waste treatment facility had to be "central," a vague concept that includes such factors as proximity to waste sources, traffic links, the availability of a trained labor force, and other considerations. In a rural state such as Schleswig-Holstein, these criteria did not leave a great number of potential sites. The site finally chosen was a location in the industrial zone of Brunsbüttel, a town of about 13,500 at the mouth of the Elbe river and near a canal linking the North Sea with the Baltic Sea.

Brunsbüttel is considered the industrial center of western Schleswig-Holstein (Johannssen 1993). In the 1960s, the local government envisioned transforming the town of then 6,500 people into an industrial and marine hub with a population of 25,000 to 30,000. This activist industrial policy eventually drew a number of chemical plants, a crude-oil processing facility, several manufacturing businesses, and two nuclear power plants to locate in and around Brunsbüttel. A local newspaper reporter, Theiss Johannssen, commented that the nuclear power plants in particular are a "monument to the faith in technology and progress that motivated political leadership in the region over the past thirty years" (Johannssen 1993). Not surprisingly, the Brunsbüttel mayor and city council in 1988 supported the siting of the hazardous waste incinerator as long as it was constructed to the existing standards and appropriately operated.

Among the businesses already in the Brunsbüttel Industrial Zone were three major chemical plants, an aluminum-manufacturing plant, and a local waste disposal company. Only the waste disposal company, which was eager to find a "grave" for collected household hazardous wastes, supported the siting proposal. None of the other companies was in need of off-site hazardous waste management services, and they neither endorsed nor opposed the incinerator. They were not dependent on the new waste disposal service and did not anticipate image problems resulting from the incinerator operations (Nussen 1993).

Opposition to the incinerator siting was widespread but diffuse. It gravitated around three unrelated groups, the Eltern für Unbelastete Nahrung [Parents for Noncontaminated Food], a farmers' organization, and the council of the neighboring town of St. Margareten. Their prime concerns were allegedly high levels of background contamination and the potential of dioxin emissions that could cause health problems and damage the agriculture-based economy outside Brunsbüttel (Johannssen 1993). The regional chapter of a national environmental organization called Das Bessere Müllkonzept [The Better Waste Concept] also got involved and used the siting controversy as a platform to promote its agenda of waste reduction. It argued that the incinerator would be unnecessary if the government forced industry to reduce, reuse, or recycle its hazardous wastes; remaining quantities should undergo physical-chemical treatment and be landfilled (Koch 1993).

The groups opposing the incinerator could build on a loose organizational infrastructure that had formed in the opposition to a controversial dump site for contaminated sludges from the Hamburg harbor (Johannssen 1993; Koch 1993). But they did not develop a decisive mobilizing capacity, and protest was evidenced only in placards and posters in front yards, a few information meetings, and a "mini-demonstration" on the first day of the hearings (Nussen 1993). Johannssen, who covered the permitting process, speculates that people had become resigned to the fact that the facility would be built no matter what they did. Early on in the permitting process Environment Minister Heydemann had stated that the incinerator project could not be stopped. According to Johannssen, such announcements must have brought back to mind the many battles against industrial projects the people of Brunsbüttel and surrounding communities have fought over the past twenty-five years, most notably the conflict over the Brokdorf nuclear power plant some 20 kilometers south of Brunsbüttel. It became a hotbed and symbol of German environmental politics—but also of state resolve to use force, if necessary, to push through with industrial projects (Bieber 1976; Stärker 1976a, 1976b).

Initially, facility proponents and state government anticipated that permitting procedures would take about two years. But by December 1993, permits still had not been issued. The reason for the delay was less popular protest than state politics and regulatory change. The application to construct and operate the Brunsbüttel hazardous waste incinerator was filed in 1988. In the judgment of environment ministry officials, the justification the applicant provided for the site was "a little superficial. But considering the absence of viable alternatives, it was deemed sufficient" (Meyer 1993). Aside from this, the application was as detailed as it should have been in its description of what was considered a state-of-the-art facility in the late 1980s (Nussen 1993). The design of the two 35,000-ton-per-year rotary-kiln incinerators with sophisticated pollution controls was similar to several other plants operating in Germany. Total construction costs then were estimated between DM 50 and DM 150 million (approximately U.S. $37–100 million). City council did not oppose the incinerator but linked its endorsement of the facility to a demand for gaining more control over local air quality and for continued access to monitoring data of the incinerator. The permitting authority, the Industry Inspectorate Itzehoe, however, deemed these questions to be "only distantly related to the project" and outside its jurisdiction (Nussen 1993). The state government made it clear that an increase of local authority over air quality control and more funding for local environmental monitoring could not be linked to the incinerator project. Despite this demonstrated resolve, local discontent and the political salience of the incinerator issue in local and state politics appear to have been sufficient to extract several symbolic but time-consuming initiatives from the state government. In 1988, shortly after the permit application was filed, a new Social Democratic Party government took power and ordered an environmental assessment of the facility.[19] Although environmental impacts would have been evaluated in the course of the permit review under the Waste Disposal Act, additional baseline studies over a period of one year were now required. The results of this review were, however, not incorporated into the decision making, since neither law nor government guidelines provided the industry inspectorate with criteria to assess the findings (Nussen 1993). Following the eighteen-month-long environmental review, the state government ordered a downsizing of the facility to one 35,000 t/a incineration line. This necessitated amendments of the application. The promulgation of the Seventeenth Federal Emission Protection Ordinance in 1990 prompted further design changes, permit application amendments, and delays. The delays in the permitting process increased the cost of the facility, now projected to be close to DM 300 million (U.S. $200 million). In 1991, cost overruns and delays forced the consortium of medium-sized companies to sell the entire project to RWE, a large utility corporation that recently expanded into waste disposal activities.

In early 1992, the complete project proposal and permit application was made public. Subsequently, some 6,000 people filed objections, to which the industry inspectorate responded individually (Nussen 1993). It also held three meetings with farmers and environmental groups to prepare for a public hearing held in August 1992. The hearing lasted five days. It was structured around sets of technical questions that were discussed on the basis of expert testimony. Experts were presented by the state government, the proponents, the town of Brunsbüttel, and environmental groups. A former mayor chaired the hearing and kept the focus on technical matters to avoid a heated atmosphere. This was the only hearing held throughout the permitting process.

Throughout 1992 and 1993, the industry inspectorate and the environment ministry continued their environmental and safety assessments of the facility, partly in response to the concerns voiced at the public hearing. More baseline studies were done, particularly to verify the background levels and dispersion pathways of dioxins in the environment. In November 1993, the industry inspectorate announced that permits would be issued unless objectors filed a court case. On December 19, 1993, an alliance of villages surrounding Brunsbüttel, environmental and farmers' groups, and individuals indicated that they would do so. According to them, their review of facility plans indicated violations of the Technical Instructions on Waste. They also charged that the environment ministry had reneged on an informal promise for a comprehensive biomonitoring program given during the hearing. There are indications, however, that the Industry Inspectorate Itzehoe will issue interim orders so that a court case would not act as a stay on the permits and so construction could commence as soon as possible. If that happens, Brunsbüttel would be the first new hazardous waste incinerator to be constructed in Germany in ten years.

Baden-Württemberg

Baden-Württemberg occupies an area of 35,750 square kilometers in southwest Germany, about half the size of West Virginia. It is renowned for its car industries and specialized machine tool manufacturing. These industries generate a substantial portion of the roughly 550,780 tons of hazardous waste that require treatment in Baden-Württemberg each year (Landtag of Baden-Württemberg 1990, 7–29). Because of the shortage of treatment and disposal facilities, Baden-Württemberg exports some 250,000 tons of hazardous waste to other *Länder*, European Community members (especially France), and (until 1990) East Germany (Landtag of Baden-Württemberg 1990, 6).[20]

In 1988, the state environment ministry initiated two *Raumordnungsverfahren* (ROV) [Regional Planning Procedures] to site two incineration facilities, one in the eastern and one in the western part of Baden-Württemberg. During the Regional Planning Procedures, various consultants were contracted to identify suitable sites. The site identification process worked iteratively along defined sets of criteria. The search concentrated on sites within industrial zones that met requirements of size, infrastructure (e.g., rail link), environmental sensitivity (e.g., outside water supply areas and nature reserves), and distance from housing areas (>500m) (Deutsche Projekt Union 1991, 7–15). Potential sites then were further investigated for geology, hydrogeology, background levels of air and water pollutants, etc. A state-owned lot in the harbor of Kehl, a city on the Rhine close to Strasbourg, France, was identified as a suitable site for the incinerator in the western part of Baden-Württemberg. Subsequently, the state government applied to the Regional District Government in Freiburg to initiate the Plan Declaration Procedures and an environmental assessment for the site. For the eastern part of Baden-Württemberg, four potential sites were identified: Westernach, Hüttlingen, Dagersheim, and Sindelfingen (Deutsche Projekt Union 1990a, 1990b).

Although municipal and regional governments and environmental groups had been consulted on site selection and environmental criteria in the later stages of the Regional Planning Procedure, local politicians and citizens' groups were outraged when the environment ministry announced the potential sites. The announcement set off a "waste revolt" with angry letter campaigns, demonstrations, and threats to blockade major highways (*Bonner Energie-Report* 1990b; Wenz 1990). Criticisms centered on insufficient consultation with potential host communities, the environmental and economic impacts, and the need for the facilities in the first place (*Bonner Energie-Report* 1990b; Landtag of Baden-Württemberg 1992a, 1992b). If industry would be serious about waste reduction and recycling, no hazardous waste incinerator would be required, critics argued.

In Kehl, the chosen site in western Baden-Württemberg, public protests focused on the "objective" site selection criteria (Müller 1990; Wenz 1989). Although the Kehl site had been studied in the context of the Regional Planning Procedure, it was primarily chosen on the basis of an earlier consultant report produced by a firm based in Hamburg—about 750 kilometers north of Kehl. Facility opponents argued, furthermore, that Kehl was among the places least central to the waste-generating industries, its background contamination levels were among the highest in the southwest, and meteorological conditions (170 days of inversion per year) were far from ideal for an incineration facility. The government's counterargument that in case of an emergency only 29,000 people would be affected in Kehl, a fraction of those in other potential locations such as Karlsruhe or Mannheim, was perceived as cynicism. Not surprisingly, the Council of Kehl, citizens' groups, and business organizations condemned the siting decision. Environment Minister Vetter countered that "St. Florians"[21] could not be tolerated and that the facility would be constructed with or without local consent (Müller 1990).

The urgency of solving the hazardous waste disposal crisis and the government's determination to have a facility sited before the next election dimmed the prospects for successful opposition. If this was the essence of a political equation applied by the state government, at least one variable was left out: Strasbourg. Strasbourg, a city of 409,000, seat of the European Parliament and sister city of Baden-Württemberg's capital, Stuttgart, had not been considered in the site selection process although it is located less than 3 kilometers west of Kehl. Within a few months of the siting announcement, Strasbourg Council voted against the "German Plan" and dozens of citizens' groups formed, collected tens of thousands of signatures against the facility, and placed banners about *"jumelage empoisonné"* and against *"le monstre allemand"* around town to greet international visitors (Wenz 1989). Soon the local Kehl-Strasbourg controversy over the siting developed into an embarrassing "German-French waste dispute" that moved on to the informal agenda of regular German and French national-government consultations (Müller 1990; Wenz 1990).

The 1992 state elections shifted the balance of power and forced the previously governing Christian Democratic Party (CDU) to form a coalition government with the Social Democratic Party (SPD). It also opened the way for a quiet solution to the almost intractable political problems of the Kehl siting. In the coalition platform, the two parties agreed to freeze the siting in Kehl and to continue the search for an appropriate site in the eastern part of Baden-Württemberg. The Kehl siting option has not been pursued since and is considered dead.

After the Kehl siting disaster the environment ministry recognized that its latitude for unilateral actions in hazardous waste management was severely limited. To break the stalemate with the alert and mobilized public in and around potential host communities, in October 1992 Environment Minister Harald Schäfer initiated a "Special Waste Forum." He invited representatives from industry, unions, environmental associations, churches, and the bureaucracy to discuss the challenges of hazardous waste management in Baden-Württemberg (Landtag of Baden-Württemberg 1992c, 388). A central objective of the forum was to bring the often emotional discussion back to a factual basis. To that end, the forum obtained the services of several consulting companies to research hazardous waste reduction and recycling potentials, as well as appropriate treatment and disposal technologies. After six months of intensive discussions, all stakeholders agreed on the data of hazardous waste generation, reduction potentials, and technology options that should guide the future hazardous waste policy of Baden-Württemberg (ECOTEC 1993; Prognos 1993; Umweltministerium Baden-Würtemberg 1993). Environment Minister Schäfer made it clear, however, that he would not call on the forum to deliberate on site selection, let alone delegate to it the actual siting decision (Landtag of Baden-Württemberg 1993, 1226). This, he said, would be too time-consuming, and irresponsible in view of the very pressing disposal crisis.

In February 1994, the cabinet of the state government announced its decision to site the hazardous waste incinerator in the eastern part of Baden-Württemberg on an abandoned military base at Dagersheim, about 25 kilometers west of Stuttgart. Residents and local governments of the neighboring towns strongly protested the decision and threatened to "sue the state government into eternity" so that the facility would never be built (Hupka 1994). Ever since, the population around the site has demonstrated its determination, opposing the facility in numerous marches, rallies, letter campaigns, and intensive lobby work. Local and regional environmental groups are, again, questioning the need for the incinerator, essentially retracting their consent to the results of the Special Waste Forum. Threats of local opposition have, so far, not deterred the state government from pursuing a permit application with the Regional District Administration in Stuttgart. The permitting process would be administered under the provisions of the 1993 Investment Facilitation and Settlement Land Act and the Federal Emission Protection Act, which entitle the applicant to a permit. Barring any procedural mistakes, only waning political will could stop the project.

PATTERNS AND PROBLEMS

Although the four case studies encompass hazardous waste incinerator sitings over more than twenty years in four different *Länder*, they show remarkable similarities and strong patterns in actors' behavior and strategies.

Strong State Intervention

In all four cases the state governments played a key role in identifying the need for a facility, providing political support for the project, and initiating the site selection process, even in those cases where the government was not the project proponent (e.g., Biebesheim, Brunsbüttel). This should not be surprising. First, the Waste Disposal Act assigns responsibility to plan for hazardous facilities to the *Länder* and leaves it up to state governments to decide whether public or private companies should manage hazardous wastes. And second, the planning and construction of a hazardous waste incinerator requires substantial investments, and project proponents, be they private enterprises, public-private partnerships, or public corporations, require some degree of certainty to raise capital and reassure creditors. State governments, therefore, necessarily function in multiple roles: as project proponents or supporters, guardians of the public good, and authorities to make political decisions that are beyond judicial review. Not surprisingly, facility opponents and the public at large have been suspicious of nontransparent decisions at the state level and their effects on the permit application review at the level of regional-district governments. The nature and extent of state involvement have undoubtedly politicized siting processes and made them vulnerable to short-term reevaluations by state politicians. Cohesion of the political elite at the state level during election time became a crucial determinant as to whether planning and siting processes continued uninterrupted for more than three or four years.

State governments have, nevertheless, proven to be *the* center of power in siting processes. By virtue of their elevated position and multiple roles, they can essentially determine the extent of nonmandatory environmental- and socioeconomic-impact assessments, they can informally negotiate with facility opponents, and they can influence the timing of permitting processes. And by flexing their political muscle, they can intimidate local governments and other opposing groups. At the same time, they cannot ignore local government or grassroots opposition with impunity.

Waning Local-Government Consent

All four cases demonstrated the strategic importance of local-government consent or support to site a facility within its boundaries and sphere of political influence. The pattern of conditions for local-government support is quite clear: as long as local politicians were not faced with an environmentally conscious electorate, they tended to support siting attempts, in part because of the prospect of increased economic activity and higher tax revenues. This has changed as voters assign more importance to environmental issues. Local politicians have become increasingly cautious about lending support to a siting proposal, if they have not opposed it outright. This development is observable throughout the case studies, although the pace of change varied across the country. Politicians in jurisdictions adjacent to a proposed facility appear to have not supported or even to have actively opposed facility sitings throughout the time period studied.

The siting attempts in Baden-Württemberg demonstrate that state governments can no longer count on local-government support, even in a traditionally conservative *Land.* As a consequence, state governments have tried to site new waste disposal facilities in either designated industrial zones or abandoned military bases. Industrial zones usually are subject to fewer land use restrictions and are outside residential areas. Abandoned and already contaminated industrial and military sites have been attractive in the past (e.g., Ebenhausen) and remain so today (e.g., Dagersheim). Former military sites fall under federal jurisdiction, but the federal government has been happy to turn over land and cleanup liability to state governments.

A Concerned but Constrained Public

The review of the permitting regulations and procedures as well as of the four case studies has shown that venues for public consultation are limited. During the early sitings at Ebenhausen and Biebesheim, the authorities did not pay much attention to concerns of the public. With the development of increasingly detailed procedural laws and regulations, formal public notice and formal public hearings have become integral parts of siting and permitting processes. Decision making, however, has remained closed. How much public input, including expert testimony at public hearings, influences decisions of permitting authorities is as difficult to assess ex post facto as it is in advance.[22] This may be one reason municipalities and environmental groups confronted with a siting proposal focus their strategies not only on input into the established permitting process but equally, if not more so, on political accountability for the "wrong" siting decision. The fact that numerous siting and permitting processes have been stopped because of political pressure and very few, if any, stopped on substantive critiques or procedural flaws certainly makes this a rational choice for facility opponents. One expression of the deliberate politicizing of the siting process by opponents is the explicit linking of issues. The city government of Brunsbüttel demanded more power over general air quality control, the town of Manching wanted compensation, environmental groups in Biebesheim and Brunsbüttel insisted on aggressive waste reduction, etc.

Politicized Processes

Three out of the four siting processes examined showed tendencies toward politicization of the site selection and permitting processes. The dynamics and consequences of this politicization can only be understood if viewed as a result of state intervention *and* counterstrategies by facility opponents. I have also noted the linking of other issues with the decision over a facility site, some of them directly, some more distantly, related. While it may be argued that issue linking has the potential of making room for political maneuvering and dialogue (e.g., informal agreements in the context of HIM expansion, the Special Waste Forum in Baden-Württemberg), it does not in itself depoliticize the process. Rather, political give-and-take and win-and-lose become even more prominent considerations in actors' strategies, perhaps at the expense of factual problem solving geared toward safe hazardous waste disposal. Further research is needed to establish whether strategies by the actors allow for compromise and whether minor and selective regulatory reforms would depoliticize the process and alter actors' conflictual behavior.

THE HAZARDS OF SITING AND THEIR POLITICAL RECOGNITION

As noted in the introduction, social-science research into the politics of hazardous waste disposal in Germany is still in its infancy. Would siting controversies have unfolded differently had social-science knowledge been further advanced and able to offer practical advice earlier? Would the development of the regulatory system have taken an alternative route? Would environmental protection have increased and been more effective? I do not know. Before I dare make some speculations about future directions in the siting of hazardous waste facilities in Germany, a brief reflection on the course and status of the debate over alternative approaches to siting processes and on suggestions for regulatory reform is in order.

Over the past twenty years there has been remarkably little change in how governments in Germany have approached the problem of facility siting. Site selection processes have been and are marked by a top-down approach: governments or contracted consultants set out for potential sites at least some criteria geared to reduce the potential interference from third parties. Pertinent regulations such as the Technical Instructions on Waste offer little guidance in site selection criteria (UBA 1992). Various administrative-court rulings have only established that site selection processes need to follow a comprehensible, objective, and logical course (Deutsche Projekt Union 1991, 9).

Over the years growing environmental awareness in the public at large and the political strength and mobilizing capacity of environmental organizations opposing hazardous waste facilities have become forces to be reckoned with. National and regional environmental groups have improved the flow of factual information. They also encourage and enable individuals and citizens' groups to speak up and closely monitor facility planning and permitting. People intervene in the permitting procedures, take to the streets, and effectively lobby local governments. The prominence of environmental issues on voters' wish lists has altered the incentive structure for local politicians, who now tend to oppose siting proposals. Local protesters increasingly succeed in driving a wedge between state and local governments. State governments appear to have been largely indifferent to these developments and did not respond with more accommodating strategies. Instead, siting processes have become politicized and have been accompanied by harsh and antagonizing polemics. Project proponents and politicians regularly brand opponents of waste disposal facilities as St. Florians.

Beyond the individual siting controversy, the inability to find politically acceptable ways to site disposal facilities has been framed[23] as an "acceptance problem" or "acceptance crisis" (Johnke 1992; Müller 1991; Noeke 1990). The terminology is indicative of the still prevalent view of the subject matter: the public has a problem with *accepting* state governments' decisions and therefore is to blame for siting conflicts. The phrase "acceptance problem" is a diffuse term that alludes to irrational fears and ignorance of, and hostility toward, modern technology. Some would also include the complex dynamics that accompany siting conflicts and the indecisiveness of politicians.

The framing of siting conflicts as an "acceptance problem" reflects the traditional view of waste management as a technological and legal challenge and indicates the dominance of the engineering and legal professions in the field.[24] Consequently, solutions were sought in stricter technological standards and increasingly detailed administrative procedures. This mode of problem solving has blessed Germany with very stringent environmental-protection regulations but cursed the country with extremely detailed and complex administrative regulations.

Over the past few years social scientists have entered the discourse,[25] and the debate over siting of waste treatment and disposal facilities shows some first signs of being reshaped. With analysts from a broader range of professions now represented in the debate, it is becoming increasingly evident that the "St. Florian" explanation is too simplistic to explain popular opposition and that earlier approaches are inadequate to explore and design processes for the successful siting of waste treatment and disposal facilities. Many authors (e.g., Hauber 1989; Herbold and Wienken 1993; Müller and Holst 1987; Johnke and Schmitt-Tegge 1993) see the main cause of siting failures in the insufficient communication between facility proponents, administering bureaucracies, politicians, and the public, as well as in nontransparent decision-making procedures. Once a decision is made, the gap between proponents and opponents is too wide, and hearings, information brochures, or suasion by a minister cannot then overcome the process failures.[26] Accordingly, reform proposals focus on the inclusion of the local population in the early stages of waste management planning, site identification, and environmental assessments of proposed facilities (Hoppe and Beckmann 1990; Johnke 1992; Jung 1988; Lahmeyer International 1992; Schenkel 1990). It is noteworthy that some authors believe improved communication and information would help to enlighten potential facility opponents and pave the way for what they would consider "rational decisions" (see, e.g., Johnke 1992; Schmitt-Tegge 1994). They maintain that the discourse should be based on "objective" scientific and technical data and that experts outside the mainstream scientific and technical communities should have no part in it. This underscores that increased public involvement is envisioned to be a limited, narrowly focused educational exercise. The public would be at the receiving end, while officials would continue to control the process.

Experiences in Germany with mediation processes akin to those tried in North America are few in number and restricted in scope and can generally be described as having the kind of educational function mentioned above. So far, four experiments are under way or have been concluded with some level of success: mediation processes over technical features of a municipal waste landfill (Herbold and Wienken 1993), the scope and technologies of cleanup efforts at an abandoned hazardous waste dump (Striegnitz 1993), and two public-forum processes geared toward a consensus over planning data and technologies prior to the siting and construction of waste disposal facilities (ECOTEC 1993; Prognos 1993). These initiatives share three common traits. First, they open administrative planning to public input and attempt to achieve a consensus that could guide (but not determine) later decisions by regional or state governments. Second, they all tend to focus on planning data for waste management or technological requirements for existing or planned facilities. And third, they provide the public with opportunities to participate and voice opinions. It remains to be carefully researched how much governments have accepted the public as equal partners in consultations and negotiations and whether governments and interest groups will adhere to compromises reached.

In the North American context, broad public participation and consultation are important features of successful hazardous waste facility sitings (Rabe 1992, 1994). So far, this feature has been absent from experiments in Germany. The legitimizing element in public consultation may, however, be central. The experience of the Special Waste Forum in Baden-Württemberg shows that a consensus over planning data does not suffice to legitimize a state government's siting decision. Consensus-oriented processes to establish planning data and technology requirements for waste disposal facilities address different types of concerns and may be necessary, but not sufficient, conditions to facilitate public cooperation. Affected communities perceive siting decisions as outside interventions and violations of their right to determine their way and quality of life. Cooperation and acceptance then become questions of legitimacy and tests for the freedom of local self-determination.[27] The issue of the legitimacy of siting decisions has not been discussed extensively in the German waste management literature. Lower Saxony is the only *Land* that has been deliberating whether to make mediation procedures mandatory in siting processes (Niedersächsisches Umweltministerium 1992). These deliberations are not far advanced, however, and the substance of such mediation processes has not been defined. Whether they would include legitimizing processes remains to be seen.

Other state governments have been even more cautious in floating ideas that would imply greater public participation in siting processes and in opening established decision-making procedures. They may feel themselves bound by federal law, which does not allow for substantial deviations from the established administrative siting and planning processes. Procedural experiments could thus be open to court challenges from disgruntled facility opponents as well as from private-sector proponents, depending on the final decision. Even more profound may be the concern that successful experiments could lead to regulatory changes that eventually might erode the power of state governments by granting of some sort of veto power to local governments or the electorate. Legal scholars and planners, most eloquently Ronellenfitsch (1989), have articulated the fear that if municipalities are empowered to veto siting decisions, state or regional governments may become incapacitated and unable to fulfill their legal obligation to provide for sufficient waste treatment and disposal capacity. Consequently, they argue adamantly against the inclusion of local governments, interest groups, or the electorate in planning decisions.

Under the current allocation of jurisdictional responsibilities, this argument has to be taken seriously. A local government with no responsibility in hazardous waste management has every incentive to oppose siting proposals. Siting failures translate into political credit for local governments, and waste management liabilities get stuck with regional and state governments. A siting strategy that presumes that the current allocation of responsibilities in waste management is cast in stone will necessarily require stronger provisions to override local resistance.

The alternative approach would be to recognize political changes, particularly municipalities' and environmental groups' demands for active involvement in siting decisions. Granted that participatory siting strategies would overcome local resistance, they may offer a viable route, and perhaps the only route, for regional and state governments to discharge their waste management functions. Chances for success are likely to be enhanced if cooperative arrangements include equal opportunities for political credit claiming and blame taking. State governments should not only share some rights in siting decisions with local governments and citizens' groups but also some responsibilities.

Increased local participation could also be blended with concepts of burden sharing. By transferring some waste-planning functions for hazardous waste management to municipalities and country governments, similar to the California approach described by Morell in this volume, and by linking the existence of plans or provision of facility sites to state transfer payments for infrastructure, local governments would have incentives to use their veto powers responsibly and to cooperate with each other.

Such reforms would require substantial regulatory changes, which take time. There appears, however, to be some space in the current system to experiment and to gain experience with what works and what does not. A recent consultant report for the Federal Ministry for Environment, Nature Protection, and Nuclear Safety detailed numerous steps state governments and permitting authorities could take within the latitude of existing procedural regulation to open up the siting and permitting process to the public (Lahmeyer International 1992). Such measures could encompass active public information by the applicant and/or the permitting authority, the formation of consultative bodies that accompany the review of applications from day one, and the mediation of conflicts with the aid of a neutral person.

CONCLUSION: FROM POLITICAL IMPOSITION TO PUBLIC INVOLVEMENT?

Germany, as most other industrialized countries, faces the challenge to dispose safely of hazardous and toxic wastes generated by industrial production. This challenge was recognized in the late 1960s, has not been solved, and is not going to go away. Continuous efforts will be required in the future. These efforts must include the siting, construction, and operation of disposal facilities, some of them highly specialized for specific wastes, others designed to accommodate a broader range of wastes. Facility siting will remain one of the central tasks in meeting the challenge of hazardous waste management.

In the past, state governments intervened in siting processes to such an extent that parts of the affected communities may have perceived siting decisions as illegitimate impositions. Governments' capacity to continue this practice has been undercut by a groundswell of environmental concern and a lasting change of the political climate.

The question is: how will governments adapt? The case studies of the sitings of hazardous waste incinerators and the general reflections on the debate over obstacles to siting have shown that at least some *Länder* governments have been experimenting with various forms of public consultation. And this needs emphasis: they are consensus-oriented consultative processes, they are experimental, and they are not uncontroversial.

What, then, are the prospects for the adoption of participatory, voluntary, or mediated siting procedures in Germany? Regulatory changes would be, if they were to occur, incremental and slow. Significant amendments to a number of federal laws and regulations would be required, including changes to the Administrative Procedures Act and the Administrative Court Regulations—changes that may have profound effects far beyond waste management. Regulatory change would have to improve public access and participation in the decision-making process and would thereby come close to a full-blown, long-term regulatory-reform project. The siting of waste management facilities, however, is a pressing problem that has to be solved within much shorter time frames.

It seems promising to study more closely the successful sitings elsewhere and evaluate whether some of the procedural elements contributing to siting success could be woven into the current siting procedures in Germany. The current regulatory framework does allow for more public consultation and even mediation processes (Lahmeyer International 1992). Initiatives in that direction would have to come from state governments. These, however, have been generally hesitant to experiment in a policy area already troubled by failure. It would be too simplistic to attribute governments' apprehension about such experiments to a lack of political will. Attitudes of bureaucrats and the public toward participatory processes, which are to a certain extent beyond government control, are crucial in determining the success or failure of innovative siting strategies. The case studies have shown that today's procedural practices and processes of facility siting are permeated by a rationalistic approach to finding and justifying policy options.[28] It may not be easy for bureaucrats to redefine their role vis-à-vis the public in making planning decisions and to accept that a participatory discourse with the public will be different (Lahmeyer International 1992, 9).[29] While this type of change should not be precluded and it may in fact be taking place, citizens' groups may be, based on earlier experiences, mistrusting and hesitant to participate in a consultative or participatory process. Likewise, their acceptance of any decision arrived at after consultation cannot be taken for granted. Nevertheless, consultative and participatory siting processes appear to be promising mechanisms to break the siting impasse that has been ruling German hazardous waste management for more than a decade and continues to rule it today. They diffuse conflicts by spreading the opportunities for credit claiming and blame taking among all actors.

The "solution" comes without guarantee and with its own risks. It would seem that a failed experiment with innovative siting processes could not yield a worse result than siting failure under current processes. I disagree. Under both scenarios a proposed facility would not be sited or constructed. But the adoption of new siting processes would imply that previous approaches to siting were perceived to be illegitimate, something already evident even if extraordinarily difficult to admit. New venues have to be explored if old means fail. Solving policy problems through experimentation requires more than a one-time effort in the wake of acute political crisis; it needs sustained commitment. Discussions over new approaches to siting hazardous waste facilities should also contemplate the possibility that a new process might fail as well, and consider what that means and what ought to be done then.

The failure of an innovative siting strategy would appear to be the ultimate nightmare for governments and the waste disposal industry. I would be less pessimistic and view such a scenario as an opportunity for a reality check. The examination of four incidents of hazardous waste incinerator sitings in Germany has shown that it took about twenty years before effective imposition of a facility site by local and state governments gradually opened to tentative opportunities for more public involvement. At the same time, dramatic changes in popular attitudes toward waste management facilities have taken place, and the power of local citizens' groups has increased. The inability to site even composting facilities, let alone hazardous waste treatment and disposal plants, can be seen as an indicator that the legal system allocates rights, responsibilities, and decision-making authority in waste management in ways that do not facilitate decision making under today's political realities. Figuratively speaking, the legal and administrative modes of resolving conflicts in waste management have reached a dead end. It will be necessary to backtrack until we get out of it, and chart new directions. If a certain innovative strategy fails, we will need to backtrack further and design a new one.

It may be difficult to detect, but a kind of backtracking and reorientation is under way. This is done in the typical German way, in which careful and cautious deliberation within government bureaucracies and a wary eye on statutory authority and bureaucratic tradition take precedence over the need to solve a problem or over consulting the public. These debates also are of fairly recent origin, and it remains to be seen whether they will affect administrative culture, procedural regulations, and governments' willingness to share power with local communities and whether they will have lasting effects. The pace of change over the past twenty years may give some clue to the magnitude of regulatory change that can be expected for the next two decades.

NOTES

1. This chapter builds on a 1993 study for the British Columbia Ministry of Environment, Lands, and Parks. Dermot Foley, Ahmed Kamel, Hans Seeliger, and one anonymous reviewer provided helpful comments on earlier drafts of this chapter. Any inaccuracies or conclusions are, however, the sole responsibility of the author. I also gratefully acknowledge the assistance of the librarians at the University of Tübingen and the information services of state legislatures in Baden-Württemberg, Bavaria, Hesse, Lower Saxony, and Schleswig-Holstein. Research for this paper would have been impossible had not numerous individuals shared information and their views on the siting and operation of incineration facilities, among them Mr. Dick (head, Engineering Department, Municipality of Manching), Mr. Frank (former general manager, Gesellschaft zur Entsorgung von Sondermüll, Munich), Mr. Gade (Special Waste Section, Lower Saxony Environment Ministry, Hannover), Dr. Hess (Hessische Industriemüll GmbH, Biebesheim), Mr. Johannssen (*Brunsbütteler Rundschau*, Brunsbüttel), Mr. Koch (Das Bessere Müllkonzept, Schleswig-Holstein), Mr. Kretschmer (Sonderabfallentsorgung Baden-Württemberg GmbH), Mr. Meyer (special-waste officer, Schleswig-Holstein Environment Ministry, Kiel), Mr. Nussen (Industry Inspectorate Itzehoe), Dr. Schweiger (Bund für Umwelt und Naturschutz in Bayern, Manching), Mr. Unger (Kommunale Arbeitsgemeinschaft "SAV Biebesheim"), and Mr. Zubiller (Head, Industrial Waste Management Section, Hesse Ministry of Environment and Nuclear Safety, Wiesbaden). The same gratitude extends to those interview partners who requested anonymity.

2. Government involvement in running hazardous waste collection and treatment systems varies greatly, however, within Germany as well as Canada and the United States.

3. Of these 15.7 million tons, about 12.1 million tons are generated in the "old *Länder*" of West Germany and 3.6 million tons are generated in the "new *Länder*" of the former East Germany. Almost 4 million tons of the total of 15.7 million tons of hazardous waste are recycled; 6.6 million tons are managed off-site, and 5.1 million tons are treated on-site. Additional volumes of hazardous wastes are generated by companies with less than twenty employees, public-transportation companies, and the German military; these are not subject to statistical-reporting legislation. Definitional differences make a comparison of these data with figures from other countries, e.g., Canada, Japan or the United States, difficult (cf. Dowling 1985). For example, Germany considers waste oils and lubricants hazardous waste; the United States does not. The United States classifies wastewaters from certain industrial processes hazardous waste, while Germany regulates such discharges under the Water

Management Act and only considers residues and sludges from industrial wastewater treatment hazardous waste.

4. Incinerator capacity depends in part on the heating value of the wastes. Generally, the lower the heating value, the more wastes can be burned within any given period of time. Over the past few years the heating values of wastes have declined as companies increasingly sold combustible hazardous wastes as secondary fuels to cement kilns. This and expansions at existing facilities explain the increase of incineration capacity to almost 1 million tons per annum (t/a) in 1992 in the western half of Germany and roughly 80,000 t/a in the eastern half as cited by Schmitt-Tegge (1994, 444).

5. Waste statistics for 1987 indicate that around 80,000 metric tons of hazardous waste are burned in municipal waste incinerators each year and possibly several hundred thousand tons end up in cement kilns and industrial boilers and furnaces (Statistisches Bundesamt 1989). The 1990 waste survey indicated that as many as 1.5 million tons of hazardous waste may be burned in industrial boilers and furnaces. While these facilities are permitted under emission control regulations, not all of them have to meet the same stringent emission control requirements as are applicable to hazardous waste incinerators.

6. The figure for 1990, however, does not include the exports to East Germany. A Greenpeace exposé estimates these deliveries to be at least 150,000 tons in 1990 (Baerens and von Arnswald 1993, 183). Substantial hazardous waste transfers from the western part of Germany to the new *Länder* in the East continue. In 1993, the *Land* of Mecklenburg-Vorpommern alone imported some 218,000 tons of hazardous waste for disposal from the western *Länder* (Landesamt für Umwelt und Natur Mecklenburg-Vorpommern 1994, 59).

7. The exception is hazardous waste landfills. Agreements between *Länder* to accept and treat each other's wastes on an exchange basis exist, but the volumes actually transferred are comparatively small. Importing large volumes of hazardous wastes from other *Länder* is, at least in the western half of Germany, a politically untenable position. The bulk of inter-*Länder* waste transports is destined for the new *Länder*.

8. The data include wastes exported for disposal only. Prior to the implementation of the *European Community Ordinance 293/93 on the Supervision and Control of Waste Transports in, into, and out of the European Communities* (EC 1993) in May 1994, wastes exported for recycling were not controlled or registered. Issues pertaining to the initial implementation of the ordinance are examined in Seeliger (1994).

9. See the federal government's reports of 1963 and 1966 on waste disposal (Deutscher Bundestag 1963, 1966).

10. See *Gesetz zur Änderung des Grundgesetzes, Bundesgesetzblatt* I, April 12, 1972: 593.

11. Substantial revisions to federal waste management legislation have passed the upper and lower houses in June 1994 but will not take effect before 1996.

12. For concise reviews of siting and permitting procedures for waste management facilities, see Müller-Heuser (1992). Müllmann (1993) discusses the effects of recent legislative changes.

13. Incinerators, for example, have to comply with regulations of the Emission Protection Act, namely the *Technische Anleitung zur Reinhaltung der Luft* (TA Luft) [Technical Instructions for the Maintenance of Air Quality] of February 27, 1986, and the *17. Bundesimmissionsschutzverordnung* [Seventeenth Federal Emission Protection Ordinance] of November 23, 1990.

14. Public notice, access to the application, and a public hearing have been standard procedure for all waste disposal facilities since the passing of the 1972 Waste Disposal Act (see §21(3)).

15. The full title is *Technische Anleitung zur Lagerung, chemisch/physikalischen, biologischen Behandlung, Verbrennung und Ablagerung von besonders überwachungsbedürftigen Abfällen* [Technical Instructions on the Storage, Chemical-Physical and Biological Treatment Incineration, and Disposal of Waste Requiring Special Supervision].

16. There is, however, a local group of the Federation for the Protection of the Environment and Nature—Bavaria. This group deals with a variety of environmental issues, including hazardous waste management at GSB.

17. The environment ministry admits that the site selection process was not followed through and that shortcuts were made. At the same time, officials point to very sophisticated site selection processes for other waste management facilities in the early 1990s that ex post facto confirmed that Biebesheim was the right choice for the HIM incinerator (Zubiller 1993).

18. In 1993, the KAG included the municipalities of Alsbach-Hähnlein, Bickenbach, Verbandsgemeinde Eich, Gimbsheim, Griesheim, Guntersblum, Oppenheim, Pfungstadt, Riedstadt, Seeheim-Jugenheim, Stockstadt, Zwingenberg, and the counties of Bergheim and Darmstadt-Dieburg.

19. The federal Environmental-Assessment Act only existed in draft form at that time.

20. In an interview with the *Stuttgarter Zeitung* in February 1990, Environment Minister Erwin Vetter stated that Baden-Württemberg exported as many as 350,000 tons of hazardous waste each year (*Stuttgarter Zeitung*, February 21, 1990, 5).

21. "St. Florians" are the German variant of the North American NIMBYs. St. Florian is the saint who protects houses from fire. St. Florians are people calling on the saint to protect their house—and light up others'.

22. Several interviewees in environmental organizations pointed out that permitting authorities usually did not provide substantive justifications for their decisions beyond references to applicable laws and regulations.

23. The use of the term "framing" does not presume malicious intent. The definition and framing of policy problems are part of every policy controversy. They are "symbolic contests over the social meaning of an issue domain, where the meaning implies not only what is at issue but what is to be done" (Schön and Rein 1994, 29).

24. The exhaustive bibliography of the Council of Experts on the Environment's 1990 report *Abfallwirtschaft* [Waste Management] reflects the lack of political and sociological understanding of popular opposition to waste management facilities (SRU 1990). The literature abounds with sophisticated works on technological, legal, and planning aspects of waste management, but social science–oriented inquiry into the problem appears to be virtually nonexistent. Among the notable exceptions are Hauber (1989) and Müller and Holst (1987).

25. This is manifested, among others, in a multiyear study by the Science Center Berlin of a mediation and negotiation process over a municipal waste disposal system in Cologne-Neuss. The study is financed by the Federal Ministry for Science, Technology, and Research.

26. Böhret (1990) and Scharpf (1991) observe similar system failures in other policy areas and have argued that decisive changes are necessary if the state is to maintain its capacity to govern into the next century.

27. Similar observations were made by Rucht (1980, 1984) in his studies of airport developments and the siting of nuclear power stations.

28. On this point see also Dyson (1982).

29. Even in the United States, where by German standards there is an openness to the introduction of new policy instruments and administrative styles, the adoption of alternative dispute resolution procedures was a slow and painful process (Manring 1994).

REFERENCES

17. Bundesimmissionsschutzverordnung [Seventeenth Federal Emission Protection Ordinance]. *Bundesgesetzblatt* I, November 23, 1990: 2545.

Abfallbeseitigungsgesetz (AbfG) [Waste Disposal Act]. *Bundesgesetzblatt I,* June 7, 1972: 873. Amended in 1986 by the *Abfallgesetz* (AbfG) [Waste Act]. *Bundesgesetzblatt* I, August 27, 1986: 1410.

Abfallbestimmungsverordnung (AbfBestV) [Waste Determination Ordinance]. *Bundesgesetzblatt* I, April 3, 1990: 614.

Amsoneit, N. 1994. A Centralized Hazardous Waste Treatment Plant–The Facilities of the ZVSMM at Schwabach as an Example. *Water Science and Technology* 29, no. 8:235–50.

Appold, W., and M. Beckmann. 1990. Ziele und rechtliche Instrumente der integrierten Abfallwirtschaft. *Verwaltungsarchiv* 81, no. 3:307–26.

Baerens, M., and U. von Arnswald. 1993. *Die Müll-Connection: Entsorger und ihre Geschäfte*. München: C. H. Beck.

Bieber, H. 1976. Bürgerkrieg im Wilster Marsch: Atomkraftwerk Brokdorf. *Die Zeit*, November 19, 2.

Böhret, C. 1990. *Folgen: Entwurf für eine aktive Politik gegen schleichende Katastrophen*. Opladen: Westdeutscher Verlag.

Bonner Energie-Report. 1990a. Ein dritter Ofen gegen den Engpaß. 11, no. 3:28–29, 32–33.

———. 1990b. Keiner weiß wohin mit dem Giftmüll. 11, no. 3:37–41.

Bundesimmissionsschutzgesetz [Federal Emision Protection Act]. *Bundesgesetzblatt* I, May 14, 1990: 880.

Bundesministerium für Umwelt, Naturschutz und Reaktorsicherheit (BMU) [Federal Ministry for Environment, Nature Protection, and Nuclear Safety]. 1992. Gemeinsame Erklärung der Umweltminister des Bundes und der Länder anläßlich der gemeinsamen Sitzung zum Thema "Abfallexporte" am 14.09.1992 [Joint Declaration of the Environment Ministers of the States and the Federal Government on the Occasion of Their Joint Conference on "Waste Exports" on September 14, 1992]. BMU, Bonn. Mimeographed.

Darnstädt, T. 1975. Chemierückstände alarmieren Umweltschützer: Casella kein Einzelfall. *Frankfurter Rundschau*, February 11: 11.

Davis, G., and J. Linneroth with B. Piasecki. 1987. Government Ownership of Risk: Guaranteeing a Treatment Infrastructure. In *America's Future in Toxic Waste Management* by B. Piasecki and G. Davis, 95–130. New York: Quorum Books.

Deutsche Projekt Union. 1990a. *Standortfindung für eine zweite Sonderabfallverbrennungsanlage in Baden-Württemberg: Flächendeckende Ermittlung von potentiellen Standorten im weiteren Einzugsbereich des Mittleren Neckarraums–Standortvorauswahl*. Stuttgart: Deutsche Projekt Union.

———. 1990b. *Standortfindung für eine zweite Sonderabfallverbrennungsanlage in Baden-Württemberg: Flächendeckende Ermittlung von potentiellen Standorten im weiteren Einzugsbereich des Mittleren Neckarraums–zusätzliche Suchschleife; Endbericht*. Stuttgart: Deutsche Projekt Union.

———. 1991. *Aktuelle Überprüfung der Standortfindung für die erste Sonderabfallverbrennungsanlage in Baden-Württemberg; Endbericht*. Stuttgart: Deutsche Projekt Union.

Deutscher Bundestag. 1963. *Erster Bericht der Bundesregierung zum Problem der Beseitigung von Abfallstoffen*. Bundestagsdrucksache IV/945, January 31.

———. 1966. *Zweiter Bericht der Bundesregierung zum Problem der Beseitigung von Abfallstoffen*. Bundestagsdrucksache VI/248, January 26, 1968.

Dick, Hr. 1993. Head, Engineering Department, Municipality of Manching. Interview with author, December 19.

Dowling, M. 1985. The International Regulation of Hazardous Waste: Problems of Definition and Classification. *Zeitschrift für Umweltpolitik*, 2/85: 141–51.

Dyson, K. 1982. West Germany: The Search for a Rationalist Consensus. In *Policy Styles in Western Europe* by J. Richardson, 17–46. London.

ECOTEC. 1993. *Neue Wege der politischen Entscheidungsfindung: Erfahrungen aus dem Sonderabfallforum Baden-Württemberg 1993*. München: ECOTEC.

European Community (EC). 1993. *Ordinance of the Council on the Supervision and Control of Waste Transports in, into, and out of the European Communities* (EC 259/93), February 1, 1993. *Official Journal of the European Communities*, L30, February 6, 1993.

Fietkau, H.-J., and H. Weidner. 1992. Mediationsverfahren in der Umweltpolitik. *Aus Politik und Zeitgeschichte*, B 39–40/92: 24–34.

Flottau, H. 1976. Hessische Umweltprobleme: Der Abfalleimer eines Riesen. *Süddeutsche Zeitung*, February 12: 3.

Frank, D. 1993. Former General Manager, Gesellschaft zur Entsorgung von Sondermüll (GSB), München. Interview with author, December 6.

Frankfurter Allgemeine Zeitung. 1994. Aufkommen von Sondermüll stark gesunken, 16–17. April 22.

Gesetz über die Umweltverträglichkeitsprüfung [Environmental Assessment Act]. *Bundesgesetzblatt* I, February 12, 1990: 205.

Hauber, G. 1989. *Abfall–Ingenieur–Bürger: Gemeinsam das Problem lösen*. Karlsruhe: C. F. Müller.

Herbold, R., and R. Wienken. 1993. Politische Strategien zur Überwindung des Durchsetzungsproblems von Abfallbeseitigungsanlagen: Der Fall der Bielefelder Abfalldeponie. *Müll und Abfall*, 5/93:317–25.

Hess, H. 1993. Public Relations Manager, Hessische Industriemüll GmbH (HIM). Interview with author, December 8.

Hessische Industriemüll GmbH (HIM) [Hesse Industrial Waste Corporation]. 1992. *Geschäftsbericht 1992*. Biebesheim: HIM.

———. 1993. HIM ändert Erweiterungspläne der SVA Biebesheim. *Im HIM-Blick*, 1/93. Biebesheim: HIM.

Hessisches Ministerium für Umwelt und Reaktorsicherheit (HMUR) [Hesse Ministry for the Environment and Nuclear Safety]. 1990. *6. Umweltbericht*. Wiesbaden: HMUR.

Hessisches Sonderabfallabgabengesetz [Hesse Special Waste Tax Act], June 26, 1991.

Hoppe, W., and W. Appold. 1990. *Umweltschutz in der Raumplanung*. Münster: Institut für Siedlungs- und Wohnungswesen.

Hoppe, W., and M. Beckmann. 1990. *Planfeststellung und Plangenehmigung im Abfallrecht: Vorschläge zur Erleichterung des Abfallrechtlichen Zulassungsverfahrens*. Münster: Institut für Siedlungs-und Wohnungswesen.

Hupka S. 1994. Böblingen und Sindelfingen kämpfen gegen Giftmüllofen. *Südwestpresse*, 1: 3. January 15.

Investitionserleichterungs- und Wohnbaulandgesetz [Investment Facilitation and Settlement Land Act]. *Bundesgesetzblatt* I, April 22, 1993: 466.

Johannssen, T. 1993. Reporter for the *Brunsbütteler Rundschau*. Interviews with author, December 13, 20.

Johnke, B. 1992. Akzeptanzproblem Abfallentsorgungsanlagen—eine Frage der Öffentlichkeitsbeteiligung. *Müll und Abfall*, 2/92:78–84.

Johnke, B., and J. Schmitt-Tegge, eds. 1993. *Akzeptanzprobleme bei Maßnahmen zur Abfallentsorgung*. Berlin: E. Schmidt Verlag.

Jung, G. 1988. *Die Planung in der Abfallwirtschaft*. Berlin: E. Schmidt Verlag.

Koch, K. 1993. Chairperson. Das Bessere Müllkonzept, Schleswig Holstein. Interview with author, December 20.

Kommunale Arbeitsgemeinschaft "SVA Biebesheim" (KAG) [Municipal Working Group "Hazardous-Waste Incinerator Biebesheim"]. 1993. *Regionales Sonderabfallvermeidungs-und -behandlungskonzept am Widerstand der Industrie gescheitert*. Press release, December 16.

Lahmeyer International. 1992. *Verbesserung von Zulassungsverfahren für Abfallentsorgungsanlagen–Leitfaden*. (Forschungsvorhaben für das Bundesministerium für Umwelt, Naturschutz und Reaktorsicherheit). Frankfurt am Main: Lahmeyer International.

Landesamt für Umwelt und Natur Mecklenburg-Vorpommern. 1994. *Daten zur Abfallwirtschaft*. Gülzow: Landesamt für Umwelt und Natur.

Landtag of Baden-Württemberg. 1990. *Mitteilung des Ministeriums für Umwelt: Sonderabfallwirtschaftsplan Baden-Württemberg*. Landtagsdrucksache 10/4545, December 27.

———. 1992a. *Kleine Anfrage des Abg. Dr. Eugen Klunzinger CDU 'Gefährdung der 64-Megbit-Chip-Produktion in unmittelbarer Nähe von Sonderabfallverbrennungsanlagen*. Landtagsdrucksache 11/1107, December 15.

———. 1992b. *Kleine Anfrage des Abg, Dr. Walter Döring FDP/DVP 'Geplante Sonderabfallverbrennungsanlage im Raum Böblingen/Sindelfingen*. Landtagsdrucksache 11/1147, December 22.

———. 1992c. *Plenarprotokoll*, October 21.

———. 1993. *Plenarprotokoll*, February 10.

Linnerooth, J., and A.V. Kneese. 1989. Hazardous Waste Management: A West German Approach. *Resources*, no. 96 (Summer): 7–10.

Mahl, R., and T. Darnstädt. 1974. Tag und Nacht rollt das Gift auf Hessens Müllkippen. *Frankfurter Rundschau*, August 1: 3.

Manring, N. J. 1994. ADR and Administrative Responsiveness: Challenges for Public Administrators. *Public Administration Review* 54, no. 2:197–203.

Meyer, H. 1993. Special Waste Officer, Schleswig-Holstein Environment Ministry. Interviews with author, December 7, 13.

Müller, A. 1990. Beim Poker mit Paris setzt Stuttgart alles auf die Karte Kehl. *uttgarter Zeitung*, March 27: 6.

Müller, H. H. 1991. Wege zur Erhöhung der Akzeptanz von Entsorgungsanlagen. *Der Städtetag*, 6/91:436–39.

Müller, K., and M. Holst. 1987. *Raumordnung und Abfallbeseitungung: Empirische Untersuchung zu Standortwahl und -durchsetzung von Abfallbeseitigungsanlagen*. Bonn: Bundesministerium für Raumordnung und Städtebau.

Müller-Heuser, G. 1992. Formaler Ablauf des Genehmigungsverfahrens für Abfallentsorgungsanlagen. In *Handbuch der Abfallentsorgung*, Ergänzungslieferung 12/92: 1–47.

Müllmann, C. 1993. Die Zulassung von Abfallentsorgungsanlagen nach dem Investitionserleichterungs- und Wohnbaulandgesetz. *Müll und Abfall*, 9/93:645–54.

Niedersächsisches Umweltministerium. 1992. *Sonderabfallkonzept Niedersachsen (Entwurf)*. Hannover: Niedersächsisches Umweltministerium.

Noeke, J. 1990. Abfallwirtschaft und Öffentlichkeit. *EntsorgungsPraxis*, 3/90:104–8.

Nussen, Hr. 1993. Permitting Offices Industry Inspectorate Itzehoe. Interviews with author, December 7, 13.

Piasecki, B. W., and Davis, G. A. 1987. *America's Future in Toxic Waste Management: Lessons from Europe*. New York: Quorum Books.

Prognos. 1993. *Abschlußdokumentation: Forum zur Sonderabfallwirtschaft Baden-Württemberg*. Basel and Stuttgart: Europäisches Zentrum für Wirtschaftsforschung und Strategieberatung.

Rabe, B. 1992. When Siting Works Canadian Style. *Journal of Health Politics and Law* 17, no. 1:119–42.

———. 1994. *Beyond NIMBY: Siting Hazardous Facilities in Canada and the United States*. Washington, D.C.: Brookings Institution.

Reststoffbestimmungsverordnung (RestBestV) [Waste Residue Determination Ordinance]. *Bundesgesetzblatt* I, April 3, 1990: 862.

Ronellenfitsch, M. 1989. Standortwahl bei Abfallentsorgungsanlagen: Planfeststellungsverfahren und Umweltverträglichkeitsprüfung. *Die Öffentliche Verwaltung* 42:737–50.

Rucht, D. 1980. *Von Wyhl nach Gorleben: Bürger gegen Atomprogramm und nukleare Entsorgung*. München: C. H. Beck.

———. 1984. *Flughafenprojekte als Politikum*. Frankfurt am Main: Campus Verlag.

Sachverständigenrat für Umweltfragen (SRU) [Council of Environmental Experts]. 1990. *Sondergutachten 1990: Abfallwirtschaft*. Stuttgart: Metzler-Poeschel.

Scharpf, W. 1991. Die Handlungsfähigkeit des Staates am Ende des 20. Jahrhunderts. *Politische Vierteljahresschrift* 32, no. 4: 621–34.

Schenkel, W. 1990. Die Planung der Abfallwirtschaft. In *Umweltschutz in der Raumplanung* by W. Hoppe and W. Appold, 132–53. Münster: Institut für Siedlungs und Wohnungswesen.

Schleswig Holstein Sonderabfallverbrennungsverordnung [Schleswig-Holstein Special Waste Incineration Ordinance], October 13, 1993.

Schmitt-Tegge, J. 1994. Incineration of Hazardous and Municipal Waste—the Situation in Germany in 1993: A Commentary. *Waste Management and Research* 12:441–53.

Schön, D. A., and M. Rein. 1994. *Frame Reflections: Toward the Resolution of Intractable Policy Controversies*. New York: Basic Books.

Schweiger, Dr. 1993. Chairperson Bund für Umwelt und Naturschutz in Bayern [Federation for Environment and Nature Protection in Bavaria], Manching. Interview with author, December 7.

Seeliger, R. 1994. Ein Lückentext aus Brüssel verunsichert die Betriebe. *Handelsblatt*, September 19: 8.

Spiegel, Der. 1973. "Hier tickt eine Zeitbombe": Heimlich wurden in Hessen Tausende von Tonnen giftigen Industriemülls auf Halden gekippt. October 1: 24–32.

———. 1974a. Giftmüll: Alles getan. May 20: 73–74.

———. 1974b. Giftmüll: "Ich darf nichts sagen." October 28: 80–84.

———. 1974c. Türkischer Honig–es kann auch Cyanid sein: SPIEGEL-Report über illegale Giftmüll-Ablagerungen in der Bundesrepublik. May 13: 54–68.

———. 1975. Prozesse: Eindruck gemacht. November 3, 54–60.

Stärker, D. 1976a. Bonn drängte auf Baubeginn in Brokdorf. *Tagesspiegel*, October 10: 3.

———. 1976b. "Soviel Brutalität gab es noch nie": Nach dem Polizeieinsatz in Brokdorf–Umweltschützer planen weitere Aktionen. *Frankfurter Rundschau*, November 2, 3.

Statistisches Bundesamt. 1989. *Abfallbeseitigung im produzierenden Gewerbe, 1987*. Stuttgart: Metzler-Poeschel.

———. 1994. *Abfallbeseitigung im produzierenden Gewerbe, 1990*. Stuttgart: Metzler-Poeschel.

Stork, W. 1991. Statement zur Eröffnung des Symposiums *Sonderabfall in Hessen*. Statement at the opening of the symposium Sonderabfall in Hessen, October 19, Riedstadt.

Striegnitz, S. 1993. Münchehagenausschuß: Mediationsverfahren zur Sanierung einer Sonderabfalldeponie. In *Akzeptanzprobleme bei Maßnahmen zur Abfallentsorgung*, edited by B. Johnke and J. Schmitt-Tegge, 183–200. Berlin: E. Schmidt Verlag.

Technische Anleitung zur Lagerung, chemisch/physikalischen, biologischen Behandlung, Verbrennung, und Ablagerung von besonders überwachungsbedürftigen Abfällen (TA Abfall) [Technical Instructions on the Storage, Chemical-Physical and Biological Treatment, Incineration, and Disposal of Waste Requiring Special Supervision]. *Gemeinsames Ministerialblatt*, April 4, 1990: 170. As amended in *Gemeinsames Ministerialblatt*, December 17, 1990: 866; February 4, 1991: 136; March 12, 1991: 139.

Technische Anleitung zur Reinhaltung der Luft (TA Luft) [Technical Instructions for the Maintenance of Air Quality]. *Gemeinsames Ministerialblatt,* February 27, 1986: 95.

Umweltbundesamt (UBA) [Federal Environment Administration]. 1990. *Jahresbericht 1989.* Berlin: UBA.

———. 1991. *Jahresbericht 1990.* Berlin: UBA.

———. 1992. *Erläuternder Bericht zur TA Abfall.* Berlin: UBA.

———. 1993a. Grenzüberschreitende Abfallverbringung (Abfall Exporte). UBA, Berlin. Mimeographed.

———. 1993b Sonderabfall–Entsorgung: Materialien zu Abfall-und Sonderabfall. UBA, Berlin. Mimeographed.

———. 1994. *Jahresbericht 1993.* Berlin: UBA.

Umweltministerium Baden-Württemberg. 1993. *Die Arbeit des Forums hat sich gelohnt! Ergebnisse sind Grundlage für die Fortschreibung des Sonderabfallwirtschaftkonzeptes für Baden-Württemberg.* Press release, June 28. Stuttgart.

Unger, H. J. 1993. Kommunale Arbeitsgemeinschaft gegen SVA Biebesheim. Interview with author, December 20.

Verwaltungsgerichtsordnung (VwGO) [Administrative Court Regulation]. *Bundesgesetzblatt* I, March 19, 1991: 1077.

Verwaltungsverfahrensgesetz (VwVfG) [Administrative Procedure Act]. *Bundesgesetzblatt* I, May 25, 1975: 1253.

Wasserhaushaltsgesetz (WHG) [Water Management Act]. *Bundesgesetzblatt* I, September 23, 1986: 1529.

Wenz, D. 1989. "Vergiftete Freundschaft" heißt es in Straßburg. *Frankfurter Allgemeine Zeitung,* August 28: 3.

———. 1990. Es raucht schon genug in Straßburg. *Frankfurter Allgemeine Zeitung,* April 9: 5.

Zielleßen, H., and T. Barbian. 1992. Neue Formen der Konfliktregelung in der Umweltpolitik. *Aus Politik und Zeitgeschichte,* B 39–40/92: 14–23.

Zubiller, C.-O. 1993. Head, Industrial Waste Management Section, Hesse Environment Ministry, Wiesbaden. Interview with author, December 15.

Part 2

Public Perceptions

Community Risk Perception and Waste Management in Three Communities at Different Stages in the Siting Process

CLYDE HERTZMAN, ALECK OSTRY

INTRODUCTION

In 1983 Vincent Covello, in an extensive review of the risk perception literature, pointed out the dearth of studies on the "community" dimension in risk perception (Covello 1983). Most empirical risk perception research has focused on understanding the psychology of risk at the individual level. As Covello pointed out, however, risk is perceived, talked about, and responded to usually within a community context. This is an awkward fact, because communities tend to be different from one another in ways that might well affect their collective perception of risk. From an epidemiological perspective, the implication is that every community must be treated as a single individual for the purposes of statistical analysis and interpretation. That is, in order to make valid inferences about the principles of community risk perception, a uniform protocol of study would need to be carried out in a statistically powerful sample of communities, sampled as though communities were individuals. In practice, this would involve a minimum of several dozen communities, and the study would still require contact with large samples of individuals in each community.

Because of the cost and time necessary to conduct even a single community study, it is unlikely that the knowledge base will develop according to such an epidemiological prescription. Instead, it is likely to proceed through an accumulation of case studies. As case studies pile up, the question "Which perceptions are unique to the particular community under study, and which have a significant element of transferability?" will become increasingly important to answer. We have begun to address this issue by studying a series of communities that are radically different from one another, anticipating that findings which have an element of consistency across such diverse sites also deserve consideration as general principles of community response.

The surveys discussed here were carried out in three communities in British Columbia that host or are slated to host a waste management facility. The initial purpose of this series of investigations was to identify determinants of facility acceptability. One of the communities was rural, while the other two were in urban locations. In one community, a medical waste incinerator (MWI) was proposed, while the other two housed operating facilities: a landfill and a municipal solid waste incinerator (MSWI). Also, each community differed in its experience of the siting process and in the number and distribution of community benefits associated with the local facility.

The first survey was conducted in neighborhoods adjacent to the University of British Columbia (UBC) in Vancouver, British Columbia. In the fall of 1991, UBC began the process of replacing a small, twenty-year-old MWI located on campus with an MWI capable of servicing the UBC hospital, as well as departments in all British Columbia universities that produce medical waste. Although a university-managed public-information process had been under way in the adjacent communities for about a year, we felt that a survey in this community would produce a baseline of largely unaware respondents among whom we could subsequently measure the impact of the siting process as it unfolded. We conducted the baseline survey in the fall of 1991. A provincial election had been called for later in the fall, and we felt that the MWI issue would be politicized, resulting in increased publicity and awareness about the MWI in the local community. Our baseline survey, in fact, showed that the community was largely unaware of the planned MWI. Twenty months after the baseline survey, in the spring of 1993, we resurveyed the UBC respondents in order to determine the impact of the siting process over this time.

The second survey was conducted in Burnaby, a city of 350,000 people within the greater metropolitan area of Vancouver. In the early 1980s, an MSWI was proposed for Burnaby's southern boundary (Greater Vancouver Regional District 1986). Publicity surrounding the incinerator siting was most intense in 1985–86. After limited opposition, the facility was constructed, and it started operating early in 1987. Since it opened, local community controversy has been minimal. Between December 1987 and July 1991, the Greater Vancouver Regional District (GVRD) Air Quality Management Division received only seventy-five complaints from residents, about 90% of which were about odors. The complaint rate fell by over 50% after the first two years of operation to approximately twelve complaints per year (Greater Vancouver Regional District 1991).

In order to characterize community attitudes toward the risks of waste disposal in an area with long experience of the issue, we conducted a survey in the British Columbia villages of Cache Creek and

Ashcroft in 1992 (Ostry, Hertzman, and Teschke 1993a). In the 1980s, two unsuccessful attempts were made to site a hazardous waste facility (HWF) near these villages. Following these attempts, the Greater Vancouver Sewerage and Drainage District succeeded in locating a landfill for Vancouver's solid waste in Cache Creek (Zeiss 1984). A chip plant was also built at Cache Creek as part of the deal to locate the landfill. Both facilities began operation in February 1989.

The results of these studies can be arranged in a crude time series (i.e., beginning with the siting and ending with a sited and operational facility) to simulate longitudinal patterns of community response to the prospect and reality of facility siting. The sequencing is roughly as follows:

1. UBC, first survey
2. UBC, second survey
3. Cache Creek responses concerning the proposed HWF
4. Burnaby responses concerning the incinerator, and Cache Creek responses concerning the landfill

Under this sequencing, the two UBC surveys occur at the beginning and at the early stages of the siting process, respectively; the Cache Creek HWF responses represent an intermediate stage; and the last two represent alternative outcomes: Burnaby is a completed noncontroversial siting, and the Cache Creek response to the landfill is a completed siting despite controversy.

This approach provides a framework for exploring the character of community responses to siting processes as they unfold over time. It is vulnerable to the claim that the communities must be comparable to one another in order to provide a single longitudinal view of the siting process. This criterion does not seem to be met, since the communities differ by socioeconomic and urban and rural characteristics. In fact, the apparent weakness is a strength. We have concentrated upon outcomes that are most likely to be coherent across communities, despite their differing character. In this way, we hope to generate hypotheses about community responses, hypotheses that underlie the siting process regardless of specific community characteristics.

METHODS

The 1990 British Columbia Provincial Voters List, which includes at least 85% of adults over the age of nineteen, was used as the sampling frame in each survey. Approximately 450 subjects were randomly selected from UBC, 350 from Burnaby, and 330 from the Cache Creek community from among all individuals whose addresses were within the postal codes closest to the facility. For the Burnaby community, subject selection differed because the shapes and sizes of postal codes adjacent to the incinerator were not uniform. Initially, 650 individuals were selected. Each subject's street address was input into a geographical information system program (MAPINFO). Those who were found to reside within 5 kilometers of the incinerator were approached to participate in the survey (350 of the original 650).

Randomly selected subjects were sent an introductory letter describing the study as an environmental survey with the purpose of determining attitudes to various waste management technologies. Approximately ten days later, each subject was phoned directly and a telephone interview requested. If consent was obtained, an interview was conducted using a structured questionnaire format. Successful interviews were completed for 224 respondents in Burnaby and 187 from the Cache Creek area, for a refusal rate of 28% in both cases. The first survey at UBC involved 265 subjects (24% refusal rate), of which 195 were still in the community and willing to provide a second interview twenty months later.

The questionnaire was based on instruments developed and tested for reliability and validity by Zeiss and Taylor in surveys of communities with waste management facilities (Zeiss 1984, 1988; Taylor et al. 1991). The instrument covered six dimensions: sociodemographic characteristics of the respondent, participation in community networks, perceptions of the neighborhood, attitudes to waste management facilities, agreement with negative statements about the sort of facility relevant to the community, and agreement with positive statements about the facility.

A major purpose of the questionnaire was to categorize subjects into subgroups based on their awareness of, and concern with, the facility. The first section of the questionnaire tested awareness by asking respondents to identify facilities they would most like to see "in their neighborhood" or "not in their neighborhood." This provided the opportunity for subjects to mention their planned or existing facility in an unsolicited manner. Subjects who were aware of the facility were asked directly, "Do you have any concerns about the site?" to determine whether or not they were concerned.

Following the categorization of subjects into the subgroups, questions modified from Zeiss (1984, 1988) and Taylor et al. (1991) were used to study attitudes toward waste management in general and the planned or existing community facility in particular. Subjects were asked to consider four hypothetical waste facility types located within 5 kilometers of their home and to rate these situations using five-point scales to indicate their familiarity with the type of facility and the desirability of each. To study attitudes to the planned or existing facility in the community, we read nine negative and ten positive statements about it and its possible impact in the community, then asked subjects to rate their agreement using a five-point scale.

RESULTS

Detailed results of the individual case studies emerging from this work have been reported elsewhere (Ostry, Hertzman, and Teschke 1993a, 1993b, 1994). In this chapter, specific findings from these case studies are presented in ways that highlight the potential for generalizability to siting processes at other, unstudied sites. These findings have two purposes. First, they can serve as hypotheses for future longitudinal studies of community responses to siting waste management facilities. Second, they have policy relevance as is. Our studies generated four main hypotheses, which follow.

Hypothesis 1: It is difficult to create awareness of proposed waste management facilities in communities.

Tables 1 and 3 present the publicity activities that were undertaken over time at the UBC site. Tables 2 and 4 present results from our questionnaire surveys, showing levels of unsolicited awareness of the facility among members of the adjacent community following each sequence of publicity activities.

Publicity about the site became visible in the community in the four months prior to our initial baseline survey. During this four-month period covered by Table 1, no special process of community involvement had yet been put in place, and so the publicity was mostly of the "news item/reaction to the news item" variety. There were, however, three public meetings held in the community, and the local ratepayers association and one environmental group (SPEC)

TABLE 1. Public-Information Activities in the Community Four Months Prior to the Baseline Survey

- 3 public meetings held about the MWI
- Newspapers and newsletters
  12 articles about the proposed MWI (5 feature articles)
    8 in local papers
    2 in UBC newsletter
    2 in major dailies
- Major mailout into the neighborhood by local environmentalist organization and feature article in its newspaper
- No major radio or TV coverage

TABLE 2. Distribution of UBC Respondents by Awareness of, and Concern about, the Proposed Facility—Original Baseline Survey

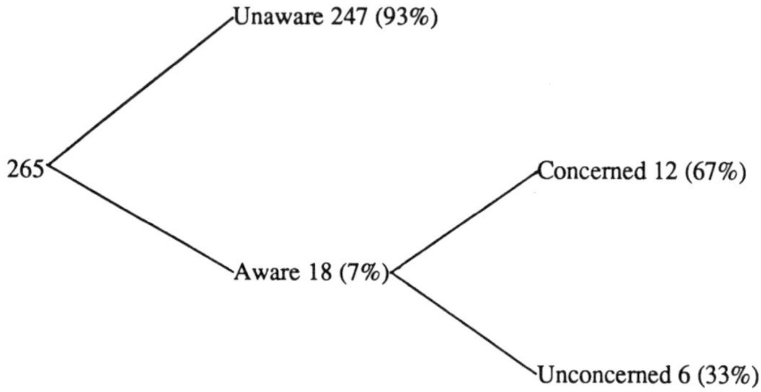

were known to be aware of the proposal. At the first survey, the level of unsolicited awareness was very low. Table 2 shows that of 265 survey respondents, only 18 (7%) were aware of the planned MWI. Although this was somewhat surprising, there had been no public controversy prior to the baseline survey.

This situation did not last. In the twenty-month interval between baseline and resurvey, the proposal became the subject of intense political controversy. It was an issue in the provincial election, wherein two local candidates ran against the site and not only won election but received jobs in the cabinet of the new provincial government. Subsequently, a community-based hazardous waste advisory committee was set up by UBC site planners, and a waste commissioner was hired by the provincial government to hold public hearings and report on biomedical waste issues. Table 3 details the large volume of articles and meetings that took place between surveys. Despite all the activity, the level of awareness in the community did not increase very much. Table 4 shows that of the 195 respondents still living in the area twenty months after the first survey and available for interview, only 23 (12%) demonstrated new awareness of the MWI siting issue. Overall, 159 respondents (82% of the sample) still demonstrated no awareness.

Although persistent unawareness of the site would be expected among some community members, 82% is a startlingly high level and calls into question the effectiveness of top-down (i.e., government- or planner-motivated) awareness campaigns. The results cannot easily be explained away by challenging the sensitivity of our survey methods based on unsolicited awareness. In Cache Creek, for example, the level of unsolicited awareness of the landfill that accepts waste from Vancouver is 96%, suggesting that lower levels of awareness found in other surveys are, indeed, valid.

TABLE 3. Public-Information Activities in the Community during the Twenty-Month Period between the Two Surveys

- Public-meeting campaign
  3 public meetings held about the MWI
- Newspapers and newsletters
  29 articles about the proposed MWI (17 feature articles)
  9 in local papers
  4 in UBC newsletters
  7 in the UBC student newspaper
  9 in major dailies
- Provincial election campaign in which opposition to the MWI was a main plank in local NDP candidates' strategy (they won the election)
- Mailout by local ratepayers association to all UEL ratepayers
- Summer 1992 appointment of a commissioner to hold public hearings and report on biomedical waste issues
- Community-based hazardous waste advisory committee set up by UBC planners, and a waste commissioner hired by the provincial government

TABLE 4. Distribution of UBC Respondents by Awareness of, and Concern about, the Proposed Facility—Resurvey

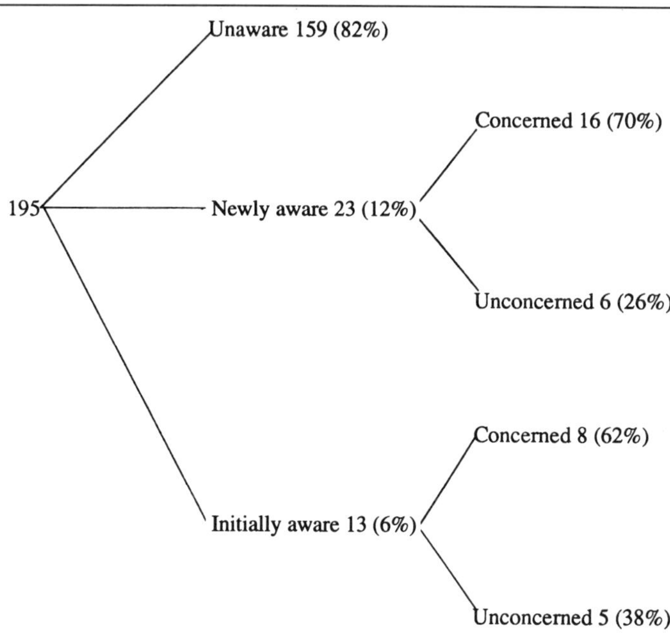

TABLE 5. Awareness and Concern in the Cache Creek Area in Relation to the Proposed but Withdrawn Hazardous Waste Facility

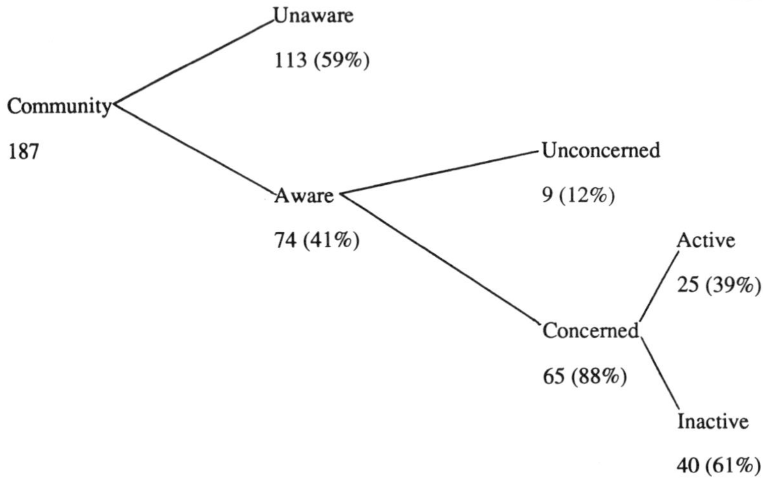

Another objection might be that the UBC community is atypical in ways that would make it specially resistant to absorbing information of community concern. One atypical feature of this community is its level of education. Eighty-seven percent of the respondents of the second survey had at least some postsecondary education. This, however, would usually be construed as a facilitator of, rather than a barrier to, awareness. A more likely candidate would be the area's transience. It included one zone (the UBC endowment lands) where graduate students live, and 36 of the original 265 subjects had left the area by the time of the second survey. Their level of awareness on the first survey, however, was not significantly different from that of those who remained. Moreover, the 82% unawareness occurred among those who had stayed in the area, of whom 62% owned their residence and can be presumed to have had a long-term interest in the community.

Hypothesis 2: People who are not easily made aware of waste management facility issues differ in important ways from those who are more easily made aware.

Table 5 shows the distribution of respondents in the Cache Creek area by awareness of, and concern about, the hazardous waste facility that was proposed, but ultimately not sited, in the area. It shows that 113 respondents (59% of the sample) remained unaware of the proposed facility despite a protracted struggle between supporters and opponents in the community. Such a level of unawareness is, once again, surprisingly high, but it is not as high as at UBC. This is not inconsistent, because the process of siting at UBC had not yet led to high levels of controversy within the community, even at the second survey. Fortuitously, the numbers of both aware and unaware respondents were reasonably large, which made it possible to explore the differences between them.

Table 6 shows how the subgroups differ with respect to sociodemographic factors and community networks. The unaware were more frequently renters than the aware, which is not surprising, but were also more frequently female, which was unexpected. More interesting from the standpoint of community process and communication is the fact that the two groups differed on four different measures of community networking: membership in a union, membership in one or more community organizations, reading the major provincial daily newspaper, and recently contacting an official. In each case, the unaware respondents were less involved in community processes than the aware.

In a more extreme case of unawareness, further differences are found. At the Burnaby site, 37% of respondents were unaware of the incinerator, even though it had been operating in their community since 1987 and the incinerator stack was clearly visible from virtually all points in the area (table 7). These individuals form a hard-core unaware group, whose existence would have been difficult to predict without empirical evidence. Table 8 shows, as in the Cache Creek case, that this subgroup is demographically different from the aware

TABLE 6. Characteristics for Which There Were Statistically Significant Differences between Those Who Were Aware and Unaware of the Hazardous Waste Facility Proposed for Cache Creek

Variable	*Aware*	*Unaware*	*(p)*
1. Sociodemographics			
Percent renters	5%	17%	.02
Sex (percent female)	31%	46%	.04
2. Community networks			
Percent in a labor union	58%	37%	.005
Percent in community organizations	38%	22%	.02
Percent who read the major daily paper	30%	17%	.04
Percent contacting community official	46%	31%	.04

*p-values based on Chi square for 2×2 contingency table.

TABLE 7. Awareness and Concern in Relation to the Operating Incinerator in Burnaby

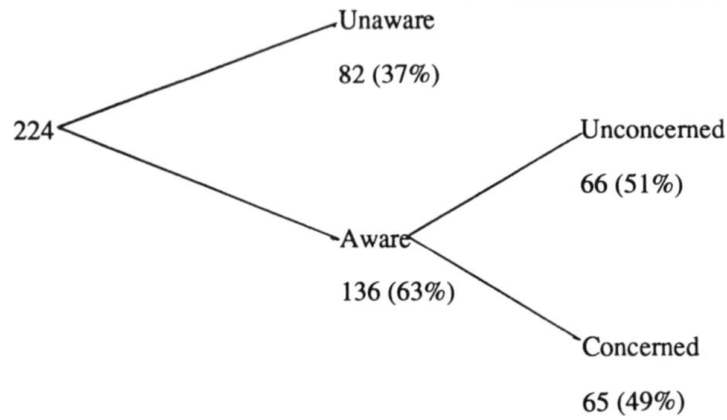

subgroup. They were more than ten times as likely to be renters; their average residence time in the neighborhood was half as long (though still over ten years); they were younger, more frequently female, more likely to be working, and less likely to be retired than the aware subgroup. Once again, their level of connection to community networks also seemed to be lower, at least according to two measures.

Much more important, however, is the information contained in figure 1. It displays the level of agreement the unaware and aware respondents (separated into the concerned and the unconcerned) have

TABLE 8. Characteristics for Which There Were Statistically Significant Differences between Those Who Were Aware and Unaware of the Neighborhood Incinerator in Burnaby

Variable	*Aware*	*Unaware*	*(p)**
1. Sociodemographics			
Percent renters	3%	32%	<.001
Mean # of years in neighborhood	21.9	11.5	<.001
Mean age (years)	54.6	45.5	<.001
Sex (percent female)	46%	65%	.006
Percent working	51%	70%	.03
Percent retired	36%	15%	.03
2. Community networks			
Read major daily paper	74%	52%	.007
Attended meetings in community	24%	10%	.008
Contacted community official	22%	35%	.04

*p-values based on Chi square for 2×2 contingency table when outcomes are expressed in proportions, and unpaired t-tests for continuous variables.

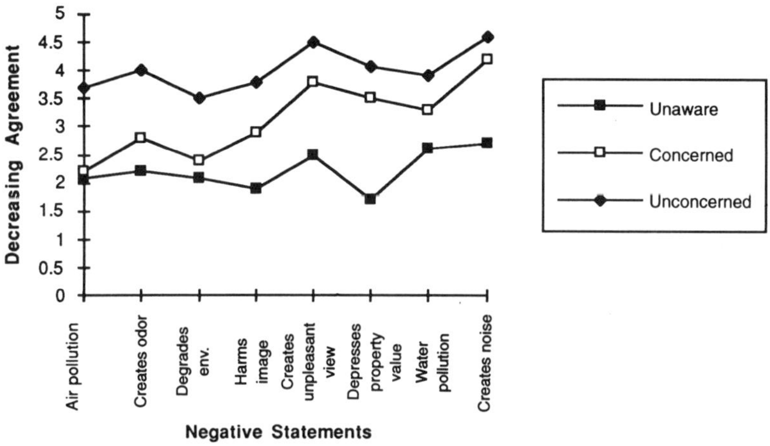

Figure 1. Agreement of aware and unaware groups with negative statements about the incinerator

with a series of negative statements about the impact of an incinerator on a community. It shows that the unaware respondents are systematically more negative in their assessment of the *potential* impact of an incinerator, compared with the assessments of those who know it is there and are, by dint of the questionnaire design, rating the actual facility in their community. Figure 1 shows that the unaware group is even more negative in its assessments than the aware and concerned group. In other words, the subgroup that is unaware of the incinerator in its backyard would hate to live in a neighborhood with an incinerator!

This demonstration of a relationship between unawareness, disconnectedness, and negativity toward a waste management facility is rudimentary but highly intriguing. We are aware of at least one facility-siting situation in another jurisdiction where what seemed like a smooth process was disrupted at the last moment by previously unaware individuals, who were suddenly forced into awareness by their neighbors and quickly became a vocal rejectionist force.

Hypothesis 3: Building majority support for facility siting means building support among the unconcerned groups rather than trying to change the opinions of the concerned groups.

Figure 2 compares the subgroups that are "aware and concerned" about the facility sited or proposed for their community, in terms of

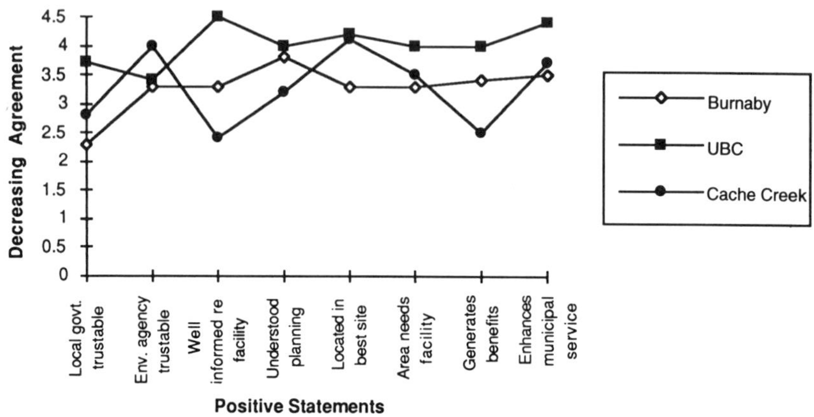

Figure 2. Concerned group ratings of equity, economic, and information issues for the facility in their backyard

their level of agreement with positive statements about the facility. All three sites are included. The UBC data come from the second survey, and the Cache Creek data relate to perceptions of the landfill site. In the latter case, we anticipate that many respondents' views are colored by awareness of the struggles over both the landfill and the hazardous waste management facility. Surprisingly, the figure shows that the opinions of the concerned groups at different sites do not differ systematically from one another, despite profound differences in the facility type, the stage of the siting process when the survey was completed, and the level of controversy about facility siting in the community.

Figure 3 makes a similar comparison among communities, but this time among those who were "aware and unconcerned." This time, a clear pattern emerges in which the unconcerned group in the community that has only begun the siting process has less positive views than the unconcerned group in the community where the facility was accepted without significant controversy, which, in turn, is less positive than the group in the community that was divided by controversy but where the facility in question was finally sited.

If these three cross sections in different communities were viewed as a time series in one community, UBC would represent an early stage, Burnaby an intermediate stage, and Cache Creek a late stage. If so, the impression is created of one group of concerned individuals whose generally negative view of facilities is static, and of another group of unconcerned individuals whose views become more positive as controversy increases and time passes. This pattern cannot easily be explained away by the hypothesis that the unconcerned group becomes more highly selected as initially unconcerned individuals

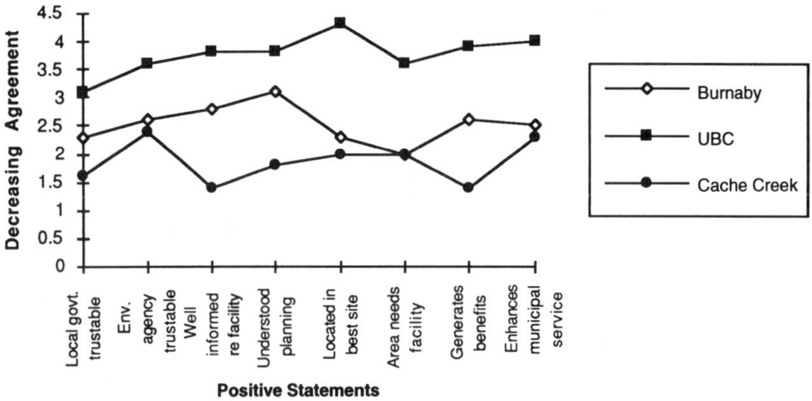

Figure 3. Unconcerned group ratings of equity, economic, and information issues for the facility in their backyard

with intermediate views change their minds and join the ranks of the concerned. If this were the case, we would expect that the proportion of unconcerned respondents would drop from UBC to Burnaby to Cache Creek. It did not.

The pattern suggests an intriguing hypothesis for longitudinal research into the dynamics of community opinion during facility siting: the success of a siting exercise depends upon building support among those who are initially predisposed to the facility, rather than on convincing those who are initially concerned to change their minds. This, unfortunately, assumes that building majority support will lead to facility siting and acceptance. In practice, it is possible that concerned opposers would be able to thwart siting even if the majority favored it. A safer hypothesis would be that building majority support for facility siting means building support among the unconcerned groups, rather than trying to change the opinions of the concerned groups.

Hypothesis 4: The credibility of the expertise of the academic community may be higher in divided communities than that of other groups.

Figure 4 shows the level of trust among aware groups, at the end of the facility siting processes, toward different agencies involved in Cache Creek at some point. In this analysis the aware group has been divided into four subgroups based on ascending levels of concern and activism: people who went to meetings to support the facility (in this case, the landfill), people who were unconcerned but inactive, people

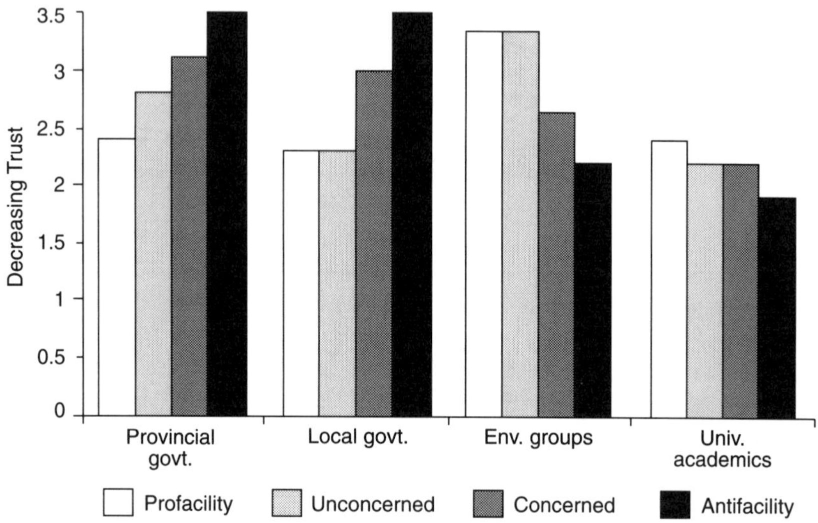

Figure 4. Trust scores for subgroups at Cache Creek

who were concerned but inactive, and people who actively did something to oppose the facility.

The figure shows, with increasing levels of activism against the landfill, decreasing levels of trust for provincial and local governments and a reverse pattern of increasing levels of trust for environmental groups. Trust scores for environmental groups were significantly lower ($p<.001$) for the unconcerned and profacility subgroups, while trust scores for the local government were significantly higher ($p<.001$) among these subgroups. All this makes intuitive sense, since the governments were supportive of facility siting and the environmental groups were critical of it. On the other hand, the average level of trust for university academics was higher than for any other agency. More important, there was no statistically significant difference in the level of trust shown the academic community as the level of concern and activism among the respondents changed. Inspection of the figure shows that there was virtually no trend toward a difference, either. These results suggest that academic experts may have more credibility in a controversial facility-siting process than do other agencies. Moreover, the results imply that academics are seen as being impartial, in the sense that they are not the particular favorite of any one of the local interest groups.

DISCUSSION

Ideally, a community's acceptance or rejection of a waste management facility in its midst (especially if that facility is meant to handle waste from other localities) would depend upon the community's collective competence in dealing with certain issues: its ability to conceive of risk as a probabilistic, rather than a dichotomous, phenomenon; its willingness to consider whether the local area has a significant responsibility to the broader community; its ability to judge whether the siting process itself is credible; its willingness and ability to appraise the technical aspects of the project; and, over time, its willingness and ability to appraise the systems established for solving the facility's problems.

The prospect of the siting process unfolding along these lines in any community would depend, to a substantial degree, upon a sense of connectedness among community members and openness and efficiency of lines of communication within the community. Our original research framework began the exploration of these factors by dividing the community into subgroups according to awareness and concern about the facility proposed or sited in their midst. Future research needs to go much further in questioning the relationships between these groups: their relative size and persistence over time, their ability to collect and synthesize information, and the quality of their communication with one another.

Consider, for example, the surprise generated by the findings underlying hypothesis 1, showing how little awareness had been engendered at UBC despite energetic efforts. This points to a misperception among planners about the efficacy and impact of their own planning and advertising activities. Future longitudinal studies could, therefore, not only survey respondents in the community but also interview planners as they go through the process to see how their understanding of their own impact on the community jibes with the actual impact of the siting program and propaganda.

From a policy standpoint, it could be that a more realistic appraisal among planners of their impact as they go through the process may help reduce "expert" versus "lay" communication difficulties regarding the potential impacts of a facility and the fairness of the siting process. It may also be true that certain communities have lines of communication that can be opened by reasonable efforts, while others simply do not. High levels of persistent unawareness, disconnectedness with lines of communication, and uninformed viewpoints with regard to technical issues may be evidence of a community that cannot competently handle a decision about siting a waste management facility in its midst.

Hypothesis 2 suggests that the unaware group in some communities may be sociodemographically distinct, particularly hard to reach, and more opposed to a facility than even the aware and concerned group. Hypothesis 3 focuses on the aware groups and suggests that it makes sense, from a policy perspective, to target the unconcerned group in order to build support for a facility. From the standpoint of further longitudinal study, it would be desirable to explore two issues.

1. Do the originally unaware, when they become aware, apportion in a different way between concerned and unconcerned groups compared with the originally aware group, and if so, what factors seem to affect the differential apportionment? This process of apportionment is important to understand because, according to hypothesis 3, the key to acceptability may lie in reaching the unconcerned.
2. In the discussion of hypothesis 3, we suggest that concerned and unconcerned individuals do not switch places, while, at the same time, unconcerned individuals become increasingly positive in their viewpoints. This initial supposition needs further evaluation. Over time, do people move from concern to unconcern and vice versa? If so, what are the factors that determine such attitude changes?

Hypothesis 4 suggests that academics may have a useful role to play as facilitators or intermediaries in the siting process. Unfortunately, our results do not help very much to understand the way in which community perceptions of various agencies change as the agencies influence the siting process over time. This issue could easily be explored in future longitudinal investigations through serial questionnaire responses supplemented with a chronicle of the activities of each agency between each questionnaire administration.

Hypothesis 1 indicates that much greater attention needs to be paid to advertising and information dissemination techniques, particularly among the hard-to-inform groups. This notion means that planners need to pay attention to information dissemination by asking questions about the routes of communication, its frequency, and who should do the communicating. The evidence underlying hypotheses 2 and 3 shows that different subgroups may need to be reached in different ways and with different kinds of information, while hypothesis 4 highlights the relevance of careful selection of the agent of communication.

REFERENCES

Covello, V. T. 1983. The Perception of Technological Risks: A Literature Review. *Technological Forecasting and Social Change* 23:285–97.

Greater Vancouver Regional District. 1986. Burnaby EFW Project of the Greater Vancouver Regional District. In *Report on Thermal Reduction of Refuse.* Vancouver.

———. 1990. *Sewerage and Drainage District First Annual Report.* Vancouver.

———. Air Quality Management Office. 1991. Personal communication to author.

Ostry, A., C. Hertzman, and K. Teschke. 1993a. Community Risk Perception: A Case Study in a Rural Community Hosting a Waste Site Used by a Large Municipality. *Canadian Journal of Public Health* 84, no. 6:415–18.

———. 1993b. Risk Perception Differences in a Community with a Municipal Solid Waste Incinerator. *Canadian Journal of Public Health* 84, no. 5:321–24.

———. 1994. Community Risk Perception and Waste Management: A Comparison of Three Communities. *Archives of Environmental Health,* no. 2:95–102.

Taylor S. M., et al. 1991. Psychosocial Impacts in Populations Exposed to Solid Waste Facilities. *Social Sciences and Medicine* 33, no. 4:441–47.

Zeiss C. 1984. Social Cost Assessment of the Proposed Waste Disposal Facilities of the Lower Mainland Refuse Project of British Columbia. Vancouver. Unpublished.

———. 1988. Siting Waste Disposal Facilities in Host Communities: Impacts and Acceptance. Ph.D. diss., University of British Columbia.

Worrying about Waste: Diagnosis and Prescription

SUSAN J. ELLIOTT, S. MARTIN TAYLOR

INTRODUCTION

Public concern over the possible toxic effects of exposure to environmental contaminants is high in the wake of such highly publicized events as Love Canal, Three Mile Island, Bhopal, Chernobyl, and St. Basile-le-Grande. For many individuals, these concerns are reinforced and given more immediacy when events occur within their own community. Recent disclosures regarding PCB disposal sites, radon gas emissions, and electromagnetic radiation also contribute to uncertainty and anxiety over risks to health, livelihood, and property values. Equally, the prospect of the future location of new waste treatment, storage, and disposal facilities compounds public worries and generates community reaction, often referred to as NIMBY (Not in My Backyard).

Thus, waste facility siting commands a high priority on the scientific and public policy agendas. The demand for waste processing, storage, and disposal facilities continues to outstrip supply. Further, despite general agreement that facilities are necessary, the process of facility siting is often impeded by community opposition, where siting is vehemently opposed when the risk becomes localized, often with cries of "not in my backyard." In this context, the public policy response often takes the form of DAD (Decide, Announce, Defend), where community residents are involved post facto in the siting process through information dissemination, public meetings, and environmental-assessment hearings. Community fears, anxieties, and worries about both real and perceived impacts, however, are not directly and fully addressed in these informational contexts for a number of reasons. First, the nature of the siting process itself tends to put the community on the defensive and tarnishes any remaining trust residents may have had in government agencies/officials. Second, there is a glaring lack of evidence to address the most often asked questions about the potential impacts of a particular facility siting on property values, environmental quality, and health. Third, the evidence that is available is often equivocal. Thus, answering the community's questions about the nature of (potential) impacts of living in close proximity to a particular facility becomes even more difficult, if not impossible.

Recent Canadian examples serve to illustrate NIMBY's impact on the siting process. In Ontario, a crown corporation (Ontario Waste Management Corporation [OWMC]) was created in the early 1970s to site a hazardous industrial waste treatment and disposal facility. After twenty years and the expenditure of several hundred million dollars, the decision on the final site was negative. Much of the delay in this case is due to the participatory nature of the siting process (e.g., lengthy public meetings, provision of intervenor funding) in anticipation of, and as a response to, substantial community opposition. In British Columbia, a plan to locate a hazardous waste treatment plant in the Cache Creek–Ashcroft area in the late 1980s failed, in large part, because of substantial community opposition (see Castle and Munton, this volume). A subsequent, short-lived crown corporation also failed in its attempt to site a facility. As a result, British Columbia is now working toward a plan to ship some hazardous waste to the Swan Hills facility, 250 kilometers northwest of Edmonton, Alberta (Watts 1993). In return, Alberta would ship waste battery acid to a location in the lower mainland of British Columbia for treatment and storage. The seeds of success for the Swan Hills facility are said to be found in the "voluntariness" of the siting process, which, in essence, eclipsed any NIMBY reaction.

Public opposition to waste disposal sites is an example of negative reaction to noxious facilities in general. It follows that past research on this topic can inform our understanding of the NIMBY response in this particular case. The literature is instructive in terms of both theoretical frameworks and empirical analyses. With regard to the former, public facility location theory (Dear 1978) and attitude theory (Ajzen and Fishbein 1980) are particularly relevant. In relation to empirical research, studies of the siting of controversial social service facilities, such as mental-health facilities, provide guidance on issues of study design and evidence on the prevalence and determinants of impacts (Taylor 1987).

As with noxious facilities generally, concerns about waste facilities vary according to a number of factors, including the type of facility (e.g., hazardous vs. nonhazardous) and the community context (e.g., urban vs. rural). Moreover, the range of impacts spans environmental (e.g., impacts on quality of air and water), economic (e.g., property values, agricultural livelihood), social (e.g., community stigma), and health (e.g., physical and psychosocial) concerns. With respect to the last, past research on the human health consequences of exposure to environmental contaminants has focused primarily on possible physical effects (e.g., cancer, adverse reproductive outcomes). More recently, psychosocial effects (e.g., worry, anxiety, emotional distress) have been the subject of increasing attention and are the focus of this chapter.

Psychosocial effects have been defined as a complex of distress, dysfunction, and disability manifested in a wide range of psychological, social, and behavioral outcomes, as a consequence of actual *or perceived* environmental contamination (Taylor et al. 1991). They have a complex etiology (Elliott et al. 1993), may occur in conjunction with—or independent of—measurable physical effects (Bertazzi et al. 1989), and may be associated with actual *or perceived* exposure (e.g., in advance of the siting of a waste facility) (Elliott 1992). They occur at various levels of social organization: individual, social-network, and community levels. At the individual level, they include emotional (e.g., worry, anxiety, guilt, loss/lack of control), behavioral (e.g., task performance, help seeking, information seeking), and somatic (e.g., headache, fatigue, depression) effects. Effects at the social network level include family disruption and social isolation as well as (potentially) positive outcomes such as the adoption of cooperative coping strategies, which may result in greater social cohesion in response to environmental stress (e.g., a proposed waste facility). Finally, at the community level, the effects of environmental contamination often include stigmatization (e.g., a "toxic community" label) and dislocation, but again, positive effects (e.g., community empowerment and enablement) may also occur.

Unlike the other chapters in this volume, the empirical focus of this chapter is on the prevalence and determinants of psychosocial effects of living in close proximity to three *non*hazardous solid waste disposal facilities in southern Ontario (Fig. 1). Two issues arise: the comparability of reactions to hazardous and nonhazardous facilities, and the implications of post-siting effects for pre-siting issues. Regarding the first, the literature indicates important parallels, while recognizing that impacts vary by type of facility (Couch and Kroll-Smith 1994; Hadden 1991; Portney 1991). For example, there is a great deal of uncertainty associated with the siting of any type of noxious facility (Greenberg 1994; Portney 1991), including both hazardous and nonhazardous waste facilities (Hadden 1991; Mazmanian and Morell 1992; Rabe 1991), that results in substantial concern for impacted populations (Couch and Kroll-Smith 1994; Greenberg 1987, 1994). Regarding the second issue, while two of the facilities reported on here have been in operation for several years, these analyses make important contributions to the siting literature by examining changing attitudes over time and the role of community context in understanding psychosocial effects. One of the three sites had only recently been ap-

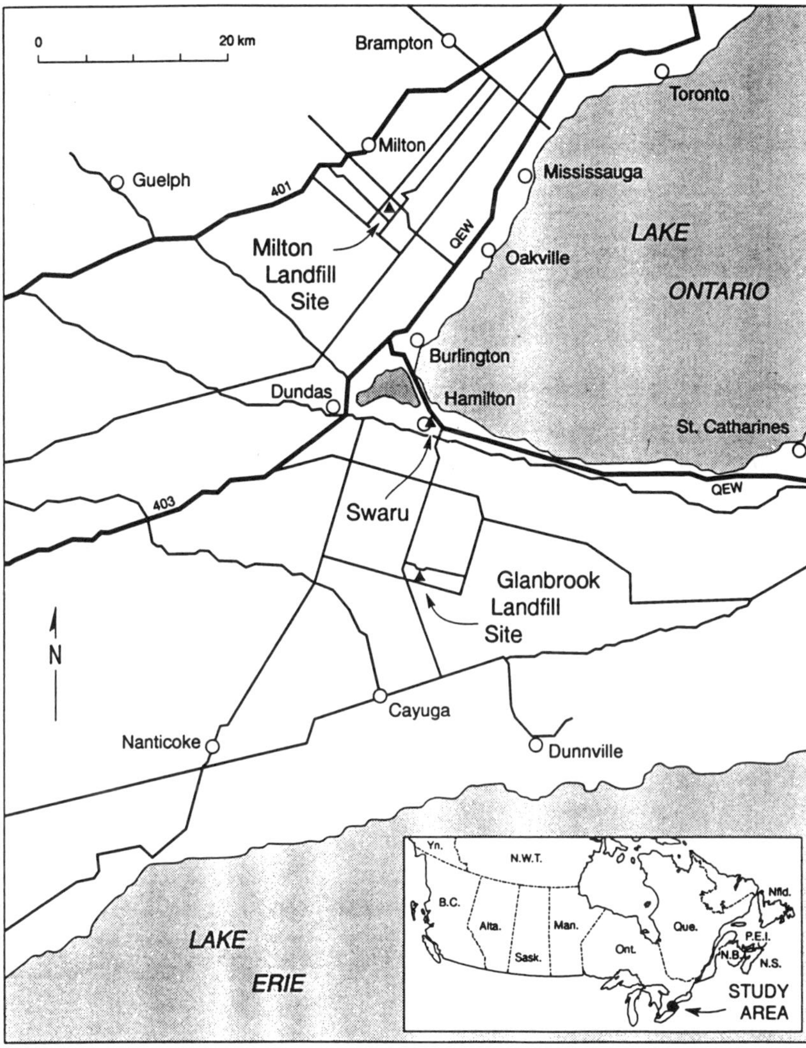

Figure 1. Location of study sites

proved at the time the initial research was conducted, but additional data were collected once it was constructed and shortly after it began accepting waste. These additional data allow analysis of changes in attitude toward the site over time. In addition, movers into and out of the area were interviewed about their attitudes toward the site and asked to report the influence of the site on their decision to move. All analyses contribute to an integrated explanation of the determinants of psychosocial effects of exposure to waste disposal facilities and the development of strategies to reduce their adverse impacts on human health and well-being.

As a prelude to examining the findings of the Ontario study, the existing evidence on the psychosocial effects of environmental contamination is briefly reviewed in the following section.

PSYCHOSOCIAL EFFECTS: A REVIEW OF THE EVIDENCE

Public facility location theory is a useful framework for research on the impacts of facility siting. It emphasizes the importance of the spatial externalities associated with the siting of facilities (i.e., the impacts of the facility on the surrounding area within the "externality field"). Externalities may affect both users and nonusers of a facility. The latter are sometimes referred to as neighborhood-associated externalities and are particularly relevant for understanding public reaction to facility siting. In the case of noxious facilities, such as waste facilities, neighborhood-associated externalities are typically perceived as negative by local residents and are fundamental in locational conflict over facility siting. Two sets of (perceived) external effects have been noted: tangible effects, which are based upon clearly recognizable, usually quantifiable, impacts of the facility (e.g., property value decline, increased traffic); and intangible effects, which refer to a wide range of nonquantitative impacts (e.g., health and quality of life concerns, neighborhood image). In the context of waste facilities, there is a tendency for the public to overstate health concerns, leaving concerns about other intangibles (e.g., neighborhood image) underrepresented (see Zeiss, this volume). The explanation may be related to the high priority our society places on health and, therefore, to the legitimacy, seriousness, and attention paid to health-related concerns.

Public-facility location theory is also useful in its recognition of the social context of siting decisions. As Dear and Taylor (1982, 16) argue, "Spatial outcomes must seek their explanations in the wider social context; there is a direct correlation between social policy and spatial outcome." The relevance of this argument is clear in light of current conflict over the siting of new waste disposal facilities in the greater Toronto area—an issue rooted in the provincial government's policy, which includes a moratorium on waste incineration, requires local municipalities to handle their own waste, and prohibits the transport of waste to other communities outside the region.

Public facility location theory is linked to attitude theory stemming from the central role of perceptions of external effects. As Dear, Fincher, and Currie (1977, 139) state, "Community perceptions of external effects determine the extent and intensity of opposition to a program and facility." In this field of investigation, terms can be used quite loosely, with the potential for ambiguity in conceptualization, measurement, and inference. Fortunately, research on public attitudes toward noxious facilities has been able to benefit from the work of Fishbein and others (Fishbein and Ajzen 1975; Ajzen and Fishbein 1980) and therefore to articulate more precisely the links between beliefs, norms, attitudes, intentions, and behavior. The last ten years have produced several studies of public attitudes toward facilities, with the result that general conclusions are now emerging with respect to the prevalence of negative and positive sentiments and their correlates at both the individual and community levels of analysis (Taylor 1987).

Studies of public attitudes toward community mental health facilities are particularly instructive regarding the determinants of reactions to facility siting. Dear and Taylor (1982) identify the importance of behavioral and normative beliefs connected with facility acceptance or rejection—specifically, beliefs related to the facility itself, the users of the facility (in their case, the mentally ill), and the local community. In the context of waste facility siting, the first and third are directly relevant. Beliefs about the facility include beliefs about impacts on the community (e.g., property values, community image and character, traffic, and noise) and on local residents (e.g., health and well-being). Beliefs about the community are based on perceptions of the suitability of the community as a host for a facility. This can be disaggregated into perceptions of the physical and social characteristics of the community and the extent to which each is congruent with the proposed facility. The concept of congruence corresponds to the notion of "fit" between form and context in the architecture and design literature (Alexander 1964).

In analyzing attitude formation and the relationships between attitudes, intentions, and behavior, it is important to go beyond antecedent beliefs and identify the external variables that in turn antecede them. In this regard, the utility of the Fishbein-Ajzen theory of reasoned action as a framework for explanation can be criticized for its de-emphasis of the role of external variables in favor of the more proximal and predictive cognitive determinants of action. Here again, research on attitudes toward community mental-health facilities is instructive in showing at the individual and community level the direct and indirect effects of external variables (e.g., facility, community, and personal characteristics) on public reactions to facilities (Hall and Taylor 1983; Taylor et al. 1984). Our recent research on the psychosocial effects of waste disposal facilities adopts this broader approach in modeling the determinants of psychosocial outcomes (Elliott et al. 1993).

A fundamental limitation of past research on psychosocial effects is the lack of conceptual clarity and the absence of well-defined conceptual frameworks. Obviously, what is measured, how it is measured, and how results are interpreted are all determined by the way concepts are defined in and of themselves as well as in relation to each other. Often, in the psychosocial literature, conceptual frameworks are left implicit, and the reader is left to grapple with multidefinitional terms such as "stress" and "depression." With a couple of exceptions in the more recent qualitative research, the majority of studies fail to make explicit the structure and properties of concepts employed (Elliott 1992; Taylor et al. 1989).

A useful theoretical framework for research on the psychosocial effects of environmental contamination comes from the literature on environmental stress and coping (Evans 1982; Evans and Cohen 1987; Lazarus and Folkman 1984). Baum, Fleming, and Singer (1985, 186) define environmental stress as "a process by which environmental events threaten, harm or challenge an organism's existence or well-being and by which the organism responds to this threat." This definition implies at least three key considerations (Evans and Cohen 1987):

1) The individual's *perception* of environmental demands and personal coping resources is the critical variable in determining the nature of the stress response (i.e., the notion of "fit").
2) Stress situations are not uniformly aversive, because personal and social mediators can ameliorate or enhance their effects.
3) Stressors may affect the individual in many ways other than the physiological, including symptoms, negative affect and interpersonal behaviors, and deficits in task performance.

A useful psychological model of response to environmental stress is that provided by Lazarus and Folkman (1984; see also Lazarus 1970; Folkman and Lazarus 1980, 1988). It contends that response to environmental stress (e.g., a proposed waste facility) is divided into two stages: *primary* appraisal, whereby the individual perceives an environmental stressor as a threat, a harm, or a challenge; and *secondary* appraisal, whereby one of two coping strategies is selected—(1) problem-focused coping (e.g., joining a citizen's action group) or (2) emotion-focused coping (e.g., adjusting attitudes toward the stressor). Feedback is provided through the process of *reappraisal,* where the stages of primary and secondary appraisal are repeated as the perception of the stressor or available coping resources change over time. It follows that the coping may shift between problem- and emotion-focused responses (Folkman and Lazarus 1980).

As indicated above, stress situations are mediated by personal, social, and other factors (Evans and Cohen 1987). The occurrence of environmental stress, the experience of psychosocial effects, and the choice of coping response are dependent upon four types of mediating factors. The first of these is *the characteristics of the environmental stressor.* For example, an invisible stressor (e.g., a proposed waste facility) may in fact lead to greater psychosocial impacts than a visible exposure (e.g., an existing landfill) (Vyner 1988). Other sensory stimuli may also lead to greater impacts (e.g., odors) (Shusterman et al. 1991). Prior experience with a stressor will usually reduce the level of impacts, while the severity and acuteness of the exposure may lead to increased impacts (Sims and Baumann 1983). Perceived economic necessity of an environmental exposure will also influence the degree of psychosocial impacts experienced; communities are much more tolerant of, for example, an industrial exposure if it provides many jobs for the community (Evans and Jacobs 1982).

The second mediating dimension that influences the experience of psychosocial impacts involves *the characteristics of the individual.* For example, an individual with a low degree of self-esteem/mastery over situations may experience greater impacts from an exposure than an individual with the opposite characteristics (Pearlin and Schooler 1978). In addition, individuals who feel they have some sense of control over exposure events may experience fewer impacts than individuals who feel less of a sense of control (Evans and Jacobs 1982; Lefcourt 1982). A final example is the access to/control over material resources; individuals with greater access tend to experience less stress and may tend toward taking action in response to a stressor more than does an individual with less access (Pearlin and Schooler 1978).

Characteristics of the social network have been shown to mediate stressful life events (Berkman and Syme 1979; Cassel 1976; Cobb 1976; Fleming et al. 1982). That is, when faced with stressful life events, people cope better when they can derive support from social relationships or social networks: "The encouragement, opinion validation, and reassurance that people get from friends and family influence their response to stress and somehow make them more resistant to its effects" (Fleming et al. 1982, 14). Indeed, previous research has demonstrated that social ties and relationships play a critical role in the determination of health status, where individuals with more social contacts have been shown to exhibit lower mortality rates than those with fewer social contacts (Berkman and Syme 1979; Cassel 1976; House, Landis, and Umberson 1988). It follows that psychosocial impacts of exposure may be lessened if one has an available and supportive social network (Edelstein 1988; Eyles et al. 1990; Fleming et al. 1982; Sorensen et al. 1987). This was shown to be the case at Three Mile Island, where higher levels of social support were associated with fewer psychological and behavioral symptoms of stress (Fleming et al. 1982). In addition, Edelstein (1988) found that in communities such as Love Canal, most residents were part of a relational web of family, friends, and coworkers and that these relationships influenced not only the extent of environmental stress experienced but also the coping strategy. Further, individuals who are very active in their communities also experience less stress from exposures and are probably more likely to practice problem-focused coping, as opposed to emotion-focused coping (Logue, Melick, and Hansen 1981; Perry 1983).

Finally, *characteristics of the wider community system* can also influence the experience of psychosocial impacts as well as the selection of a coping response. These characteristics include, for example, the nature of the political and economic systems and their influence on an individual's access to institutions. The portrayal of environmental exposure events in the media is also a major factor (Eyles et al. 1990; Roht et al. 1985; Sandman et al. 1987; Sims and Baumann 1983). The media is typically the public's main source of information about an environmental exposure. Oftentimes exposure events are sensationalized in the media, so that they are made to seem more severe than they truly are. Thus, the stage of primary appraisal is made even more difficult. In addition, communities are often provided with misinformation or conflicting information that makes decision making (for example, the decision whether to evacuate) an even more difficult process than it already is (Eyles et al. 1990; Sandman et al. 1987). This categorization of mediating factors provides the basis for the conceptual framework (Fig. 2) that guided the research reported in the next section.

Empirical research on the psychosocial effects of environmental contamination reveal a mix of significant and nonsignificant associations across a broad range of exposure-outcome combinations (Elliott 1992; Taylor et al. 1989). While most studies address direct environ-

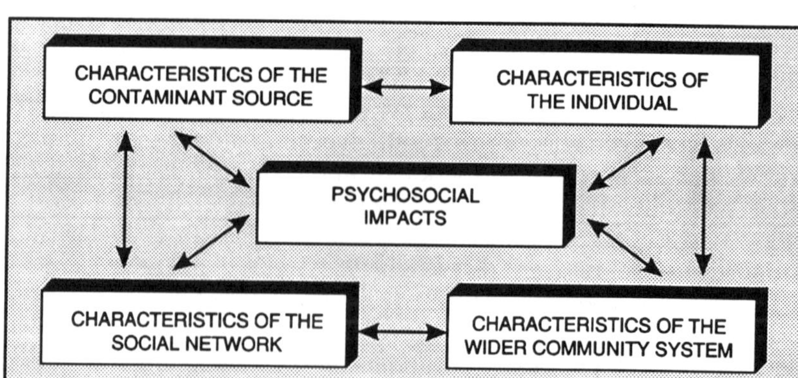

Figure 2. Conceptual framework

mental exposure-outcome relationships, many also examine the role of mediating factors (e.g., the coping abilities of exposed individuals) (Bachrach and Zautra 1985). The vast majority of studies have been conducted at the individual level, providing little, if any, evidence of the community context within which environmental stressors are situated or of impacts at other levels of social organization (e.g., the family or social network levels). With respect to study design, the majority are cross-sectional, thus providing weak evidence for causation.

The most frequently examined source of contamination is nuclear power, especially related to the Three Mile Island accident. Other sources examined, in descending order of occurrence, are ambient exposures (e.g., lead and smog), hazardous waste disposal, heavy industry, solid waste landfills, and power transmission lines. The health outcomes addressed can be categorized under five headings. About two-thirds of the studies discuss *diagnosed disease,* which includes stress, psychiatric somatoform outcomes, and affective disorders such as depression. Indicators of *objectively measured dysfunction* include both physical and psychological functions, assessed using clinical examination, physiological measures, laboratory tests, and psychological/behavioral measures. Psychological functions are most often measured using prevalidated existing instruments such as the Beck Depression Inventory. *Illness stigmata* include physical, emotional, or cognitive complaints. Anxiety, fear, depression, helplessness, and loss of control are the most commonly assessed emotional complaints. *Illness behavior* refers to behavior resulting from the presence (or the belief in the presence) of disease. Only about a third of the studies examine this outcome. *Disability* is addressed in very few studies. Outcomes assessed are related to vocational and recreational activities, activities of daily living, domestic and interpersonal relations, and social activities.

Overall, therefore, the literature provides equivocal evidence for relationships between environmental exposure and psychosocial impacts across a wide range of exposure-outcome relationships. With the exception of a small number of well-conducted studies of the (perceived) health impacts of exposure to (hazardous) waste facilities, few studies deal directly with issues of siting, and the differences among them preclude a generalization of findings. The majority have, however, found some relationship between exposure to existing or proposed facilities and psychosocial effects, including mood symptoms (Hertzman et al. 1987); demoralization (Horowitz and Stefanko 1989); symptom reporting (Lipscomb et al. 1991); stress, anxiety, and poorer levels of perceived health status (Dunne et al. 1990); and perceived health risks from a proposed solid waste incinerator (Wiedemann, Schutz, and Peters 1991) and a hazardous waste treatment facility (Portney 1991).

THE ONTARIO STUDY

The Ontario study examined the psychosocial effects of living in close proximity to nonhazardous waste disposal facilities (Fig. 1). Three facilities in southern Ontario were included. The Solid Waste Reduction Unit (SWARU) is an incinerator for nonhazardous municipal solid waste from the City of Hamilton, operated for the Region of Hamilton-Wentworth by a private corporation twenty-four hours per day, five days per week. It has been operating since the early 1970s. The Glanbrook landfill has been in operation for approximately twelve years and accepts nonhazardous municipal solid waste from the Hamilton-Wentworth Region as well as ash from the incineration process at SWARU. It is a 400-acre (162-hectare) site, of which approximately 250 acres (101 hectares) are active for landfilling. The third facility is a new landfill site in Milton. Approval for construction was given in June 1990, when an appeal to the Supreme Court of Ontario by a local residents' group was denied. The site is approximately 400 acres (162 hectares), has a life expectancy of thirty years, and will take municipal solid waste from the Halton Region. The site began operating in November 1992.

These sites were judged to be similar on the basis of type of exposure (chronic), source of exposure (point), and nature of contaminants (nonhazardous municipal solid waste). It was recognized, however, that they differed sufficiently to limit both comparison between sites and the validity of across-site analyses. Therefore, a parallel case study design was used in which each site formed an independent study. The primary analyses reported below are therefore by site. Selected combined analyses are also reported to examine variation in the pattern of effects between sites.

The complex nature of psychosocial effects requires an integrated set of methods to address effects that may be real or perceived, span a range of outcomes, and occur across various levels of social organization. Thus, a suite of quantitative and qualitative methods was employed. First, a survey was administered by telephone to a stratified random sample of approximately 250 residents living in four distance zones around each of the three sites. It was introduced as a general quality-of-life survey. Instrument development, content, and format are discussed elsewhere (Elliott 1992; Taylor et al. 1991). Essentially, the instrument consisted of five sections: (1) attitudes toward the neighborhood or home area, including factors liked and disliked and ratings of satisfaction; (2) social networks and their possible mediating influence on psychosocial effects; (3) measures of general psychosocial health and well-being, i.e., the twenty-item version of the General Health Questionnaire (GHQ-20) (Goldberg 1972), the somatic-symptom checklist from the Symptom Check List-90 (SCL-90) (Derogatis, Lipman, and Covi 1973), and a subset of items from the Critical Life Events Scale (Holmes and Rahe 1967); (4) a combination of closed- and open-ended items to determine residents' levels of awareness, knowledge, concern, and action regarding the site or more general environmental problems, with emphasis on health concerns; and (5) standard sociodemographic variables, as a check on the representativeness of the sample and as potential mediators of psychosocial effects. At the three sites 696 interviews were completed, representing response rates of 57% (SWARU), 84% (Glanbrook), and 75% (Milton). Interviews averaged twenty-seven minutes in length.

The samples included 53% females and differed little on reported education level. Respondents at Milton tended to be older than at the other two sites. Overall, respondents at Glanbrook and Milton were quite similar and contrasted in several respects with respondents at SWARU. This is not surprising given the rural/urban contrast. Compared to respondents at SWARU, respondents at Glanbrook and Milton were more likely to have total household incomes over $30,000, have a partner, have children in the home less than five years old, be employed full-time, have a higher number of persons per household, speak English at home, own their own home, and have lived longer in both the current residence and the local area.

The timing of the research facilitated a longitudinal component at the Milton site. The initial data were collected immediately after the site was approved (summer 1990). Two years later (summer 1992), when landfill construction was nearing completion, a follow-up survey was conducted. There was a second follow-up survey in January 1993, two months after the site began accepting waste. The follow-up surveys were conducted with different subgroups of the same respondents interviewed in the 1990 baseline study, and used the same questionnaire. This design yielded two cohorts for a within-person analysis of reappraisal of the environmental stressor over time. The longitudinal study at Milton also included telephone interviews with households that had moved into or out of the area subsequent to the baseline survey. A checklist of topics was used to determine awareness and concerns about the site as well as its influence on the decision to move.

In the overall research design, qualitative methods were used to complement and extend the survey data with the purposes of examining the wider community context within which psychosocial effects occur and the role of social and cultural mediating factors. A combination of qualitative methods (i.e., depth interviews [thirteen per site], focus and discussion groups) was used to reconstruct the ways in which subgroups of the survey respondents experienced the waste facilities and their impacts in the context of their everyday lives. While the study results presented here are based primarily on the survey data, they are contextualized by the findings from the qualitative analysis (for more details, see Eyles et al. 1993). All analyses contribute to an integrated explanation of the determinants of the psychosocial effects of exposure to waste disposal facilities, and to the development of strategies to reduce their adverse impacts on human health and well-being.

Results from the analysis of the baseline survey data (1990) are reported for site-specific psychosocial effects: *concern, health concern,* and *actions*. In addition, analyses combining data from all three sites are reported for outcomes *concern* and *actions* to illustrate the influence of site on psychosocial effects. Throughout these analyses, attempts were made to model the exposure, individual, social network, and health status determinants of psychosocial effects. These outcomes are related to the processes of primary (concern, health concern) and secondary (actions) appraisal of an environmental stressor. Longitudinal results for the Milton site address the process of reappraisal by documenting changes in outcome levels reported over time. This is supplemented by the results of a Milton movers study, which documents awareness and concerns of those who moved into or out of the area subsequent to the baseline epidemiologic survey.

Overall, the percentage of respondents in the baseline survey reporting *concern* at SWARU was low (28%) relative to the other two sites (Milton, 74%; Glanbrook, 67%). There was some evidence at each site of a gradient of concern, with the highest levels of reporting within the closest zones. Further, the concern most frequently mentioned varied by site (SWARU, air pollution; Glanbrook, site operation; Milton, water quality/pollution).

Particular concerns reported via the depth interviews were similar and again indicate the importance of site and situational factors. At SWARU, where air quality (specifically stack emissions and smells) was a major concern, pollution seemed to be accepted as a fact of living in a highly industrialized area characterized by heavy traffic flow; in fact, the discussion group participants identified SWARU as simply "a drop in the bucket." SWARU, however, was singled out as an identifiable contributor to the poor air quality of the area; that is, respondents complained specifically about a black "soot/ash," attributed to SWARU emissions, that settles on cars, outdoor furniture, laundry, and so on. Key informants noted (fall 1990), however, that complaints about SWARU had declined since a recent retrofit.

At Glanbrook, the concern about site operation was related to the nature of wastes being disposed of on-site as well as other safety and security measures. This is not surprising given that site access is closed to all but Regional employees. Restricted access makes area residents naturally suspicious about the nature of activity at the site, particularly for those who recall previous contamination events at another site in the local area (Hertzman et al. 1987). The most important theme emerging from the media analysis of Glanbrook area newspapers (analyzed for September 1, 1988, to October 1, 1991) was site monitoring (Taylor et al. 1993). The concerns most frequently mentioned in the depth interviews at this site were traffic and seagulls, although, for those living nearest the site, reduced property values, blowing debris, odors, and noise were of greater concern even though they felt their properties were far enough away from the site not to be seriously impacted. The depth interview data also indicated that residents placed a great deal of confidence in the authorities that monitored and operated the site. This contrasts quite sharply with the survey data, as well with as the Milton experience, where residents expressed extreme distrust in the opinions of "experts" and authorities who have assured them there is nothing to worry about. The same dichotomy of opinion was documented in the Glanbrook discussion group data. For some, there was a sense that "things are being hidden" from area residents. In fact, two of the men in the group wanted to be able to enter the site unannounced to see for themselves what types of wastes are being dumped. Despite these suspicions, several people in the discussion group expressed confidence in the local landfill committee.

The situation was quite different at Milton, where the concern reported most frequently in the survey was water pollution. Given that the vast majority of respondents rely on private wells for their domestic water supply, this finding is not surprising. In fact, in addition to traffic and property values, it was also a major concern raised in the depth interviews and discussion groups. Fears were expressed about increased health risks as a result of problems with leaching and runoff. These fears were often exacerbated by the knowledge that some harmful substances will unavoidably enter the site and by a lack of trust in the opinions of the "experts" or authorities who have assured them that there is nothing to worry about. This lack of trust seems to stem primarily from the perception that experts have been proven wrong in the past and that strict monitoring procedures will be relaxed over time. Again, these thoughts were reiterated by members of the discussion groups at Milton. Related to this were feelings of resignation: the dump was going to be there regardless, and what could residents possibly do about it? In the depth interviews, these

TABLE 1. Profile Characteristics of Respondents More Likely to Report Site-Related Concerns

SWARU	*Glanbrook*	*Milton*
satisfied with current health	living closer to site	older
GHQ-20 score above cut point	more close friends	living closer to site
union/commercial-group member	not satisfied with social activities	SCL-90 score above cut point
owner living close to site	environmental-group member	more past community involvement
	fewer helping activities with neighbors than before	younger resident with less past community involvement
	more helping activities among residents living longer at current address	

statements were often combined with expressions of frustration and anger.

At all three sites, residents in the depth interviews reported latent concern about the dumping of toxic wastes (e.g., PCBs), which were perceived as a much greater and more immediate risk to human health.

The importance of community context for understanding differences in site impacts can also be seen in profiles of respondents at each site more likely to report site-related concerns (table 1). There are several points to be made from these profiles, which were generated using logistic regression analysis (see Elliott et al. 1993 for details of the statistical analyses that produced this profile and the profiles described in subsequent tables). The first is to acknowledge the heterogeneity of the determinants of concern between sites. There is no single "profile of concern" across the three study sites. Differences in site and population characteristics influence psychosocial effects. Second, there is no simple cause-and-effect relationship between exposure and concern. While distance from the site was an important determinant at all sites, several other characteristics related to the individual, social network, and health status also contribute to the profiles. Third, the evidence confirms that psychosocial effects occur whether exposure is actual or perceived. The highest levels of concern reported were at Milton, a site only recently approved at the time these data were collected. Finally, social network factors were stronger predictors of concern in the rural study areas (Glanbrook and Milton)

TABLE 2. Characteristics of Respondents More Likely to Report Health-Related Site Concerns

SWARU	*Glanbrook*	*Milton*
environmental-group member	living farther from site	living closer to site
household income $60–70,000	SCL-90 score above cut point	more past community involvement
	environmental-group member	
	woman who had not experienced a recent serious illness	
	SCL-90 score *below* cut point among residents living closer to the site	

than in the urban area (SWARU). For example, Glanbrook residents with more close friends as well as neighborhood connections and longer lengths of residence were more likely to report concerns about the landfill in their area.

The frequency of reporting *health concerns* at each site shows the same pattern as concern in general: SWARU (21%), Glanbrook (49%) and Milton (52%). The depth interview data confirm that few people at SWARU had concerns about health, although those who did mentioned that they or one of their family members suffered from asthma, recurring bronchitis, or other sorts of respiratory problems. These individuals firmly believe that emissions from the SWARU stack contribute to the poor quality of the air and thus have an impact on their health. Several other interviewees stated that although they have not experienced any immediate somatic effects, they worry about the possible long-term effects of exposure to "toxins" from SWARU and other local industries. Despite the high levels of health concerns reported in the survey, depth interview respondents at Glanbrook were more concerned about the health of their livestock than their own health. A possible explanation is that Glanbrook residents do not rely on private wells for their domestic water and so, in contrast to Milton residents, are not as concerned about leachate or runoff except as it might affect the watering of livestock.

The profiles of respondents more likely to report health-related site concerns again show distinct differences across sites (table 2). Distance from the site was a significant factor at two of the sites.

TABLE 3. Characteristics of Respondents More Likely to Take Actions in Response to Site Concerns

SWARU *(1 or more actions)*	*Glanbrook* *(2 or more actions)*	*Milton* *(2 or more actions)*
living closer to site	living closer to site	living closer to site
owners vs. renters	has a spouse/partner	longtime resident at current address
not employed full-time	no children < 5 years	some university education
not satisfied with area	longtime resident of area	GHQ-20 score above cut point
belongs to several community groups	greater past community involvement	satisfied with social activities
	belongs to church/ religious group	belongs to children's group
	belongs to both a social/recreational and an environmental group	

Although the direction of the relationship at Glanbrook may at first seem counterintuitive (farther rather than closer respondents reporting more health-related site concerns), it may simply be an indication that respondents living closer to the site were worried about other things (e.g., property values, blowing debris, seagulls). One common feature of these profiles is the consistently important role of community involvement factors (e.g., environmental group membership, greater past community involvement).

The frequency of reporting site-related *actions* was: SWARU (21%), Glanbrook (60%), and Milton (71%). In addition, there was clear evidence of a gradient of actions with distance from each of the three sites, with the highest levels of reporting in the closest zones. Documenting the actions taken in response to site concerns moves the analysis from the stage of primary appraisal to the stage of secondary appraisal, in terms of problem-focused coping responses. The profiles of action-related characteristics again differ across the three study sites (Table 3). Generally, three points can be made from these profiles. First, given the low level of reporting of the outcome variable at SWARU relative to the other two sites, different cut points of activity (one vs. two or more actions) were used in the analysis of determinants; one action could be incidental, two or more actions are more likely deliberate. Second, closeness to the site emerged as an important

TABLE 4. Across-Site Profiles of Respondent Characteristics for Outcomes Concern and Actions (One or More)

Concern	*Actions*
living closer to site	living closer to site
Milton resident	Milton resident
longtime resident of area	longtime resident of area
SCL-90 score above cut point	SCL-90 score above cut point
employed full-time	greater past community involvement
recently lost spouse/partner through death	belongs to church/religious group
talks with neighbors very often	nobody to count on for help when in need
rates neighbors as (very) friendly	belongs to environmental group
participates in fewer helping activities with neighbours	moderate self-perception of past community involvement
belongs to several community groups	Milton resident with greater past community involvement
older resident with greater past community involvement	
Milton resident with greater past community involvement	

characteristic in each of the action-related profiles. Finally, the profiles are dominated by social-network characteristics, which indicate an individual's connectedness to both primary (friends and family) and secondary (neighborhood and community) social networks as well as their wider community system. This finding supports the use of a socioecological conceptual framework when investigating psychosocial effects.

Although the parallel case study design made within-site analysis the primary focus, there was good reason to explore the direct effects of site on psychosocial outcomes. This is consistent with the conceptual framework, which includes characteristics of the contaminant source as determinants of psychosocial outcomes. To this end, profiles were developed for *concern* and *actions (one or more)* with data from all three sites simultaneously, including site as a factor (Table 4). They indicate the importance of site-related characteristics, given that site emerged as a profile characteristic in both models. At the same time, however, site did not dominate in either profile. The number and diversity of significant variables are further evidence of the socioecological determination of psychosocial effects and the importance of community context. Given that the study design focuses on an independent analysis of three separate sites and situations, any across-site comparison of outcomes must be made with caution. There are, however, some similarities in the profiles. Respondents more likely to report both concern and action were longtime Milton residents living closer to the site with SCL-90 scores above the cut point as well as relatively high levels of past and present community involvement. Finally, a point that clearly emerges again from these particular profiles is the role and importance of social-network factors.

What are not apparent in these quantitative findings are the effects on individual lives, families, and community. Substantively, the findings from the quantitative and qualitative approaches used in this research are similar (Eyles et al. 1993) but tended to be expressed more in terms of anxieties, worries, and impacts in the qualitative data than in the responses to the survey questions. For example, in one of the Milton discussion groups, participants noted that while the community fight against the siting of the landfill did tend to bring the community together, it also placed a great deal of stress on individuals and families. This stress and anxiety were thought to have contributed to the strain on one couple's marriage, which soon after ended in divorce, and to have hastened one woman's death from cancer. This illustrates the need for enhancing quantitative findings with qualitative data collected via other approaches to the research questions as well as the importance of placing psychosocial effects within a broad community context.

As previously noted, the timing of the research project facilitated a longitudinal analysis of changes over time in the prevalence of individual-level psychosocial effects and actions at the Milton site. The sizes of the follow-up samples are small (i.e., 70 in June 1992, immediately prior to site opening, and 44 in January 1993, two months after site operation began, compared with the initial 187), but this presents less of a problem, as the data come from the same respondents, therefore allowing within-person analysis of change. The survey instrument was virtually identical to the original and, overall, 61% of the original respondents were reinterviewed. The samples from the first and the second follow-up surveys were representative of the original sample and well matched on most characteristics, but there was a larger percentage of females in the second group, and mean age was higher in the first group, as was the median length of residence.

With respect to site- and situation-specific effects, the data on dislikes about the area provided a measure of the frequency with which respondents volunteered the site as a concern. The frequency of mentioning the site as a dislike increased significantly for the first group (9% vs. 38%) but remained approximately the same for the second group (14% vs. 11%). In contrast, the overall prevalence of concern decreased over time for both sample groups (81% vs. 71%; 71% vs. 61%). There was little change in the types of concern reported, although they were focused on water quality/quantity, traffic, and proximity of the landfill to respondents' homes. These results are not surprising given the timing of the research in the context of site history. The presence of traffic associated with site construction and waste disposal, as well as associated nuisance effects, would have become increasingly apparent by the time both follow-up surveys were conducted. The concern about water quality and quantity suggests an intensification of anticipatory fears, given the prematurity of potential water problems associated with the landfill. Health concerns form a major component of overall concern, and that pattern has persisted over time, with only slight decreases in reporting (56% vs. 51%; 46% vs. 39%). The percentage of respondents reporting that they had taken one or more actions decreased over time for both sample groups (80% vs. 71%; 71% vs. 50%). In addition, the distance gradient of actions away from the study site evident in the initial survey data largely disappeared over time, as the largest decreases in the percentage of people taking action occurred in the zone(s) closest to the site.

The Milton follow-up study also included open-ended telephone interviews (July 1992) with respondents who had moved into (n=4 of 7) and out of (n=7 of 7) the study area. A checklist of interview topics included length of residence, tenure status (own vs. rent), type of property (farm vs. nonfarm), and feelings about the impacts of the landfill on their decision to move and on property values. Among the out-migrants, the most profound impact was reported by the two respondents who had been expropriated from the zone closest to the site. They had very different experiences from those who had moved by choice. They commented that "our lives are finished"; that "the Region took everything away"; and that investment was lost and property values were "down the drain." The experiences of those who moved out by choice contrast sharply with those who were expropriated. The former reported that the landfill had only a relatively small influence on their decision to move, with the exception of one former resident who said that the landfill was the primary reason for moving, specifically citing concerns about water shortages, seagulls, and property value decline as well as frustration with regional government.

Of the four new residents who agreed to be interviewed, three were aware of the site before they decided to move to the area. The other maintained that prior knowledge would not have significantly influenced the decision to move despite the feeling that the site had adversely affected property values in the area. In general, property values were not considered to be a direct concern. In fact, the two renters in this group moved into the area because of the reasonable rent, which was perhaps a function of proximity to the site. The factors influencing appraisal of the "threat" posed by the landfill site included anticipated duration of residence in the area and the perceived distance from the landfill. For example, the two renters anticipated that they would soon be in the position to purchase a home elsewhere, while one of the owners planned to move away in a few years. In addition, both homeowners felt unthreatened because they lived outside the zone (within one-half mile of the site) that they perceived as impacted by the landfill. These generally positive responses may reflect the (emotion-focused) coping responses of new householders who may feel the need to justify their decision to buy/rent a residence near an impending landfill, especially if they had no prior knowledge. It is not surprising, perhaps, that they would claim that they "do not intend to live there long," that they are "only renting," or that it is their neighbors living closer to the site who are at risk, not them. As one resident expressed it, there is a prevalent "wait and see" attitude, implying that primary appraisal of the landfill may change over time through reappraisal on the basis of the experience of living in close proximity to the site.

APPLICATIONS AND IMPLICATIONS

Potential applications of the research findings are linked to the purpose of the larger research program: to determine the effects of exposure to environmental contaminants on human health and well-being and to develop strategies to reduce their adverse effects. Three issues can be addressed from the results of the initial survey. These are related to the levels of concern and action reported at each site, the types of concern reported, and the determinants of concern and action.

The higher levels of concern and action at Milton and Glanbrook compared with SWARU confirm that the recent controversy and debate over the introduction of the Milton site has generated widespread reaction throughout the community. At Glanbrook, where the site has been operating for over ten years, concern persists at quite a high level even though the site has not been a high-profile issue in the recent past. This implies that concern does not necessarily decrease steadily over time. The much lower levels of concern and action at SWARU, which has been operating for over twenty years, suggests counterevidence. It is very doubtful, however, whether the SWARU situation is comparable, given that it is a very different type of facility (incinerator vs. landfill) and is situated in a very different setting (urban industrial vs. rural). In short, context is very important for understanding differences in level of concern.

Variations in the types of concern reported in the survey and interviews reinforce the importance of site and community characteristics in any effort to address and alleviate community concerns. At SWARU, the most frequently mentioned concern was air pollution. At Glanbrook, it was site operation (e.g., type and nature of wastes being disposed of and safety concerns), perhaps related to the fact that the landfill is closed to public access, thereby arousing the suspicions of area residents. At Milton, the most frequent concern was water pollution, which reflects the importance of agriculture in the local economy. For the majority, these concerns were linked to worry about health implications. The longitudinal results at this site underscore the relationship between pre-siting issues and post-siting effects, as potential long-term health impacts of water pollution continues to be the major concern at this site.

Profiles of the characteristics of respondents more likely to report site- and situation-specific effects reveal that a combination of factors is significantly associated with concern and action but that the specific factors involved vary by outcome and site. This latter finding is consistent with previous studies, which show no strong consistency in the characteristics of individuals and groups reporting psychosocial effects (Taylor et al. 1989). Again, the implication is that strategies aimed to address and alleviate psychosocial effects need to be specific to the characteristics of the populations in particular situations.

The longitudinal component of the Milton study provides insights about the ongoing process of reappraisal. The higher percentage of respondents volunteering the landfill site as a dislike, combined with the lower percentage reporting concerns about the landfill, suggests that anticipatory fears about the potential negative impacts of the landfill have diminished over time and that concerns have been replaced by a general dislike. It remains to be seen, however, how further reappraisal will influence perceptions of the landfill one or two years hence as residents begin to experience living with the landfill. Interviews with residents who had moved from and into the area since the initial survey (1990) indicated that movers were more likely to report that the site had influenced their decision to move if they lived closer to the site. Not surprisingly, households that were expropriated were much more profoundly impacted, both emotionally and financially, than individuals who had moved by choice. The appraisal of new residents was influenced by a combination of factors related to anticipated duration of residence in the area, to housing tenure, and to the perception that the landfill is too far away to affect them. This primary appraisal may change over time with the experience of living in close proximity to an operating landfill.

What is the prescription for this diagnosis of the psychosocial effects of living in close proximity to these waste facilities? It is naturally as complex as the outcomes themselves. In addition, the prescription is only a partial one, given a number of gaps in our knowledge.

Recommendations for addressing psychosocial effects can be organized around three key themes: prevention and reduction of psychosocial effects, protection of at-risk groups, and mitigation of impacts through public involvement in facility operation and management. First, the most effective approach to prevention and reduction would be to work toward the elimination of the problem in the first instance (i.e., government agencies should continue to promote, develop, and implement strategies to reduce the need for new waste disposal facilities through the processes of recovery, reduction, re-use, and recycling). Related to this, agencies should solicit from community groups their perceptions of waste disposal technologies that might be more acceptable in particular communities. For example, in the SWARU sample, there was a perception that landfills are a greater source of concern than incinerators, despite the fact that at the time of the initial survey, a newly elected New Democratic government placed a moratorium on incinerator development in the province of Ontario. SWARU residents' preference for incinerators over landfills, however, may be related to the nature of the community context (i.e., an urban industrial community). Indeed, a recent survey of residents on Vancouver Island, British Columbia—a nonindustrial community—indicated a much stronger preference for landfills over incinerators, although the priority identified by this sample group was to reduce the need for facilities in the first instance (Malatest and Associates 1991).

Further, there remains considerable uncertainty and distrust regarding the types of waste being disposed of and the possible future contamination of water, air, and soil. Such uncertainty fosters concerns and anxiety, which could be alleviated to a considerable extent if the public felt they had fuller and easier access to information on the composition of the waste and on the measures in place to prevent and control contaminants. In this vein, the public should be given the opportunity to observe management practices firsthand. For example, as shown by the Glanbrook data, respondents had a substantial amount of trust and confidence in a site-monitoring committee made up of area residents, despite concerns they had about the site itself. In addition, the emphasis should be on a proactive approach designed to address concerns as they arise, so as to reduce potential psychosocial distress due to anxieties that are either unheard or unattended. For example, at SWARU there is a very quick and comprehensive complaint-response protocol, which involves personal visits to the complainant as well as offers of site tours. This may be one of the factors contributing to lower levels of reported concern at this site. Actual or perceived community exclusion from site management and planning leads to a sense of disempowerment, distrust, and resentment (Hadden 1991; Rabe 1991). This was very apparent among Milton respondents at all stages of data collection. It promotes and perpetuates adversarial relations between the community and the agencies responsible for the facility. Genuine involvement of community members in site management and planning can substantially prevent and reduce associated psychosocial distress.

Second, with respect to the protection of at-risk groups, the findings show that the determinants of psychosocial effects vary in relation to the combined characteristics of the site and of the local population. No clear profile of at-risk groups, which might be generalizable across sites, emerges from these analyses. It follows that facility planners and operators need to be sensitive to issues and concerns arising from the conditions that are to some degree specific to each situation. Although the findings do not support a simple distance-decay relationship for psychosocial effects, proximity to the site emerges consistently as a significant and important determinant of impacts. It follows therefore that psychosocial effects will be reduced by minimizing the impacts of the facility on those living closest to the site (i.e., within the first 2–3 kilometers). Psychosocial effects related to solid waste disposal facilities are typically a response to perceived risk. This should not be grounds for dismissing or minimizing their importance and, thereby, for failing to recognize and address them as legitimate. To do so potentially increases distrust and uncertainty and in turn compounds psychosocial distress.

Finally, as noted above, public involvement in facility operation and management can go a long way toward mitigating psychosocial effects by reducing distrust and uncertainty. For example, advisory committees provide a vehicle for monitoring and promptly responding to community concerns about facility siting and/or operations. The involvement of all stakeholders facilitates mutual understanding of respective positions and lessens the potential conflict arising from adversarial relations between the community and facility operators. The representation and participation of community members (as in the case of Glanbrook) promotes empowerment, reduces misperceptions, and fosters trust—all of which potentially decreases psychosocial effects. As Hadden (1991, 50) suggests, "If the people accept the ways in which decisions are made, they will accept the decision itself."

While the focus of this chapter has been on psychosocial effects of non-hazardous waste facilities, both existing and proposed, a number of parallels with hazardous waste facility siting can be identified. These include the uncertainty inherent in the siting process; individual and community levels of distrust in the opinions of "experts" and authorities who have assured them there is nothing to worry about; suspicion regarding the nature of wastes proposed for disposal at the site(s); individual and community lack of faith in "safe" technologies; fear of latent health effects; frustration and anger resulting from a lack of genuine involvement in the siting process; substantial psychosocial effects on expropriated individuals and households; and, finally, the apparent persistence of concerns over time. Further, some or all of the recommendations made here can be applied to both hazardous and non-hazardous waste facilities but are limited by the gaps in our knowledge of psychosocial effects of exposure to waste facilities, existing and proposed. These gaps can be addressed through additional research, in several ways. First, as was attempted in this study, ongoing research should use designs that combine quantitative and qualitative approaches to examine the effects of social and cultural context on psychosocial outcomes. As seen in these findings, explanation and understanding are substantially enhanced. Further, research should be guided by existing conceptual models but should also aim to advance theory through the clarification and integration of key concepts of environmental stress and risk.

Second, more longitudinal research is needed to address the issue of reappraisal in the context of individual and community response to environmental stress. While the research reported here provides data for three points in time at one site, at least one additional stage is required to address the process of reappraisal during the period that the site has been fully operational. Ideally, prospective cohort studies should be conducted that have the power to detect changes in psychosocial effects over time in response to environmental exposures, risk appraisal, and coping responses. Another important methodological issue is the development of appropriate outcomes measures. It is not clear that the combination of general (e.g., GHQ and SCL-90) and site-specific measures used in our work provides the sensitivity and specificity desirable for detecting psychosocial effects in exposed populations.

In addition to longitudinal research, more emphasis needs to be placed on comparative (case) studies. This can take two forms. First, the research reported on in this chapter takes place in one area of southern Ontario. Given that one of the primary conclusions of the work is that psychosocial impacts cannot be divorced from the wider community context within which they occur, it makes sense to add to the inventory of communities and contexts. In addition, using similar survey instruments and qualitative approaches across communities and contexts provides a foundation for the meta-analysis of site and situational factors influencing the determinants of psychosocial effects of exposure to waste disposal facilities and the development of strategies to reduce their adverse impacts on human health and well-being.

REFERENCES

Ajzen, I., and M. Fishbein. 1980. *Understanding Attitudes and Predicting Social Behavio*r. Englewood Cliffs: Prentice Hall.

Alexander, C. 1964. *Notes on the Synthesis of Form*. Cambridge: Harvard University Press.

Bachrach, K. M., and A. J. Zautra. 1985. Coping with a Community Stressor: The Threat of a Hazardous Waste Facility. *Journal of Health and Social Behaviour* 26:127–42.

Baum, A., R. Fleming, and J. E. Singer. 1985. Understanding Environmental Stress: Strategies for Conceptual and Methodological Integration. In *Advances in Environmental Psychology*, edited by A. Baum and J. E. Singer, 5:185–205. Hillsdale, N.J.: Erlbaum Associates.

Berkman, L. F., and S. L. Syme. 1979. Social Networks, Host Resistance, and Mortality: A Nine-Year Follow-Up Study of Alameda County Residents. *American Journal of Epidemiology* 109, no. 2:186–204.

Bertazzi, P. A., C. Zocchetti, A. C. Pesatori, S. Guercilena, M. Sanarico, and L. Radice. 1989. Ten-Year Mortality Study of the Population Involved in the Seveso Incident in 1976. *American Journal of Epidemiology* 29, no. 6:1187–1200.

Cassel J. 1976. The Contribution of the Social Environment to Host Resistance. *American Journal of Epidemiology* 104, no. 2:107–23.

Cobb, S. 1976. Social Support as a Moderator of Life Stress. *Psychometric Medicine* 38:300–314.

Couch, S., and J. S. Kroll-Smith. 1994. The Chronic Technical Disaster: Toward a Social Scientific Perspective. In *Environmental Risks and Hazards*, edited by S. Cutter, 297–309. Englewood Cliffs, N.J.: Prentice Hall.

Dear, M. 1978. Planning for Mental Health Care: A Reconsideration of Public Facility Location Theory. *International Regional Science Review* 3:93–111.

Dear, M.J., R. Fincher, and L. Currie. 1977. Measuring the External Effects of Public Programs. *Environment and Planning A* 9:137–47.

Dear, M. J., and S. M. Taylor. 1982. *Not on Our Street*. London: Pion Books.

Derogatis, L. R., R. S. Lipman, and L. Covi. 1973. SCL–90: An Outpatient Psychiatric Rating Scale—Preliminary Report. *Psychopharmacology Bulletin* 9:13–28.

Dunne, M. P., P. Burnett, J. Lawton, and B. Raphael. 1990. The Health Effects of Chemical Waste in an Urban Community. *Medical Journal of Australia* 152:592–97.

Edelstein, M. R. 1988. *Contaminated Communities: The Social and Psychological Impacts of Residential Toxic Exposure*. Boulder, Colo.: Westview Press.

Elliott, S. J. 1992. Psychosocial Impacts in Populations Exposed to Solid Waste Facilities. Ph.D. diss., Department of Geography, McMaster University, Hamilton, Ont.

Elliott, S. J., S. M. Taylor, S. Walter, D. Stieb, J. Frank, and J. Eyles. 1993. Modelling Psychosocial Effects of Exposure to Solid Waste Facilities. *Social Science and Medicine* 37, no. 6:791–804.

Evans, G. W., ed. 1982. *Environmental Stress*. Cambridge: Cambridge University Press.

Evans, G. W., and S. Cohen. 1987. Environmental Stress. In *Handbook of Environmental Psychology*, edited by D. Stokols and I. Altman, 2:571–610. New York: Wiley and Sons.

Evans, G. W., and S. V. Jacobs. 1982. Air Pollution and Human Behaviour. In *Environmental Stress*, edited by G. W. Evans, 105–32. Cambridge: Cambridge University Press.

Eyles, J., D. Sider, J. Baxter, S. M. Taylor, and D. Willms. 1990. The Impacts and Effects of the Hagersville Tire Fire: Preliminary Analysis and Findings. Paper presented at the 1990 Ontario Ministry of Environment Technology Transfer Conference, November 19–20, Toronto.

Eyles J., S. M. Taylor, N. Johnson, and J. Baxter. 1993. Worrying about Waste: Living Close to Solid Waste Disposal Facilities in Southern Ontario. *Social Science and Medicine* 37, no. 6:805–12.

Fishbein, M., and I. Ajzen. 1975. *Belief, Attitude, Intention and Behavior*. Reading, Mass.: Addison Wesley.

Fleming, R., A. Baum, M. M. Gisriel, and R. J. Gatchel. 1982. Mediating Influences of Social Support on Stress at Three Mile Island. *Journal of Human Stress* 8, no. 3:14–22.

Folkman, S., and R. S. Lazarus. 1980. An Analysis of Coping in a Middle-Aged Community Sample. *Journal of Health and Social Behavior* 21:219–39.

———. 1988. The Relationship between Coping and Emotion: Implications for Theory and Research. *Social Science and Medicine* 26, no. 3:309–17.

Goldberg, D. P. 1972. *The Detection of Psychiatric Illness by Questionnaire: A Technique for the Identification of Non-psychotic Illness*. London: Oxford University Press.

Greenberg, M. 1994. Health Effects of Environmental Chemicals. In *Environmental Risks and Hazards*, edited by S. Cutter, 297–309. Englewood Cliffs, N.J.: Prentice Hall.

Hadden, S. 1991. Public Perception of Hazardous Waste. *Risk Analysis* 11, no. 1:47–62.

Hall, B., and S. M. Taylor. 1983. A Causal Model of Attitudes toward Mental Health Facilities. *Environment and Planning A* 15:525–42.

Hertzman, C., M. Hayes, J. Singer, and J. Highland. 1987. Upper Ottawa Street Landfill Site Health Study. *Environmental Health Perspectives* 75:173–95.

Holmes, T. H., and R. H. Rahe. 1967. The Social Readjustment Rating Scale. *Journal of Psychosomatic Research* 11:213–18.

Horowitz J., and M. Stefanko. 1989. Toxic Waste: Behavioral Effects of an Environmental Stressor. *Behavioral Medicine* 15:23–28.

House J. S., K. R. Landis, and D. Umberson. 1988. Social Relationships and Health. *Science* 241:540–45.

Lazarus, R. S. 1970. Cognitive and Personality Factors Underlying Threat and Coping. In *Social Stress*, edited by S. Levine and N. A. Scotch, 143–64. Chicago: Aldine.

Lazarus, R. S., and S. Folkman. 1984. *Stress: Appraisal and Coping*. New York: Springer.

Lefcourt, H. M. 1982. *Locus of Control: Current Trends in Theory and Research*. 2d ed. Hillsdale, N.J.: Lawrence Erlbaum Associates.

Lipscomb, J. A., L. R. Goldman, K. P. Satin, D. F. Smith, W. A. Vance, and R. R. Neutra. 1991. A Follow-Up Study of the Community near the McColl Waste Disposal Site. *Environmental Health Perspectives* 94:15–24.

Logue, J. N., M. E. Melick, and H. Hansen. 1981. Research Issues and Directions in the Epidemiology of Health Effects of Disasters. *Epidemiologic Reviews* 3:140–62.

Malatest and Associates. 1991. *Community Survey: Solid Waste Alternatives.* Victoria, B.C.: Robert A. Malatest and Associates.

Mazmanian, D., and D. Morell. 1992. *Beyond Superfailure: America's Toxics Policy for the 1990s.* Boulder, Colo.: Westview Press.

Pearlin, L. I., and C. Schooler. 1978. The Structure of Coping. *Journal of Health and Social Behavior* 19:2–21.

Perry, R. W. 1983. Environmental Hazards and Psychopathology: Linking Natural Disasters with Mental Health. *Environmental Management* 7, no. 6:543–52.

Portney, K. E. 1991. *Siting Hazardous Waste Treatment Facilities.* New York: Auburn House.

Rabe, B. 1991. Beyond the NIMBY Syndrome in Hazardous Waste Facility Siting: The Alberta Breakthrough and the Propects for Cooperation in Canada and the United States. *Governance* 4, no. 2:184–206.

Roht, L. H., S. W. Vernon, F. W. Weir, S. M. Pier, P. Sullivan, and L. J. Reed. 1985. Community Exposure to Hazardous Waste Disposal Sites: Assessing Reporting Bias. *American Journal of Epidemiology* 122:418–33.

Sandman, P. M., D. B. Sachsman, M. R. Greenberg, and M. Gochfeld. 1987. *Environmental Risk and the Press.* New Brunswick, N.J.: Transaction Books.

Shusterman, D., J. Lipscomb, R. Neutra, and K. Satin. 1991. Symptom Prevalence and Odor-Worry Interaction near Hazardous Waste Sites. *Environmental Health Perspectives* 94:25–30.

Sims, J. H., and D. D. Baumann. 1983. Educational Programs and Human Response to Natural Hazards. *Environment and Behaviour* 2:165–89.

Sorensen J., J. Soderstrom, E. Copenhaver, S. Carnes, and R. Bolin. 1987. *Impacts of Hazardous Technology: The Psycho-social Effects of Restarting TMI–1.* Albany: State University of New York Press.

Taylor, S. M. 1987. Community Reactions to Deinstitutionalization. In *Location and Stigma*, edited by C. J. Smith, 223–46. Boston: Unwin Hyman.

Taylor, S. M., S. J. Elliott, J. Eyles, J. Frank, M. Haight, D. Streiner, S. Walter, N. White, and D. Willms. 1991. Psychosocial Impacts in Populations Exposed to Solid Waste Facilities. *Social Science and Medicine* 33, no. 4:441–47.

Taylor, S. M., S. J. Elliott, J. Eyles, J. Frank, M. Haight, D. Streiner, S. Walter, D. Willms, and N. White. 1993. *Psychosocial Impacts in Populations Exposed to Solid Waste Facilities.* Report prepared for the Ontario Ministry of Environment. Department of Geography, McMaster University, Hamilton, Ont.

Taylor, S. M., J. Frank, S. Walter, M. Haight, N.F. White, D. Streiner, D. Willms, S. Birnie, and S. J. Elliott. 1989. *The Psychosocial Impacts of Exposure to Environmental Contaminants in Ontario: A Feasibility Study.* Report prepared for the Ontario Ministry of the Environment.

Taylor, S. M., G. B. Hall, R. C. Hughes, and M. J. Dear. 1984. Predicting Community Reaction to Mental Health Facilities. *American Planning Association Journal* 50:36–47.

Vyner, H. M. 1988. *Invisible Trauma: Psychosocial Effects of the Invisible Environmental Contaminants*. Toronto: D. C. Heath.

Watts, R. 1993. Sihota Eyes Alberta for B.C. Waste. *Victoria Times-Colonist*, November 14, A2.

Wiedemann, P. M., H. Schutz, and H. P. Peters. 1991. Information Needs Concerning a Planned Waste Incineration Facility. *Risk Analysis* 11, no. 2:229–37.

Part 3

Options and Strategies

Using Co-management to Build Community Support for Waste Facilities

Alun Richards

The "perfect site" concept for locating potentially unwanted facilities seems to be difficult for society to let go. Perfection, like beauty, is in the eye of the beholder. Proponents (individuals, corporations, and institutions that propose facilities) have to deal with communities, self-defining groups that anticipate being affected by the projects. *Proponents need communities, not sites.* Communities need a significant level of control over their destinies, particularly when dealing with potential risks or unknowns. Proponents should, therefore, be prepared to share some of their decision making. In the end, giving away power gains power—for both parties.

The purpose of this chapter is to suggest some general principles that should apply to most siting projects. Principally, it advocates the use of "co-management" as a central theme of the siting, development, and ultimate management process. The approach outlined here is based on the experience of siting a central hazardous waste management facility in Manitoba.

It is possible to get demonstrated support for a typically difficult-to-site project, but this is unlikely to be achieved without having the public, primarily the affected community, be part of the process. With the community demonstrably and vocally supporting a project, its chance of success is immeasurably increased. Offering communities a major part in investigating the project and in making decisions, including the siting decision, is one means of developing this support. Developing a longer-term relationship based on a mutually accepted agreement will help maintain the relationship.

The above is broadly covered by the term "co-management," with the community and, potentially, other stakeholders having a major decision-making role in a project. This role could include, but may not be limited to, project design, siting, engineering/architectural design, development, operation, decommissioning, or any other issues that are considered locally to be important. An invitational, co-managed approach requires the project proponents, and even the regulators, to reconsider many of their traditional attitudes and bring stakeholders into the process at any early stage prior to making many of the most significant decisions (Connell 1992). While the early stages of the project are likely to be slower and require a great deal of flexibility on the part of the proponent, there is a reasonable expectation that the later formal regulatory process will be much less controversial and therefore faster.

MANITOBA'S SITTING PROCESS

The Manitoba Hazardous Waste Management Corporation (MHWMC) is a provincial crown corporation established to develop and operate a hazardous waste management system. The system will provide for the handling of provincially regulated hazardous waste (about 3,100 substances are listed as "dangerous goods," excluding federally regulated radioactive and explosive materials). The corporation does not have a monopoly and is regulated as any private developer or operator. The system will include waste management at source, transportation, storage, treatment or access to treatment, education, and other aspects. Among other components, the system requires a combined physical-chemical treatment plant and a storage/transfer facility (see Fig. 1) to handle up to 30,000 tonnes per year of organic and inorganic materials. The facility will also bioremediate up to 40,000 tonnes per year of hydrocarbon contaminated soil. The siting process for this facility sought an area with not only a high level of environmental and technical capability but also demonstrated community support.

The four-year invitational siting and joint review process began in late 1988. It proved very effective, with about sixty municipalities, singly or in groups, expressing some interest. Many of the detailed aspects of the process, including communication and decision making, were developed from the experience of working with those communities. Four rural communities and the City of Winnipeg took the process through to a level of serious discussion. Two of these ended their investigations with close losses in premature referenda. Another was rejected because it was judged by the MHWMC not to have a suitable level of natural environmental protection. By 1991, the siting had become focused on the Rural Municipality (R. M.) of Montcalm (population 1,700) and the City of Winnipeg (population 650,000). The sites under consideration in these two municipalities were excellent from environmental, technical, and business viewpoints. The MHWMC chose the Montcalm site (see Fig. 2), 75 kilometers south of Winnipeg and 15 kilometers north of the U.S. border, because of the community's demonstrated support and the clearly articulated terms and conditions of that support.

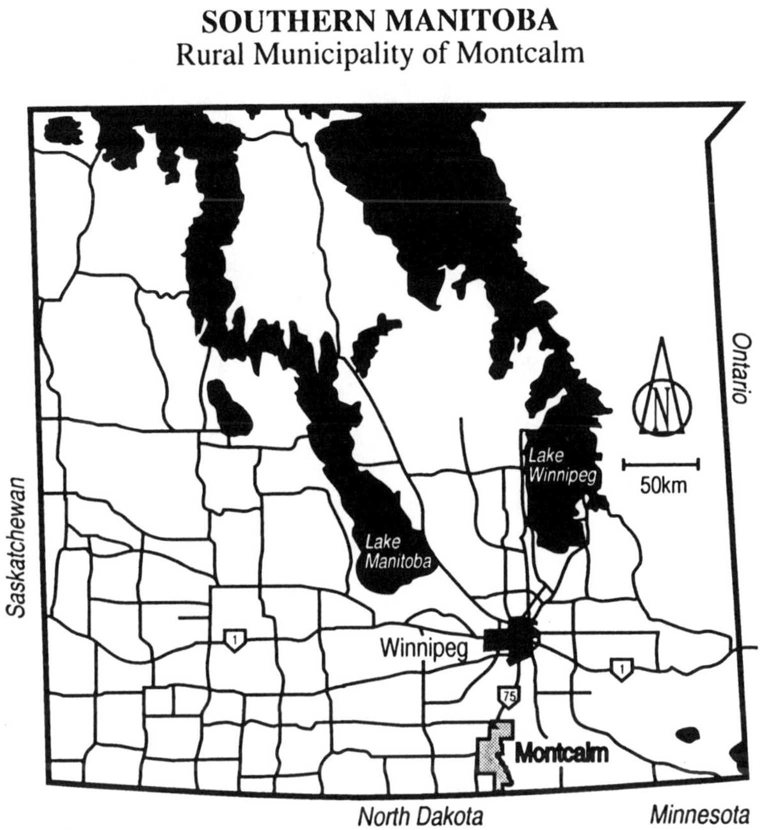

Figure 1. Location of the MHWMC facility

The MHWMC filed its application to construct and operate its central hazardous waste management facility in March 1992. The environmental hearings were held in Montcalm in June and lasted three evenings. Of the fourteen mostly local interventions, thirteen supported, while one opposed, the project. Construction began in September 1992, almost exactly four years after the siting process began and less than three years after the initial contact with Montcalm.

During 1993, a soil remediation facility and a transfer station were constructed at the Manitoba Environment Centre (MEC). At the same time, a private-sector investor/partner was being sought to construct the remaining MEC facilities and to operate the system developed by the MHWMC. At the end of 1994, the corporation moved its head office and related formal and functional activities to leased Montcalm office space in the expectation of reaching an agreement with a partner

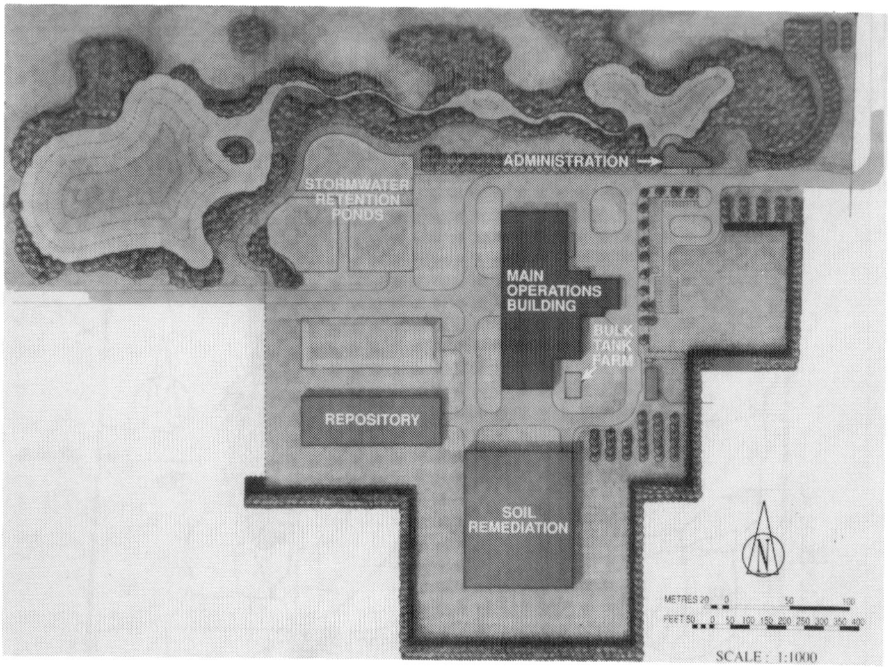

Figure 2. Conceptual design of the Manitoba Environmental Centre in the Rural Municipality of Montcalm

and beginning construction of the main treatment, transfer, and support facilities in 1995.

Based on extensive discussions on the siting issue, the assumption was made that no community should have a facility that it does not want. To begin the process, the MHWMC publicized its siting search on the basis that it would not even consider an area unless invited by the community to do so (Latonas 1988; Latonas and Kucera 1988). Typically, a municipal council would request information by means of a presentation. Following this, the MHWMC would usually hold one or more open houses to inform as many people as possible and assess, through a simple questionnaire, their level of interest in continuing the discussion. The corporation learned from experience that a large attendance and a very high level of support from those attending were necessary to consider continuing the discussion in that area. The MHWMC staff learned the hard way that apathy meant NO. If there was enough interest and no significant backlash, a joint local review process would be implemented. This process would only continue as long as local support and corporate interest in the area were both present. The community was not asked to support the project but, rather, to support the joint review on the basis that it had a final decision.

In this voluntary (invitational) approach, the community could opt out of the process at any time. No site would be proposed by the MHWMC without the community's demonstrated support, which could be based on community-proposed terms and conditions. Co-management was a major part of the process, with the potentially affected communities having a significant level of power, including the final say in some decisions. One test of a real co-management process is the ability, on some issues, for the community's wishes to overrule the corporation's.

The main steps in the siting process included

- developing proposed environmental and technical siting criteria to establish computerized maps of areas of greatest interest on a provincial scale,
- soliciting interest from communities on an invitational basis (i.e., the MHWMC would not consider any community unless invited to do so),
- the communities' and the corporation's jointly investigating the issue, project, and potential siting, with either party being free to withdraw at any time.

THE COMMUNITY'S ROLE IN A VOLUNTARY PROCESS

The joint reviews involved community advisory committees (CACs) chosen by the municipal council(s), ideally representing a range of interests and opinions, and supported financially either directly by the MHWMC or indirectly through the municipality. Each CAC independently reviewed and amended the corporation's draft siting criteria and siting process as was locally appropriate and reviewed design, operation, impact, and any other issues that it felt were appropriate. The committee then publicly reported its findings.

Montcalm's CAC conducted a very public review with the ongoing involvement of the MHWMC. The CAC produced a report, with its conclusions including proposed terms and conditions, and recommended a municipal referendum. That vote subsequently established a 67% level of local support for the project. The community terms and conditions were eventually developed into legally binding contracts between the MHWMC and the municipality, and between the Province of Manitoba and the municipality.

A referendum had been successfully used to site a fully integrated hazardous waste management facility at Swan Hills, Alberta. (See Castle and Munton, this volume.) It was MHWMC policy to have the community choose the form of its decision, but siting a facility such as this would have been extremely difficult without a referendum. While a difficult hurdle to cross, it is the most equitable form and most reliable proof of a community's decision. Once established, the referendum results proved invaluable in the public regulatory-review process.

The concept of a voluntary process receives wide, almost universal, support but must be based on a number of principles to have enough substance to survive a siting process. Some of the principles of the community's role that should be agreed upon in advance are that

- a community's NO will be honored,
- the process will be public and have no major surprises,
- the community will have the ability to negotiate terms and conditions of support with the understanding that the proponent may also be negotiating with other communities,
- the co-management process will continue past the permit stage through to operation and eventual decommissioning,
- since trust can be personal and temporary, promises or guarantees must be given legal or regulatory substance,
- a clear understanding must exist on the form, definition, and implications of the community's decision.

THE PATTERN OF ISSUES

The issues around which the discussions are centered are the same and in the same order of priority as arise in general discussions with members of the public. In the MHWMC experience of over 500 committee and public meetings, open houses, presentations, and other venues, the pattern of occurrence and priority of concerns are as follows: (1) environment/health/safety, (2) site selection and the decision-making process, and (3) (once the first two are answered) economic opportunities. People will not accept real or perceived environmental damage in return for economic benefits. In our experience, where economic incentives were raised in discussion, the response was often a fear of being perceived to be taking something negative in return for economic gain. The search for general economic benefits certainly was the "carrot" that triggered most of the initial community interest, as would be expected in the case of any industrial facility. The "perception is everything" aspect of hazardous waste management, however, held true regarding the offering of inducements or compensation. Certainly, it would not have increased the project's chances.

This order of importance for issues generally applied to all of the communities in the Manitoba process. It is particularly significant during the development of a comanagement process and the eventual agreement(s). The MHWMC-Montcalm agreement, other than reimbursing the R. M. of Montcalm for its direct costs related to the comanagement activities and providing some agricultural-product and property value guarantees, has only one paragraph relating to local economic benefits and employment. It is a fairly loosely worded "all things being equal" reference. Even the process of negotiating the agreements reflected the pattern of issues, with economic benefits being one of the last thoughts.

On the issue of economic benefits, once Montcalm and the MHWMC approved each other, the corporation implemented a policy of hiring and purchasing from the area. This has since encompassed some major expenditures. The local response to even the small initial hiring and purchasing activities was positive because, as with the rest of the project, it showed intent and established a track record. Later construction delays increased local pressure for doing more in the area, including moving the MHWMC head office to Montcalm.

STAGES, ISSUES, AND CO-MANAGEMENT STRUCTURES

In the Manitoba experience, the structure of the co-management activities developed from the main issues of concern that were, in part, linked to the stages of the project's development. These stages can be described as follows:

- *Initial stakeholder and community discussions* centered on environment/health/safety and voluntary-siting issues that required a corporate response of formal policies and involved a very short-term corporate track record. The co-management was an extremely loose process of public discussions, with the concerns raised and the corresponding MHWMC actions taken being well documented to show responsiveness.
- *The joint community-based review* of local aspects of environment/health/safety and siting required a structure of much more formal information sharing, investigation, and joint decision making. The corporate response had to be extremely flexible, following the issues as they arose through a formal local co-management process. Specific guarantees, such as letters from the CEO, or specific studies requested by the community representatives were occasionally required.
- *During the environmental-impact assessment (EIA) process,* an agreement should be negotiated between the facility proponent and

the host community, dealing with such issues as the ongoing joint power-sharing processes, limits to facility development, local environmental monitoring, and property/crop value protection. In the Manitoba case, once the siting became focused on Montcalm, an agreement in principle was negotiated between the MHWMC and a community advisory committee. While it had little legal basis, it was approved by the corporation's board of directors and the Montcalm municipal council. The agreement was included in the EIA to show intent.

- *Prior to construction,* a final formal contract should be concluded, involving details of community controls, facility operation, protection measures, financial support for the process, and local benefits. This should provide for a level of flexibility. The MHWMC-Montcalm contract, guaranteeing operational issues, was concluded first, followed by a Province of Manitoba-Montcalm contract. Since there was no guarantee that the MHWMC would not be partly or wholly privatized, the obligations of the corporation or its successor had to be established separately from the regulatory and shareholder obligations of the province.
- *Once the site has become operational,* the co-management process consists of formally implementing the terms of the agreement, with both parties helping each other along a learning curve. The negotiations never actually end. The mutual trust, while considerable, can never be taken for granted. In the Manitoba case, the operational stage began with the construction of a small transfer facility for the full range of hazardous wastes and a bioremediation building for hydrocarbon-contaminated soil.

THE MONTCALM AGREEMENTS

The agreements negotiated between the R. M. of Montcalm and the MHWMC and between Montcalm and the Province of Manitoba were based on proposed terms and conditions applied by the CAC to its recommended support for the project. These were then negotiated by the various parties and linked to any licenses issued on the site. They are also public and available to anyone interested.

The Montcalm-MHWMC agreement, which would also apply to any purchaser or leaseholder of the facility, covers a number of major issues, including

- ongoing co-management, including community representatives on the MHWMC board, a senior-level plant co-management committee, and a liaison committee made up entirely of local representatives with access to all environmental data;

- required community support for any major changes and expansions at the facility or significant changes to the license;
- community protection programs such as land and agricultural-product sale value guarantees, environmental monitoring and reporting (with funds for the community to acquire technical assistance), and training and emergency response undertakings;
- standards for facility operation, with a limit being placed on the annual capacity for the engineered landfill for stabilized nonhazardous residues from the treatment process;
- the right of the MHWMC to handle all provincially regulated hazardous wastes except PCBs;
- a brief reference to local economic opportunities given price, quality, and other relevant considerations;
- linkage of the agreement to any licenses and permits related to the facility;
- the understanding that wastes may come from other jurisdictions, with Manitoba wastes being given priority;
- financial support from the MHWMC for the community to operate its part of the co-management program.

The agreement between Montcalm and the province guaranteed a number of issues, including

- ownership of the land on which the facility is constructed always to be held either by the province directly or through a crown corporation;
- the financial obligations in the case of operator default of the property, in product value protection programs, and in long-term environmental monitoring;
- appointment of at least two board members chosen by the municipality;
- complete and free access to all government-generated information relating to the facility and its operation.

The negotiation of these agreements, while including aspects that had been discussed for some time, began formally with the Montcalm Advisory Committee's report on its activities and findings (Montcalm Advisory Committee 1991). The report suggested a referendum on the siting issue, based on a "tentative list of recommended terms and conditions." Following the positive referendum vote and prior to the EIA submission, MHWMC management negotiated an agreement in principle. This agreement, accepted by the MHWMC board of directors and by the Montcalm council, was included verbatim in the EIA.

There were some issues that had to be addressed by the government of Manitoba and for which the corporation could only promise support.

Following the EIA submission and prior to the license being issued, the formal agreements between the community and the corporation and the community and the provincial government were legally drafted and ratified. It was at this stage that legal counsel for all of the parties became involved. The community/corporation co-management agreement and the environmental license were linked through references in their texts. This made the community an indirect part of the regulatory enforcement and the regulators somewhat responsible for ensuring that the agreement is not contravened. It also made the regulators responsible for keeping the community informed.

A NEED FOR FLEXIBILITY

Three major aspects of the siting process are necessary to create a level of general public comfort with the process and, ultimately, with the proposal (Richards 1993):

- *siting flexibility* at the regional as well as local levels, allowing the choice of a variety of communities and sites within a particular area;
- *regulatory capabilities* as perceived locally to be effective, whether they are local, provincial, or federal;
- *co-management* at the local level.

Not all of the above can be as evident or as effective as might be desired. It is, however, possible to make up for limitations in one by increasing another. For example, a shortfall in the actual or perceived regulatory capabilities may be compensated by increasing local monitoring and control through the co-management process. Many of the demands on either regulatory controls or co-management may be offset by more flexibility in the proponent's siting requirements. Limits to siting flexibility may be compensated by tougher local controls. There appears to be a direct relationship between these three issues and the ability to partially compensate one with another.

Urban and rural responses to the Manitoba siting process were significantly different, reflecting different values, communication, and decision-making processes. Communities define their own social boundaries, and these are more likely to match municipal boundaries in rural areas than in urban neighborhoods. Rural areas, with a culture of consensus, a generally higher level of personal and community self-reliance, and potentially more interest in economic benefits, were found in the Manitoba experience to be more interested in the concept of co-management than were urban neighborhoods (Richards 1991, 1992).

While most of the Manitoba experience was with rural communities, the siting activities within the city of Winnipeg indicated the need for a different process that would recognize the split between neighborhood concerns and city-level decision making. An urban process would likely rely less on co-management and more on siting flexibility with specific, expanded regulatory controls.

Siting process decisions need to be responsive to a variety of stakeholders, including NGOs and the community. A small, flexible, relatively autonomous project management team is more likely to be able to respond to a small, flexible, relatively autonomous group of people such as a rural community or urban neighborhood. This allows the proponent to be visible and definable as a group of people who can be consistent and credible and who can build up personal trust with NGOs and residents. One key to the success of the Manitoba siting project is the fact that the MHWMC consisted of about twenty staff who, to varied extents, were all involved in siting, design, ongoing waste management, public education, and other aspects of the issue. This provided the public with a clear proponent that was both approachable and flexible enough to respond to changing situations.

Finally, proponents have rights too! They have the right to be tough negotiators and to withdraw from a community if they cannot achieve what they feel is reasonable. To be able to do this, they should be working with more than one community at a time. This gives the proponent negotiating flexibility while taking much of the pressure off the communities involved.

LESSONS LEARNED

The project team in the Manitoba siting experience went through an intensive learning process as part of the siting process. While many decisions were made early on, including the need for community support, most of the actual details of the process were learned through experience. There is no magic formula or code to break in achieving community support for a potentially undesirable facility. There are, however, a number of general lessons learned that may apply to other siting processes. The following are some of them.

The Courtship

A voluntary siting process is much like a courtship, with all of the parties involved, particularly the community and the proponent, requiring the ability to conduct their investigations on a no-strings-attached basis. Any decisions, the final approval in particular, must be jointly made and may involve conditions. The proponent should be out in public being seen and communicating. From a practical viewpoint, to build grassroots support, the proponent's "show" should go "on the road." A high level of community-proponent familiarity prior to community leaders' expressing formal interest in the project will help to avoid an instant local backlash. From a practical standpoint, the MHWMC found that holding well-advertised public open houses was the best means of estimating the potential level of support and of getting enough information out into the community to minimize the backlash that can result from a municipality's indication of interest.

Stakeholder Support

There are a number of stakeholders, other than the community and the project team, who should be prepared to indicate at least some level of support, if only for the process:

- the proponent's senior management, directors, and financial supporters,
- politicians of all parties at the regulatory level and at the community level,
- environmental groups,
- industry groups,
- the press, insofar as trusting the technical expertise of the project team.

The Public Venue

Public meetings are often referred to as the last of the blood sports. They may be popular, but they provide more opportunities on both sides for preaching, grandstanding, and creating a media opportunity than they do for constructive two-way communication. Come-and-go open houses or other face-to-face or small group meetings provide much better venues for discussing controversial issues (Connell 1992).

Community Priorities

Most projects do not attempt demonstrably to make community priorities their priorities. To assume that a community will trust the government(s), the EIA process, the proponent, or anyone else to ensure their health and safety is almost certainly a guarantee of failure. The proponent needs a means of dealing with the health and safety concerns of the local residents in such a way that they will feel comfortable with the conclusions. Better yet, the residents could internally address the issue. Since this is likely the major issue in any hazardous waste siting project, offering financial inducements prior to dealing with it may simply widen the credibility gap.

Choosing Communities

Any proponent in a voluntary siting process needs the ability to be selective with respect to the communities. Technical, environmental, and economic considerations also have major roles and may supersede a particular community's apparent willingness. Given the significant economic and political costs of working with communities that do not successfully complete the process, every effort must be made to screen out areas that are likely to fail. Environmental and other traditional forms of site screening are straightforward and can be done early in the process, prior to significant community discussions. Care must be taken to avoid unnecessarily ruling out areas because of nonessential "technical" criteria. Assessing a community's interest in the process and its ability to weather it is more difficult and requires extensive local public discussions so that the proponent can feel comfortable with the investment necessary to follow up a municipality's formal expression of interest.

Successful Failures (When a Community Says NO)

During the joint investigation, the proponent retains the right to drop the community as well as vice versa. Saying "no thanks" to one community because of local opposition or lack of interest will enhance the proponent's credibility elsewhere. To assist in its choice of interested communities, the proponent should always be working with more than one at a time.

A siting project can benefit from a rejection in two ways. First, it proves the concept of a community's right to refuse, and it establishes the proponent's credibility when the process in that area is dropped despite time and money invested. This is particularly important, since the public is suspicious that a proponent will not leave once significant money has been spent investigating a local area. The second benefit is that the level of the rejection may show the possibility of success. (This was the case, for example, in the Manitoba experience, where two close referenda losses, in Rossburn and Hamiota, suggested that other communities might well vote favorably.) In any case, after a referendum defeat, the proponent must formally and quickly honor the request to cease its activities in that area.

When to Proceed

Many of the more difficult decisions throughout the siting process are related to the timing of the decision to proceed to the next step, for example, to form an advisory committee or begin site testing. Some of the judgment needed can only be gained through trying and, in many cases, failing. In general, the mistakes in proceeding are often from moving too fast or taking too big a step. The process can only proceed at the community's pace, which, given communication needs, may be seen by the local advisory committee or municipal council to be slow. Making the community drive the pace and making the steps as small and methodical as possible will minimize the risk of proceeding too quickly or in the wrong direction.

The Community Advisory Committee—Proponent Relationship

While discussed generally elsewhere in this chapter, it is worth emphasizing that in the experience of the Manitoba project, advisory committees that had no elected officials, that conducted all of their discussions in public and communicated their activities to the public, and that proceeded methodically had the best chance of successfully completing their mandate. Corporate assistance for this community review was by reimbursement, usually through the municipality, for expenditures incurred by the CAC, rather than by up-front cash payments. Any CAC will need some advice on how it might proceed, including draft terms of reference, items to do, timing, and the critical paths of issues. This advice could be provided directly by the proponent or indirectly by an independent facilitator on the understanding that it is advice only and that the CAC makes its own decisions on how to proceed.

The Vote

A referendum is the only form of decision making that will be given serious local credence. (A referendum, or plebiscite, as used in the Swan Hills, Alberta, experience, was instantly the local choice in every community with which the MHWMC worked.) There are a number of benefits from a referendum. From the local perspective, it settles suspicions about politicians and manipulation. From the proponent's perspective, provided the question is clear and the vote is fair, the support is indisputable as evidence for the EIA process, as documentation for a financial investor, and as a reason for seeking political support. The last point is especially significant, since, whether the proponent is public or private, the expectation of a referendum decision allows all of the politicians to state their intention of letting the people decide. A political dispute, especially if it is partisan, is likely to destroy the project (Cooke 1993).

Cost-Effectiveness of Voluntary Siting and Co-management

The Manitoba experience, from the public siting discussions in late 1988 to the successful Montcalm vote in late 1991, was extremely fast compared with other high-profile siting processes. It was also successful, something that puts it well ahead of most siting attempts. This is not to say that all voluntary processes are more cost-effective than all other processes, but there is certainly no indication that they are either more risky or more costly. Community support, in the Manitoba experience, resulted in significant savings in land costs, a straightforward EIA process, financially and operationally reasonable community terms and conditions, and a complete lack of litigation. Support also provided the benefits of locally based lobbying for proceeding expeditiously and for ensuring that servicing, zoning, and otherwise supporting the site's development proceeded without a hitch. Demonstrated local support significantly enhanced the value of the asset—in this case, the site—to potential investors. In short, a voluntary process, while not necessarily more costly, will almost certainly provide a more valuable result if successful, and a less damaging result if unsuccessful.

Corporate Commitment

When corporate and other decision makers are in a room full of angry citizens, there is no problem getting them to support a voluntary process as well as any terms and conditions that may go with it. Once the process is complete or even smoothly under way, there is a great temptation to slip back into previous thinking patterns. Promises of many levels of safety or other forms of expensive undertakings are natural when one is confronted by an angry grandmother, but difficult to implement in the cold hard light of the bottom line. Informed commitment to a voluntary, co-managed process has to be achieved and constantly reaffirmed through all management levels, including the board of directors and the shareholders. Certainly there can be time, financial, or other limits put on this. The proponents have the responsibility to be tough, principled negotiators, but there is no point in a project such as this if the co-management process ends.

There will be times when the government, the management, or the ownership changes and the continuation of co-management activities is questioned. Communities and project staff can anticipate this with binding agreements, regulations, dispute-resolution planning, and other means to prepare a process through which to deal with the problem. Communities should also understand the significant time and legal responsibilities placed on them through this role and not take the process for granted, no matter how pleased they may be with the development.

The Issue of the 1990s

Much of the siting discussion will be centered on the community's concern that the project may just be the foot in the door to a much larger development. In fact, more recent experience indicates a greater potential for the opposite. A community, to its surprise (less so to the proponent), may be faced with the possibility that the facility, even if approved, may not be built or staffed as extensively as proposed. In fact, it may not be built at all. A community may wish to seek guarantees that its work and, sometimes, anguish in considering the facility will be awarded by the development as proposed once local and regulatory approvals have been received.

CONCLUSION

Whatever the community, the actual siting of a facility is less of an issue than its presence in, and acceptance by, the community. A successful long-term relationship between a proponent and a community requires the proponent to

- acknowledge local concerns and match these with appropriate corporate responses;
- effectively communicate its (i.e., the proponent's) needs, goals, and limitations;
- demonstrably fulfill its promises, commitments, and policies with the understanding that trust must be earned;
- encourage a level of community control over those issues considered locally to be significant, including, for example, the siting discussion/decision or major technology decisions but not necessarily detailed design or day-to-day operations;
- facilitate a process that reflects openness, local control, and a lack of surprises.

In the end, the structure of any co-management arrangement should develop and continue to change following the interests of the affected parties. A successful siting project will involve a co-management structure that fits the community and will be wary of attempting to make a community fit any preconceptions.

As mentioned at the beginning of this chaper, giving up power can actually gain power for the proponent. Siting has been shown to be a social process with technical aspects, rather than the other way around. In fact, the term "siting" may be misleading, since a supportive community, not necessarily the site, is the necessary achievement. There is no sign that this will change through the 1990s or beyond.

REFERENCES

Connell, B. E. 1992. The Siting Approach for a Hazardous Waste Management Facility in Manitoba: Comment and Critique. In *Innovative Approaches to the Siting of Waste Management Facilities: A Guide to Nonconfrontational Siting Procedures*, edited by Resource Futures International. Pittsburgh: Air and Waste Management Association.

Cooke, R. J. 1993. Some Observations on Voluntary Siting Processes: "The Upside and Down Side." Paper presented at the Ontario Ministry of Environment and Energy Technology Transfer Conference, Toronto.

Latonas, G. P. 1988. *Discussion Document: Proposed Site Selection Process for the Development of a Hazardous Waste Management System*. Manitoba Hazardous Management Corporation, Winnipeg.

Latonas, G. P., and E. Kucera. 1988. *Discussion Document: Proposed Site Selection Criteria for the Development of a Hazardous Waste Management System*. Manitoba Hazardous Management Corporation, Winnipeg.

Montcalm Advisory Committee. 1991. *Final Report—Hazardous Waste Management Facility Investigation, May 1990–July 1991*. Montcalm, Man.

Richards, A. B. 1991. Implementing a Voluntary and Responsive Siting Process in Rural and Urban Settings. *Impact Assessment Bulletin* 10, no. 1. International Association for Impact Assessment, Belhaven, N.C.

———. 1992. Contrasting Public Participation and Co-management Experiences in Hazardous Waste Management Facility Siting in Urban and Rural Communities. Paper presented at the twelfth annual meeting of the International Association for Impact Assessment, at the World Bank, Washington, D.C.

———. 1993. Using Co-management in a Voluntary Hazardous Waste Facility Siting Process. Paper presented at the International High Level Radioactive Waste Management Conference, Las Vegas, Nev.

Voluntary Procedures for Siting Noxious Facilities: Lotteries, Auctions, and Benefit Sharing

Howard Kunreuther

INTRODUCTION

There is a growing recognition among practitioners and researchers that new approaches are needed for siting facilities that are perceived to be noxious and potentially hazardous by the public.[1] The traditional "Decide, Announce, Defend" (DAD) philosophy, where a developer makes a series of technological choices as to the type of facility that is needed and then announces and defends these decisions, has generally failed to find sites for facilities. By not consulting the public in advance, the process generates a set of negative reactions by concerned stakeholders, and those opposed to the facility find legal ways of preventing it from being sited (O'Hare, Bacow, and Sanderson 1983).

A voluntary choice siting procedure is one of the more promising approaches for finding homes for noxious facilities, since it overcomes many of the objections posed by the DAD philosophy. Under this approach, the burden for finding a home for a facility is placed on the applicant to convince one or more communities that it will be desirable for them to house the facility because it is perceived to be acceptably safe and the benefits package is sufficiently attractive to the residents (Kunreuther, 1995). In fact, voluntary procedures have been successfully utilized in locating a hazardous waste facility in both Alberta (Rabe 1991; Zeiss, Castle, and Munton, this volume) and Manitoba (Richards, this volume). Rabe (this volume), looking at different approaches to siting low-level radioactive waste (LLRW) repositories in Canada and the United States, concludes that a voluntary process for site selection is more likely to be successful than processes where residents in the community or region have a limited say in the final outcome.

This chapter proposes a voluntary procedure for siting, with the following features. *Well-specified standards* are used to deal with the risks that the facility creates to the community itself as well as to areas beyond its political boundaries (i.e., the *transboundary-risk* problem). Some form of *compensation* to the host community should be considered so that the benefits from constructing the facility are shared with those who agree to have it in their backyard. *Formal voting procedures,* such as a referendum, should be used to determine which communities are willing to accept the facility. Suppose N communities vote to host a specific facility and there is only a need for $M<N$ communities to house the facility. A lottery-auction procedure is then proposed whereby communities can specify the benefits package they require for having the facility in their backyard.

The next section develops a simplified example that illustrates the importance of having well-specified standards for dealing with risks. In particular, we focus on the divergence between private and social costs when there are transboundary risks and indicate the importance of some type of public siting authority in specifying safety standards and in monitoring procedures. The third section examines different types of compensation for encouraging communities to support a facility. The details of a voluntary process, highlighting the lottery-auction procedure, is then presented in the fourth section and illustrated for the case of finding a home for an LLRW facility in Ontario. The concluding section raises a set of open questions for future theoretical and empirical research.

WELL-SPECIFIED STANDARDS FOR DEALING WITH TRANSBOUNDARY RISKS

The literature evaluating siting processes to date has focused almost entirely on the impact that a facility will have on a particular community or region with a well-defined set of political boundaries.[2] Today there appears to be general agreement that well-specified standards coupled with monitoring and control procedures are critical for a community even to consider hosting a facility.

In one study a few years ago, Carnes et al. (1983) found that opposition to hazardous facilities decreased significantly if residents felt that the facility was subject to local control in dealing with risks posed by the facility. A survey of the attitudes toward the proposed high-level radioactive waste repository at Yucca Mountain, Nevada, revealed that over 80% of those Nevadans who claimed they would vote against the facility felt it was very important to form a local committee with the power to shut down the repository if it did not meet standards. Of particular note is the difference in attitudes toward this measure by those who claimed they would vote in favor the facility; only 41% of this group consider such a measure to be very important (Kunreuther and Easterling 1992).

There has been only limited discussion to date on how one deals with *transboundary risks,* that is, the negative potential impacts of the facility on areas outside the political jurisdiction in which it is located. Not surprisingly, transboundary risks reinforce the need for strict safety standards that are well enforced. Unless the risks to areas outside the facility's political boundaries are explicitly considered, there may be a large divergence between the social and private costs from any proposed facility. Furthermore, by explicitly considering transboundary risks as an integral part of the siting problem, it may be possible to overcome facility opposition that may otherwise limit the ability of the developer to find a volunteer host community.

An Illustrative Example

Consider the challenges facing a private firm or developer that is trying to find a home for a solid or hazardous waste facility. For concreteness, suppose that the facility is an incinerator. Community Y has expressed an interest in hosting the incinerator right near its political boundary in a part of town that has relatively few homes and businesses. Residents in Community Z are also subject to health and environmental risks from the facility.[3]

A voluntary siting process has been proposed whereby all the residents in Y over eighteen can vote on a referendum as to whether an incinerator should be located in their backyard. Suppose that if a certain percentage (e.g., two-thirds) of Y's residents support the facility, then it will be deemed approved and construction will begin in 1996. Those residing in Z have no official vote on whether the incinerator should be located in Y, but they can publicly protest the facility in order to encourage residents in Y to vote against it.

Each resident i in community Y will determine whether to vote in favor of a particular facility by considering the benefits to him or her (Bi) and the perceived risks associated with the facility. The benefits can be direct compensation to an individual, such as a reduction in property taxes, or they can take the form of communitywide or regional improvements, such as additional health-related services, higher salaries to attract more and better teachers for the schools, and/or new recreational facilities. A benefits package may also contain contingent arrangements, such as guarantees against property value declines due to the facility if a family tries to sell its home, and reimbursement for any health and/or environment impacts from the facility.

The risks associated with the incinerator for each individual i in Y are characterized by a perceived probability that some type of loss will occur to him or her. These risks can be mitigated (but not necessarily eliminated) through enforcement of safety standards and regulations. If the benefits package is attractive enough and/or the perceived risks associated with the facility are sufficiently small to the resident in community Y, then he or she will vote yes to constructing the incinerator.

The developer has no economic incentive to provide residents in Z with any benefit package or reduce the risks facing this group. From the firm's perspective, the only votes that count are the ones from community Y. Thus, it is conceivable that the majority of the residents in Z may face certain risks from the new facility for which they will not be sufficiently compensated by the developer, so that they would disapprove of the facility even though it was approved by those living in Y.

The above example illustrates the divergence between private and social costs due to transboundary risks. The private costs to the developer only revolve around residents in Y, while the social costs include the impact on individuals in both Y and Z. Unless some steps are taken to protect community Z against possible economic, health, and safety losses, the above voluntary siting process can prove costly at a societal level.

The presence of transboundary risks is a form of externality that is normally associated with public goods or bads. Thus, a hazardous waste facility poses risks of different degrees to all individuals within a certain radius at the site, and there is little that a person can do to alleviate this risk once a facility is built, short of moving out of the area. Individuals can engage in collective action to lobby against having a facility in the first place, but this involves costs to them that they may not be prepared to incur. Hamilton (1993) has shown that private firms will want to locate facilities in communities or regions that generate the least political opposition, and provides empirical evidence that the host commuities will not necessarily be the ones that generate the lowest externalities.

Role of a Public Siting Authority

The presence of externalities suggests a clear role for a public siting authority (PSA) or government agency to play as part of the siting process. More specifically, such a group would need to impose strict mitigation measures and standards that reduce the risks to both the host community and its affected neighbors before any developer or firm engages in the search for a site.

To reflect adequately the concerns of all the affected residents, the siting authority or government agency would have to have jurisdiction over a geographic region that encompasses both the host community and those areas subject to the transboundary risks from a facility. Furthermore, it would need to be empowered by a governmental body that was concerned with the welfare of a wide area and did not reflect just the narrower interests of citizens from certain jurisdictions.[4]

In the case of waste disposal plants that are on the border of two countries, this may require international cooperation in the development of appropriate standards and regulations. The measures imposed by the siting authority or government agency are likely to be at least as stringent as those recommended by the private developer, because they reflect risks over a broader area than just the host community. Of course, if local land use decisions could be appealed by surrounding communities to some higher court of law, there could be a much lower incentive to locate on borders.

An additional reason for imposing strict safety standards by a public authority or governmental authority before the search for a site begins is the reduction of conflicts that are otherwise likely to emerge between the developer that relies on scientific experts for characterizing risks and the residents in the community who have their own perceptions of the risks.

While the experts normally measure risks in quantitative terms (e.g., the probability and the anticipated consequences of an accident), the public takes other factors, such as dread, unfamiliarity with the technology, and catastrophic potential, into account when evaluating its concerns. Residents in Z who oppose the facility can feed into the fears of those residing in Y, encouraging them to vote against the proposal.

Evidence on how the public's perceptions of risk differ from the scientists' views is illustrated by a study demonstrating that a layperson's willingness to pay (WTP) is particularly high for risks, such as hazardous waste and air pollution from sulphur, where there is considerable dread and uncertainty in the public about their exposure to the risk. Scientists, on the other hand, do not consider factors such as dread or lack of familiarity to be relevant in characterizing the degree of risk from a particular activity (McDaniels, Kamlet, and Fischer 1992). Another study of laypersons and toxicologists revealed large differences between these two groups in their assessments of chemical risks (Kraus, Malmfors, and Slovic 1992).

In addition to standards and regulations imposed at the time the facility is sited to deal with risk perceptions of both the experts and the public, there is a need for monitoring and control procedures to assess the performance of the facility once it is in place. If the facility does not meet the specified standards, then the PSA should have the ability to shut down the facility until there is convincing evidence that standards will be met in the future. Thus, the Shoreham nuclear

TABLE 1. Percentage of U.S. Residents Voting for Different Facilities under Alternative Mitigation Packages

	Landfill	*Incinerator*	*Repository*
No measure	18	14	13
Inspection	53	42	32
Inspection and shutdown	67	58	43

Source: Jenkins-Smith et al. 1993.

power plant in Long Island was ordered to be permanently closed in 1991 after a study revealed that it was impossible to evacuate the residents from the area should an accident occur (Lyall 1991).

When stringent standards that address public concerns with the risk are required, there is a greater likelihood that a positive vote will be forthcoming if a siting referendum is instituted. In a survey, conducted by the University of New Mexico in February 1993, of over 1,200 American households selected at random, individuals were asked about their attitudes toward alternative mitigation measures as a condition for siting different types of facilities (Jenkins-Smith et al. 1993).

Table 1 depicts the percentage of individuals who would vote in favor of a landfill for municipal waste (Landfill) or an incinerator for hazardous waste (Incinerator) within either 1 or 10 miles of their homes, and for a repository for disposal of high-level nuclear waste (Repository) within either 10 or 50 miles of their homes, under the following conditions:

> No mitigation measure is explicitly mentioned (No measure). Regular inspections will ensure that all regulations are being met (Inspection).
>
> Local officials have the ability to close down the facility if they detect any problems (Shutdown).

Table 1 shows that in the case of the landfill only 18% approved the facility with no measure. The percentage increased to 67% when both inspection and shutdown were made part of the mitigation package. These two measures had a positive impact on attitudes toward both an incinerator and a repository, but the percentage favoring either of these facilities was somewhat less than for a landfill. In fact, in the case of a repository only 43% percent of the respondents would have voted in favor of hosting the facility, even if the mitigation package included both measures. The data do suggest the importance of instituting stringent safety measures if one wants to gain support for a facility.

ROLE OF COMPENSATION OR BENEFIT SHARING

For residents to support a facility in their backyard or in surrounding areas, the benefits associated with having it need to be greater than maintainance of the status quo. One way to satisfy this condition is to provide compensation to communities that site the facility as well as to those nearby. Compensation, however, will be viewed as a bribe unless the affected groups feel that the facility satisfies rigorous safety standards that will be well enforced.

Six types of compensation have been identified by Gregory et al. (1991) for facilitating siting decisions. The most common form of compensation is *direct monetary payments,* which can take the form of guaranteed annual payments or tipping fees on the waste that is stored. In Charles City, Virginia, the developer of a landfill collects a tipping fee, which it pays to the city, amounting to about $1 million in revenues annually. This has lowered property taxes and allowed for the rebuilding of the city's ailing school system.

In-kind awards take the form of grants to communities or regions for improving health care facilities, housing, education, or other services that enhance the citizens' well-being and reduce risks that they face. In Swan Hills, Alberta, subsidized housing was provided for thirty-five housing units in conjunction with the siting of a hazardous waste facility (Rabe 1991). In Charles City, the operator collects the county's garbage free of charge. This is also the case in Grandview, Idaho, where Wes-Con provides free garbage pickup to residents as part of a package associated with siting a waste disposal facility (O'Hare, Bacow, and Sanderson 1982).

Contingency funds are used to cover losses from accidents or other adverse effects of the facility. For example, trust funds could be established to cover the damages and health-related costs incurred by victims of an accident.

Property value guarantees protect residents of the host community and surrounding areas against any decline in the resale value of their homes because of the location of the facility. This type of compensation has been offered by Champion International Corp. as part of a program for siting an industrial landfill. The company monitors the changes in the sale prices of homes in the county over a ten-year period and pays residents if there are any adverse changes in the property value due to the presence of a landfill (Ewing 1990). Kodak initiated a similar program to assure homeowners near its production facility that they would not lose equity (Lonsberry 1990).

Benefit assurances guarantee direct or indirect employment for community members either during construction of the facility or during its operational phase. This type of benefit has positive externalities that will make the facility more attractive to neighboring communities as well as to the host site. The hazardous waste treatment center in Swan Hills promised fifty-five new jobs and convinced town leaders that other new developments, such as a new hospital, would now be feasible.

Economic goodwill refers to contributions to local organizations and expenditures for projects that are important to the community and the surrounding area. The private corporation responsible for the Swan Hills facility planted 400 trees for town beautification, provided $65,000 to support local activities, including golf course development, and made charitable contributions, such as sponsoring a hockey school and donating a bear rug to the town council chambers (Rabe 1991).

Several surveys have investigated the impact that compensation has on people's willingness to accept a facility in their backyard. A survey of 844 Tennessee residents in 1989 (Bacot, Bowen, and Fitzgerald 1994) revealed that approximately 30% were willing to accept the construction of a landfill at a nearby site with no mention of benefits. When benefits were provided in the form of rebates on property taxes or for schools, over 60% of these same individuals supported the landfill. In the University of New Mexico national survey (Jenkins-Smith et al. 1995), respondents were asked whether they would vote in favor of a landfill or incinerator when economic benefits were provided to residents within 50 miles of the facility. The percentage supporting either of these facilities doubled when some form of compensation was introduced into the picture.[5]

A PROPOSED VOLUNTARY PROCEDURE

At a National Workshop on Facility Siting in 1990, a group of practitioners and researchers developed a set of guidelines for siting noxious and/or hazardous facilities. These guidelines, which were formalized in a "Facility Siting Credo," focused on developing not only a workable and fair procedure for locating a facility but also an outcome that satisfied distributional (equity) and benefit-cost (efficiency) considerations.

A study of twenty-nine siting cases, both successful and unsuccessful, across the United States and Canada confirmed the importance of two features of the process for finding a community that agrees to host a facility: a broad-based public participation process, and the perception by host community residents that the facility was the best solution to their waste problem (Kunreuther, Fitzgerald, and

Aarts 1993). Both of these elements should be considered in designing a siting process.

The Facility Siting Credo also emphasized the desirability of a voluntary-siting process. This section proposes a three-stage siting process that explicitly addresses the issues of community and transboundary risks while using compensation to address the concerns with equity. The implicit assumption is made that a new facility is viewed as socially desirable. If a volunteer site cannot be found, however, then the status quo will be maintained rather than force a community or region to site a facility. Two key questions are what type of facility to construct and where to locate it.

Stage 1: Screen Appropriate Sites and Specify Standards

In this first stage of the process, the PSA determines a set of sites that meet prespecified technical criteria. At the same time, the PSA specifies a set of safety standards that a proposed facility will have to meet at the local, regional, or national levels, depending on the nature of the risks. For example, if there are transportation risks associated with shipping the material from different sources to their final resting place, then this factor should play a role in determining what sites are suitable candidates. If the facility has the possibility of causing air pollution to neighboring areas, then this risk needs to be considered when specific performance standards for the facility are being set.

Swallow, Opaluch, and Weaver (1992) suggest a procedure for screening all sites so that they satisfy a set of technical constraints associated with health, safety, and environmental effects. For example, landfill sites must meet certain hydrologic and geologic conditions so as to make it highly unlikely that pollutants will migrate off-site.

One issue that should be addressed in screening acceptable sites is whether to exclude certain communities or regions on equity or fairness grounds. There are two extreme views normally taken with respect to this question. If a voluntary siting process is to be utilized, then one can argue that any community can decide for itself whether it wants the facility. This is the position that will be taken in this chapter. On the other hand, there may be concern by the siting authority that communities that already have their share of noxious facilities should be excluded from consideration. Then a siting map should be drawn that excludes these places from being considered, even though they may be technically suitable areas.

In Alberta, fourteen communities were initially interested in hosting a proposed hazardous waste facility. Nine of these were subsequently eliminated either on environmental grounds or because of strong public opposition (McGlennon 1983). In Pennsylvania, legislation was passed that requires the Department of Environmental Resources to screen out potential sites for an LLRW facility using technical criteria. At the same time, citizens have been invited to provide information and articulate concern about the siting process.

Stage 2: Engage in a Referendum

Each community or region that is considered an acceptable candidate on both technical and equity grounds has the opportunity to decide whether it wants to host a facility. To facilitate this process and encourage public participation, planning grants could be given to communities that are seriously interested in considering having the facility in their backyard. These funds would be used for feasibility studies and public-information efforts. The acceptance of such a grant does not imply a commitment to accept the facility. Rather, the funds would be designated to initiate a process whereby the community or region could specify conditions, including compensation arrangements, that would make the site acceptable. Planning grants were utilized in Illinois in an attempt to find a home for an LLRW repository (English 1992) and by the nuclear waste negotiator in an attempt to find a state or Indian tribe willing to host a monitored retrievable storage (MRS) facility for the temporary storage of spent nuclear fuel (Easterling and Kunreuther 1995).

Once the concerned public has an opportunity to consider the proposed site, some type of formal voting procedure needs to be instituted to determine whether there is sufficient support for the facility by community residents. Switzerland has a tradition of using referenda for exactly this purpose and they have been successfully used in finding homes for the disposal of municipal waste (Renn, et al. 1994) and for an underground repository for storing low- and mid-level nuclear waste (Frey and Oberholzer, in press).

In the United States, the most successful example of the use of referenda for finding homes for waste disposal facilities is the Community Partnership Program instituted by Browning Ferris Industries (BFI). Under this program, which was initiated in 1990 in New York State, each town had an opportunity to consider hosting a solid waste landfill in return for a share of the benefits from its operation.

From the outset, BFI went to great lengths to ensure that the process was entirely voluntary. The company required an official invitation by the community even to begin the process, at which time it began an educational campaign of open houses and tours of other BFI facilities. Eagle, a community with 1,300 residents, agreed to this process and eventually held a referendum in which the citizens turned down the facility. True to its word, BFI left town. Only after a grassroots movement organized another referendum that overwhelmingly favored community partnership did the company return. Today a site has been identified and a benefits package worth between $1 and $2 million will go with the new landfill. Hence, most residents feel that they are better off than under the status quo (Browning Ferris Industries 1993).

Stage 3: Lottery-Auction Procedure

Suppose that some type of formal voting procedure has been conducted in each of N communities to determine whether there was sufficient interest by the residents in hosting the facility if a prespecified set of conditions were met (i.e., clearly defined safety standards, monitoring and control procedures, and a well-defined compensation package worth C*). If there are no volunteers for the facility, then either the statue quo will be maintained or some default procedure specified in advance will be instituted.[6]

The siting process and benefits package should explicitly consider the impact of any transboundary risks that may still exist even after safety standards are imposed. To address this problem, it may be necessary to consider the residents both within the community and in affected areas. For example, compensation could be based on the number of citizens exposed to a particular risk, whether they are in the community where the facility is located or in the neighboring locales.

Thus, if a facility was placed near a political boundary, then citizens in the abutting political jurisdiction would be compensated for this risk, and a certain predetermined percentage of voters in both communities would have to approve the facility in some type of referendum.[7]

To be quite specific, suppose that any community where a majority of the residents favored the facility would be a viable candidate for a facility where only a single location was required. Assume that N>1 communities are comfortable with the proposed standards and regulations developed by the PSA in stage one of the process to support a compensation package in a referendum. In the case of Alberta, five communities held plebiscites in 1982, and all supported having a hazardous waste facility in their backyard (Rabe 1991).

The siting authority would now be asked to institute the following two-step procedure to determine the actual host site (see Kunreuther and Portney 1991 for more details):

Step 1. *Lottery to Select Tentative Site.* Each of the candidate sites will be assumed to have an equal chance of being chosen as the tentative site by lottery. If the siting process ends with the lottery, then the winner of the draw knows that it will receive C*.

Step 2. Competitive Process to Determine Final Site. Each of the communities (including the one selected by lottery) now has an opportunity to host the facility by bidding an amount less than C*. The process for specifying a final amount would involve discussions between the developer and each of the host communities. For example, a community that was not selected by lottery might propose to its citizenry a benefits package worth $C<C^*$. If this proposal passes, then the developer would decide to locate the facility in this area unless the lottery winner was willing to accept a similar benefits package.

The most important feature of this process is that if no one decides to submit a bid lower than C*, then the lottery winner hosts the site. At the end of this phase of the process, everyone will know that a tentative site has been chosen to host the facility without any reference to income levels relative to the other candidate sites. Furthermore, this site may well host the facility if the other communities do not want to make a counterbid for compensation lower than C*.

Historically, lotteries have been utilized for selecting individuals to undertake onerous or burdensome tasks such as serving in the army, so this proposed procedure is not unusual. Draft lotteries were utilized to select men for the Army in the United States during World War II and in the Vietnam War (1970, 1971). In the Norwegian draft between World Wars I and II, people were able to sell their right if selected by a draft lottery, an analog to the proposed two-step procedure for siting the facility (Elster 1989).[8]

The equity argument for the lottery is a subtle but important one. When all communities are considered equally desirable for having the facility, there can be no complaints that poorer towns are being forced into taking the facility. Only if a poor community really wants to host the facility will it enter a bid below C*.

To illustrate this point, suppose one of the more affluent communities was chosen by lottery. A low-income community may decide not to enter the bidding even though it was willing to host the facility. For example, it may have agreed to be a candidate site because residents felt that the status quo was unacceptable and a facility was needed somewhere in the region. By volunteering as one of the candidates, public-spirited citizens may have felt that they were satisfying a greater societal need. In offering themselves as a candidate, however, they recognized that others were doing the same. In this case they would have little interest in bidding for the facility once another community was selected by lottery.

From a psychological perspective, there are good reasons to have a lottery as a first step in the sitting process. Following the random drawing, the status quo has been changed from "no facility" to a "potential site" with the candidate having substantial benefits (C*). Let

C_i represent the minimum amount of compensation that a community would have to receive to be indifferent between hosting and not hosting the facility.

Any community not chosen by lottery may now feel that C^*-C_i reflects a loss to them for not having hosted the facility. In other words, the problem is now reframed in their minds, so that if community i decides *not* to enter the bidding in step two, it is likely to feel that it has lost C^*-C_i in potential benefits. In other words, loss aversion (see Kahneman and Tversky 1991) may stimulate an interest in bidding for the facility by those not chosen by lottery.

Of course, there will always be some dissatisfied residents in each of the candidate sites. For the host community, there will be the minority group who opposed the referendum but was unsuccessful in preventing the facility from being sited. For the other communities, there will be a segment of the population that would have liked to have had the facility in its backyard given the available benefits package. By instituting referenda, one automatically disappoints a segment of the public. The challenge is to design a siting agreement so that relatively few individuals would prefer the status quo to hosting a facility.

STRATEGIES FOR SITING LOW-LEVEL RADIOACTIVE WASTE IN ONTARIO

The voluntary process in Ontario for siting of a low-level radioactive waste (LLRW) storage facility offered an opportunity to utilize a lottery-auction approach if the two communities that were seriously interested in the facility had been willing to accept this procedure.[9] As it turned out, only one community volunteered to store the waste, so the issue of introducing such a process will have to await another occasion where there is competition for a waste disposal facility.

Historical Background

The waste to be stored is mainly low-level waste generated by a crown corporation in southern Ontario and dumped in three sites, two of which are temporary depositories. When the decision was made to decommission the storage facilities, decision makers used traditional engineering principles to identify sites that would be the most suitable according to several measures, including environmental impact. This is an example of the classic DAD philosophy of siting, since it did not allow for public participation until after the proposed site was located, and even then usually in a confrontational setting.

While the waste to be moved was low-level and represented a relatively low risk to human health, the "engineering" problem beame political as enraged citizens in prospective host sites took their concerns to national politicians. The process ended not with the selection of a site but, rather, with a campaign promise by Prime Minister Brian Mulroney to block the proposed sites.

Current Procedure

This siting failure illustrated the need to reform the selection process in Ontario so that it would be more attractive to potential hosts. The revised procedure was administered by three federal task forces: one recommended a voluntary-siting program after a two-year study, the second was responsible for implementing phases one to three of the five-phase process, and the third has completed phase four of the process. Assisting the siting task force and municipal councils throughout these stages have been community liaison groups (CLGs). Intended to be strong but impartial partners alongside the task force, the CLGs are composed of interested volunteers representing various local interests from the potential host communities.

One encouraging sign that the voluntary process is an improvement over traditional approaches is the level of trust that has been generated. Although a number of residents of prospective sites disapprove of the construction of a low-level nuclear waste storage facility in their community, there seems to be a consensus that both the task force and the CLGs have been fair throughout the process.

Initially, fourteen communities expressed an interest in learning more about the voluntary-siting process through a series of public meetings. Four communities then voted to become involved in the site search procedure, although only two remained under active consideration through the fourth phase—Geraldton and Deep River. Geraldton, which is located in north central Ontario, had jurisdicational problems that exacerbated an already complicated siting process. This small town is surrounded by crown land, meaning that land would have to be annexed for the new facility. Despite the community's willingness to proceed to complete studies of potential sites and hold a referendum, a nearby Indian tribe was opposed to the annexation, making the required provincial approval of it all the more contentious. Ultimately the decision on land annexation was given too late to permit the necessary site selection technology to be completed and for a referendum to be held within the time frame set by the municipal council. Hence, Geraldton was forced to withdraw as a candidate site.

In Deep River, located in eastern Ontario, the task force completed site selection and negotiated with the town a "Community Agreement in Principle" (CAP) containing mitigation measures, community control, and compensation provisions. A referendum on the CAP was held; 60% of those voting supported the project.[10] The community was already host to a facility run by Atomic Energy of Canada, Ltd., so the citizens were relatively sophisticated about nuclear technologies and their associated risks.

Conventional wisdom would say that a public well-informed of the risks is more likely to vote yes in a referendum. The town's nuclear experience, paradoxically, was thought to be working against the siting of the facility in this case. The general attitude in Deep River was that the waste to be moved was indeed low-level and, hence, relatively low risk. Rather than being willing to accept a low-risk facility in their community, some in the town questioned why the waste needed to be moved from southern Ontario at all. These residents wondered why so much time and money has been spent on trying to find a new home for the waste that required extensive transportation costs and possible risks. These concerns obviously did not carry the day when the referendum was held.

The progress made by the task force has highlighted two issues associated with a voluntary-siting process. The first challenge is the required referendum. A voluntary process, by definition, is one in which communities can and should be able to opt out if they believe the facility is no longer in their interest. The referendum, though, is at the end of phase four, just prior to actual implementation. This raises the question of how much public money should be invested in project assessment if the community will simply opt out in the end.

Second is the question of compensation. The siting task force is not authorized to make any binding offer of compensation to the host community; it can, however negotiate the details of any agreement. The task force can guarantee two things. First, any damage caused by the facility, such as damage to roads or sewers, will be remediated. Second, the community that accepts the facility will receive a net benefit or reward (i.e., equity compensation) for its willingness to help solve a long-standing environmental problem facing the province. Equity compensation, while definitely a part of the overall package, is a matter that can be negotiated in principle and then recommended to the federal government. The government can then decide whether the price to be paid to the voluntary community is appropriate. This raises the question of how many communities might have chosen to stay in the process had they known the size of future compensation packages.

A Proposed Lottery-Auction Procedure

Had both Geraldton and Deep River decided that they were interested in hosting a facility, the lottery-auction procedure described above might have offered an opportunity for both communities to determine whether they would be willing to host the facility to satisfy their own needs as well as broader societal concerns.

Under the proposed procedure the siting authority would first specify the types of monitoring and control procedures that would be followed so that the community and surrounding areas would feel that the facility will be constructed and operated safely (stage one). Furthermore, if there were problems, the residents would know that specific steps would be taken to correct these defects or close the facility.

In order for such a procedure to work, some type of compensation or benefit-sharing package would have to be put on the table. More specifically, the Ontario Siting Task Force should present both communities with a package that could be part of their decision on whether to support the facility. The Indian tribe next to Geraldton would also have to be given compensation for its land in order for it to support the facility.

If a lottery is instituted between these two communities, both would view the hazardous waste treatment plant as a trade for some benefits that at the same time helps the province of Ontario and Canada. Whether there would be some type of bidding system or other type of negotiation between the communities and the siting task force would have been a topic for discussion prior to the lottery.

As it turned out, Deep River was the winner of the "lottery" because Geraldton withdrew from consideration. The use of a referendum in Deep River highlighted the importance of a voting procedure where a proposed mitigation and compensation package was designed by the siting task force for residents to consider. In essence, the referendum had the same features as the lottery procedure: 1) it put the key issues of mitigation, monitoring, and compensation on the table for everyone to consider, and 2) it did not commit either community to take the facility without due process.

OPEN QUESTIONS

If the voluntary siting process is utilized, one of the issues that deserves further discussion is the role that environmental groups and organizations can play. Suppose that one of these public-interest groups feels that it would be inappropriate for community Y to host a proposed incinerator because it feels that the technology is unsafe.[11] What are acceptable ways that information on the public interest group's views of the risk can be presented to the affected residents?

More studies need to be undertaken not only to understand what the affected public considers to be an intrinsically fair process but also to address issues of what constitutes an acceptable outcome. How can appropriate safety standards be developed and implemented so that the citizenry feels it is protected? What type of specific compensation programs are able to address the range of concerns that arise in response to siting proposals? Are there any generic compensation strategies that have been successful across different types of facilities, or does the developer and siting authority need to tailor the strategy to a particular facility? How can trust between the different interested parties in the siting debates be restored so that there is a greater chance that facilities will find homes in the future?

There are no easy answers to these questions. It is my view, however, that a voluntary process is a very promising framework for incorporating procedures and policies that will be viewed as reasonable by the relevant participants. The contributions and case studies in this volume attest to the potential of voluntary siting for turning "Not in My Backyard" (NIMBY) into "Maybe in My Backyard" (MIMBY) and then into "Yes in My Backyard" (YIMBY).

NOTES

1. This chapter reflects helpful discussions with Doug Easterling, Kevin Fitzgerald, Paul Kleindorfer, Roger Kasperson, Paul Slovic, and John Villani. Partial support for the paper is provided by the William and Flora Hewlett Foundation and NSF Grant #88-90299.

2. For summaries of the literature, see Portney 1991; Kleindorfer and Kunreuther 1994.

3. Decisions to locate near political boundaries often occur in practice. For example, Tulleytown, Pennsylvania, chose to locate a hazardous waste facility right near its borders, and this created considerable controversy with its neighbors. See Faulhaber and Ingberman (1995) for a discussion of this and other examples, as well as a rationale for locating at the political boundary.

4. If a public siting authority solely reflected the interests of those who elected government representatives, then it may be unconcerned with the safety of residents in other communities who may not have supported them in the last election.

5. For more detail on the results of other surveys and the role that compensation plays in siting hazardous facilities, see Kunreuther and Easterling 1995.

6. For example, in the case of storing high-level radioactive waste, the default option could be on-site interim storage at nuclear reactors in special-purpose canisters (Easterling and Kunreuther 1995).

7. I am grateful to John Villani for suggesting this alternative procedure for determining compensation if there are transboundary risks.

8. In contrast to the successful Norwegian experience, a lottery with an option to buy out of the Army was used during the U.S. Civil War, and it led to draft riots. This suggests the importance of making sure that every community understands the implications of the two-step lottery-auction procedure before a referendum or plebiscite is conducted to determine whether residents want to participate in it.

9. I am grateful to Dennis Wood, head of the Ontario Siting Task Force, for providing the background material on the siting process in Ontario. He participated in a Wharton Risk Management and Decision Processes Center siting workshop at the University of Pennsylvania in January 1994, where he described the challenges Ontario faced in utilizing a voluntary procedure. He also has provided an updated status of the situation in a personal communication to Don Munton on October 11, 1995.

10. More than 70% of the eligible residents in the community voted on the referendum.

11. Organizations such as Greenpeace have strongly opposed the construction of incinerators in communities because they feel the incinerators produce environmental hazards. They have developed a set of guidelines for preventing these facilities from being built (*Greenpeace* 1991), but there is limited empirical evidence as to what impact these efforts have had on community attitudes toward these facilities.

REFERENCES

Bacot, H., T. Bowen, and M. Fitzgerald. 1994. Managing the Solid Waste Crisis: Exploring the Link between Citizen Attitudes, Policy Incentives, and Siting Landfills. *Policy Studies Journal* 22:229–44.

Browning Ferris Industries, Inc. 1995. Personal communication from Rob Raynham, BFI, Inc., to author. April.

Carnes, S. A., E. D. Copenhaver, J. H. Sorenson, E. J. Soderstrom, J. H. Reed, D. J. Bjornstad, and E. Peele. 1983. Incentives and Nuclear Waste Siting: Prospects and Constraints. *Energy Systems and Policy* 7, no. 4:324–51.

Easterling, D., and H. Kunreuther. 1995. *The Dilemma of Siting a High Level Radioactive Waste Facility*. Boston: Kluwer.

Elster, J. 1989. *Solomonic Judgments: Studies in the Limitations of Rationality*. New York: Cambridge University Press.

English, M. 1992. *Siting Low-Level Radioactive Waste Disposal Facilities*. New York: Quorum Books.

Ewing, T. F. 1990. Guarantee near a Landfill. *New York Times*, July 8.

Faulhaber, G., and D. Ingberman. 1995. Markets vs. Governments: The Political Economy of NIMBY. In *The Political Economy of Environmental Protection: Analysis and Evidence*. Ann Arbor: University of Michigan Press.

Frey, B., and F. Oberholzer. In press. Falt Siting Procedures: An Empirical Analysis of Text Importance and Characteristics. *Journal of Policy Analysis and Management*.

Greenpeace. 1991. How to Stop the Incinerators: A Community Owner's Manual. January–February, 20–21.

Gregory, R., H. Kunreuther, D. Easterling, and K. Richards. 1991. Incentives Policies to Site Hazardous Facilities. *Risk Analysis* 11:667–75.
Hamilton, J. 1993. Politics and Social Costs: Estimating the Impact of Collective Action on Hazardous Waste Facilities. *RAND Journal of Economics* 24:101–25.
Jenkins-Smith, H., H. Kunreuther, R. Barke, and D. Easterling. 1993. UNM Mitigation/Compensation Survey. Unpublished report.
Kleindorfer, P., and H. Kunreuther. 1994. Siting of Hazardous Facilities. In *Handbook of Operations Research* by S. Pollock, A. Barnett, and M. Rothkopf. Amsterdam: Elsevier/North-Holland.
Kraus, N., T. Malmfors, and P. Slovic. 1992. Intuitive Toxicology: Expert and Lay Judgments of Chemical Risks. *Risk Analysis* 12:215–32.
Kunreuther, H. 1995. Voluntary Siting of Noxious Facilities. In *Novel Models of Participation,* edited by O. Renn and T. Webler. Boston: Kluwer.
Kunreuther, H., and D. Easterling. 1992. Gaining Acceptance for Noxious Facilities with Economic Incentives. In *The Social Response to Environmental Risk,* edited by D. Bromley and K. Segerson, 151–86. Boston: Kluwer.
———. 1995. The Role of Compensation in Siting Hazardous Facilities. Wharton Risk Management and Decision Processes Center Working Paper 02–01–95. Philadelphia: University of Pennsylvania.
Kunreuther, H., K. Fitzgerald, and T. Aarts. 1993. Siting Noxious Facilities: A Test of the Facility Credo. *Risk Analysis* 13:301–15.
Kunreuther, H., and P. Portney. 1991. Wheel of Fortune: A Lottery/Auction Mechanism for Siting of Noxious Facilities. *Journal of Energy Engineering* 117, no. 3:125–32.
Lonsberry, Robert. 1990. New Kodak Home Incentives. *Rochester Times-Union,* 1A.
Lyall, S. 1991. U.S. Appeals Court Clears Way to Dismantle Shoreham A-Plant. *New York Times,* April 20, 1.
McDaniels, T., M. Kamlet, and G. Fischer. 1992. Risk Perception and the Value of Safety. *Risk Analysis* 12:495–504.
McGlennon, J. 1983. The Alberta Experience . . . Hazardous Wastes? Maybe in My Backyard. *The Enviornmental Forum* 2:23–25.
O'Hare, M., L. Bacow, and D. Sanderson. 1983. *Facility Siting and Public Opposition*. New York: Van Nostrand Reinhold.
Portney, K. 1991. *Siting Hazardous Waste Treatment Facilities: The NIMBY Syndrome.* New York: Auburn House
Rabe, B. 1991. Beyond the NIMBY Syndrome in Hazardous Waste Facility Siting: The Alberta Breakthrough and the Prospects for Cooperation in Canada and the United States. *Governance* 4:184–206.
Renn, O. 1994. Fairness and Competence in Siting a Landfill: A Case Study from Switzerland. Paper presented at the IIASA Workshop on Fairness and Siting, May 21–22, at the International Institute for Applied Systems Analysis, Laxenburg, Austria.
Swallow, S., J. Opaluch, and T. Weaver. 1992. Siting Noxious Facilities: An Approach that Integrates Technical, Economic, and Political Considerations. *Land Economics* 68:283–301.

Tversky, A., and D. Kahneman. 1991. Loss Aversion and Riskless Choice: A Reference Dependent Model. *Quarterly Journal of Economics* 106:1039–62.

Zeiss, C. 1991. Community Decision-Making and Impact Management Priorities for Siting Waste Facilities. *Environmental Impact Assessment Review* 11:231–55.

Directions for Engineering Contributions to Successfully Siting Hazardous Waste Facilities

CHRISTOPHER ZEISS

INTRODUCTION

Engineering expertise often drives the linear, goal-oriented process of siting waste facilities and efficiently seeking approval under short deadlines and stringent cost limits. The design is conducted without input and is set to meet government regulations at least cost. This efficient approach, however, often contributes to community resistance because it ignores elements, and violates key principles, of community decision making. Failed siting attempts are inefficient because they do not provide the necessary service, often at great cost.

Therefore, a careful analysis of the engineering design process, on the one hand, and of the community decision-making process, on the other, are presented to contrast the differences and develop successful engineering siting intervention measures.

Engineering design has been defined as a process consisting of six sequential steps (French 1971; Ullman 1982; Love 1986):

1. Identifying and defining the need for the facility as precisely as necessary or possible in order to achieve the required production or service; waste facilities basically provide waste treatment and disposal services
2. Analyzing the problem with a statement of the limitations placed on the facility (standards, market requirements, time lines, etc.) and with the criteria of performance (costs, reliability, safety, environmental quality, fairness, etc.)
3. Developing conceptual designs as broad solutions, such as different treatment and disposal methods (e.g., landfilling, physical-chemical treatment, incineration, etc.), with an outline of the means of performing each major function but without elaborating on the details of established methods

4. Evaluating and selecting a solution, based on the performance criteria and limitations developed under the problem analysis (point 2) and through development of the solution in detail
5. Building and operating the facility as designed
6. Monitoring the performance of the facility to ensure that the criteria are met

These generic engineering design steps are shown in sequence in Figure 1. The engineering design of a waste facility is often presented and seen from the outside as a linear process with the final solution as the single outcome. In contrast, however, the engineering design process involves frequent and intensive iteration from subsequent steps (evaluation of designs) back to preceding tasks, such as the definition of need, problem analysis, and concept designs. The key characteristic of engineering design processes is this iterative search for solutions with more flexible, lateral, and open thinking than is evident from the final solution. This chapter identifies and develops engineering intervention methods to consider, integrate, and address environmental impacts and host-community concerns within the engineering design process.

The premise of the proposed design approach is that engineering expertise is essential to the proper design and operation of waste facilities and, in particular, to the reduction and management of facility impacts. Engineering efforts must therefore not only provide proper

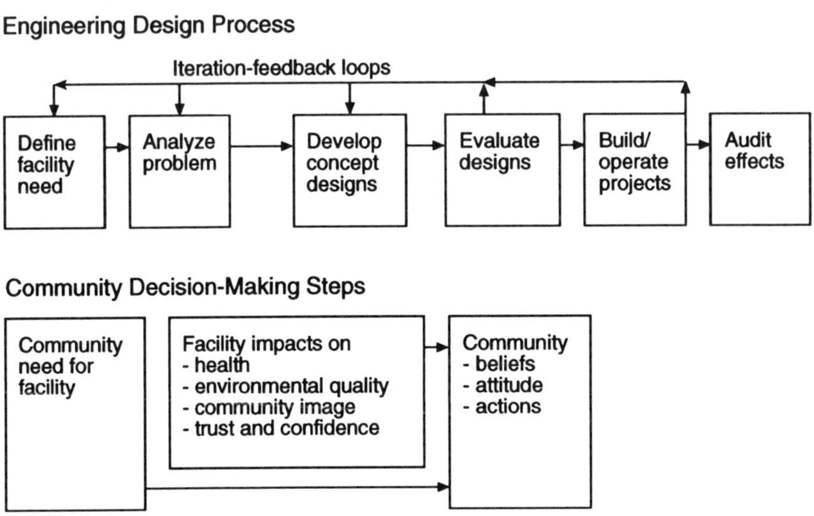

Figure 1. Engineering Design Process and Community Decision Process

design and operation to meet health, safety, and regulatory requirements but also effectively address host community concerns about facility effects. Siting agreements may contain a large range and variety of "incentive" measures to avoid, reduce, or redress actual or possible impacts caused by waste facilities. Because the possibilities are so diverse, it is extremely important for engineers to choose the most effective approaches and sequence to avoid the use of expensive, ineffective measures and, maybe more important, to avoid unintended negative effects caused by the incentive measures themselves. The selection of substantive interventions for siting packages must be based on valid siting principles to ensure positive and effective results. This chapter presents the theory and application of effective engineering intervention measures so that they fit into the engineering design and the waste facility siting processes.

The following section presents a failed siting attempt and identifies the elements of a prospective host community's attitude formation process. These elements are contrasted with the engineering design process to develop siting intervention measures. Then the principles of effective siting interventions are developed from the theory and applied in the discussion of siting cases. At the end of the chapter, the siting interventions are synthesized into a proposed design and siting process.

SETTING THE STAGE—A CASE STUDY OF A MAROONED SITING ATTEMPT

With the opening of the Alberta Special Waste Treatment Centre (ASWTC) in Swan Hills, Alberta, in 1987, the siting, construction, and operation of a comprehensive, integrated hazardous waste treatment center had been accomplished. While the access to treatment was provided for generators in the province's two large cities, most of the other, smaller communities were still without access points. As a result, the owners of the treatment facility initiated a program to distribute five to six smaller transfer stations to capture hazardous waste from smaller regional centers. Private companies were to build, own, and operate these transfer stations (Zeiss and Paddon 1992).

The anticipated size of transfer stations could range from 5,000 to 20,000 ft^2 of building floor space on 2- to 5-acre lots. The total investment is anticipated to range from $500,000 to $1 million per transfer station. This type of operation would likely employ two to three persons, possibly on a part-time basis. Spill containment measures include secondary containment (curbing, pallet trays, retention sumps), as well as gas and fire detection and extinguishing systems. Safety and emergency procedure plans are required. The most likely

TABLE 1. Waste Facility Impacts and Benefits

Physical Impacts	*Nonphysical Impacts*	*Facility Need and Benefits*
Health Risks	**Economic**	Protection of health and environmental quality
Air quality	Property values	Convenience
Water quality	Infrastructure costs	Employment
Fires, explosions	Slow development	Tax revenue
Accidents	**Social**	Development and diversification
Nuisances	Image, status	Energy and resource recovery
Odor	**Political**	
Noise	Control	
View	Fairness	
Environmental	Confidence	
Quality		
Air		
Water		

areas for location of facilities are industrial parks with light to heavy industrial or commercial land uses. These land uses are likely to be least sensitive to the types of risks and impacts that are commonly expected to occur from transfer stations (see table 1).

An oil and gas field service company headquartered in Calgary was very interested in the concept. During the spring of that year, the company identified potential locations. By the summer, the firm had decided to establish a transfer station to handle primarily oil and gas hazardous wastes. The community identified was Morinville, a small (population 5,400 residents) farming community that also served as a bedroom community to Edmonton. Morinville is 60 kilometers north of Edmonton on a major highway with access to Swan Hills. The community's industrial park, situated 1 kilometer from the nearest residence, had considerable available property, and the company had several employees residing in Morinville.

Morinville's economic development officer was enthusiastic about the business venture, and a development permit was granted, subject to the advertising of the project with a review at a public meeting. The town councilors who supported the business venture saw an opportunity for Morinville to diversify its economy and to add to its tax base. Morinville had suffered considerably during the recession of the early 1980s, with a number of businesses leaving the industrial park. Councilors on the economic development committee were satisfied that provincial regulations offered the necessary safeguards for the operation of the facility, and many were excited to have the chance to participate in a new industry that was believed to have considerable potential.

When the advertisement was published in the *Morinville Record*, it failed to explain that the transfer station would be handling hazardous waste. Prior to the public meeting, however, word got out in the community that the firm's plans involved collecting hazardous waste from around the province and consolidating it into large consignments for shipment to Swan Hills. By the time of the public meeting, many people had become concerned about the proposal.

Under the terms of the development agreement, the public meeting came with less than one week for the community to appeal the permit. At the appeal hearing, several hundred people packed the hall carrying placards expressing their rejection of the proposal. The quotes in Table 2 reflect the concerns in the community about the transfer station's impacts. The lack of trust and confidence in the process, the government, and the proponent; the absence of understanding for the need for the facility; and the concern with a new and unfamiliar type of facility are all nonphysical impacts, while community image and health risk impacts are physical impacts. No nuisance impacts were mentioned at this point.

As a result, the appeal board rejected the application. The proponent firm's president said that his company had learned quite a bit

TABLE 2. Community Concerns about a Hazardous Waste Transfer Station

Quote	*Impact Concern*
Why do we need it, and why here?	Need
I'm worried about health risks to me and my children.	Health to current and future generations
A nice rural community is no place for industry and urban waste.	Community image
We don't know if the transfer station will fit into an industrial park.	Unfamiliar new facility
Why should we take their waste?	Fairness, responsibility
Why is the transfer station being sited here? Is someone taking bribes?	Confidence, trust
We didn't have enough time!	Confidence in process
Why didn't they say the transfer station was for hazardous waste?	Trust

from the experience. "We will be more cautious in our approach. We will try to educate the public more than we did this time, before we apply for a permit," he said. The few who supported the development advocated economic benefits—jobs and tax revenues. The opposition was of two camps: those who did not see the development "fitting" into their community (a rural town that is not the place for industry) and those who were fearful of the pollution and potential health risks. Finally, confidence in the proponent and the process was severely diminished.

In summary, while the economic benefits were recognized, the need for the facility from the community's perspective was questionable. The facility's impacts on health and community image and the residents' distrust of the facility design and the siting process resulted in negative beliefs and created strong negative attitudes among many residents and stakeholders. The benefits were presented as economic gains but were outweighed by the physical risks and intangible impacts that the hazardous waste transfer station may have generated. This is a familiar cause for similar outcomes of siting attempts (Zeiss and Atwater 1987).

The community considers the need for the facility and the impacts on the community from its perspective. Community need and impacts on health, environmental quality, community image, trust, and confidence influence community beliefs and attitude about the facility. These elements have been noted elsewhere (Bord et al. 1985; Kunreuther et al. 1990; Zeiss 1991; Zeiss and Paddon 1992). Community need, facility impacts, and the resulting community beliefs, attitude, and actions are shown in Figure 1 in the sequence they are considered by host community residents (Fishbein and Ajzen 1975; Zeiss and Lefsrud 1995). The arrows show the cause-effect interactions. Community need directly affects beliefs and attitude early in the process. Then facility impacts are considered (Zeiss and Lefsrud 1995).

In the conventional siting process (as outlined in the marooned siting attempt), the community's consideration of these elements usually proceeds after the engineering solution has been presented. Some of the community decision-making steps, however, resemble those of the waste facility design process. These elements are aligned with the corresponding engineering design steps in figure 1. Just as engineering defines facility need, the community identifies need from its perspective. The community analyzes the problem of hosting a facility by identifying impacts the facility might cause and evaluates the solution by forming an attitude and deciding what actions to take. The similarity suggests three categories of siting interventions to mediate between the engineering design and the community decision-making processes: (1) need identification, (2) impact assessment and management, and (3) benefits and offsets.

In the next section, engineering siting intervention measures are developed from the theory and are applied to practical siting cases.

ENGINEERING SITING INTERVENTIONS

Theory and Principles

Relevant theoretical principles are identified and developed into guidelines for designing engineering siting interventions. Need definition, impact assessment and management, and facility benefits, compensation, and incentives are analyzed.

Need Identification

The large societal need for waste facilities are often apparent and undisputed in the provision of public health through proper and efficient waste disposal. A waste facility's capability to fulfill a locally apparent need, however, is scarcely analyzed. While this point may appear intuitive, several studies in Canada and the United States support the importance of determining the local need in addition to the broader societal need for the facility (Gregory et al. 1991; McQuaid-Cook and Simons 1989; Camerer and Kunreuther 1989; Zeiss 1991; Zeiss and Paddon 1992; all for waste facilities).

"Need" is defined, for the purposes of siting, as the recognition of a deficiency in the status quo. Therefore, need, by this definition, is present in a host community when the community believes its status quo condition to be below a desired or expected level, the reference point. Therefore, the fulfillment of a need redresses a current loss and contrasts with the approach of adding new benefits to a siting agreement. The reasons for the difference are provided by prospect theory (Kahneman and Tversky 1979, 1984).

Prospect theory predicts that losses weigh more heavily than benefits (Figure 2 (a); Kahnemann and Tversky 1979). The larger absolute values of losses are reflected in the steeper slope of the loss curve in the prospect theory. The values of losses have been shown to exceed those of equal gains by two- to fifteenfold. This asymmetry of loss and gain values is exhibited for decisions about money, environmental quality, fishing licenses, lottery tickets, and other units. It seems to apply for many goods (Knetsch 1983). Therefore, residents are predicted to respond more sensitively to impact losses from waste facilities than they are likely to be attracted by benefits. As a result, reducing the facility impact loss L by an impact reduction ΔL is predicted to be more valuable in improving the residents' attitude about a facility than offering compensating benefits as additional gains ΔG from the reference point (Zeiss 1991; Figure 2).

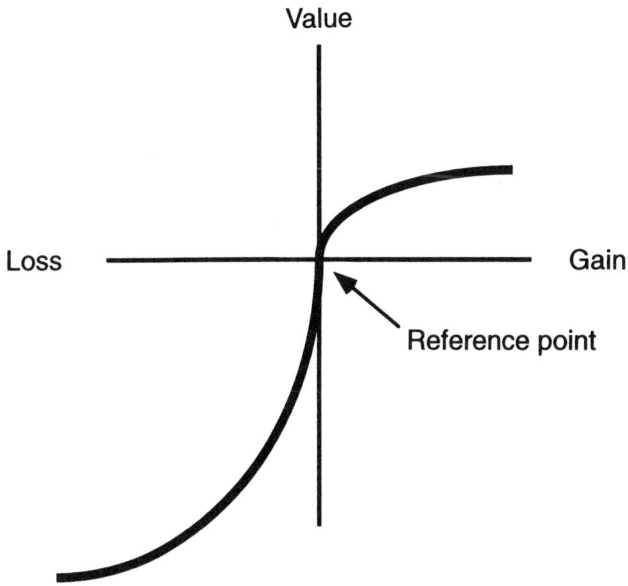

a: Prospect Theory Value Curve (source: Kahnemann and Tversky 1979)

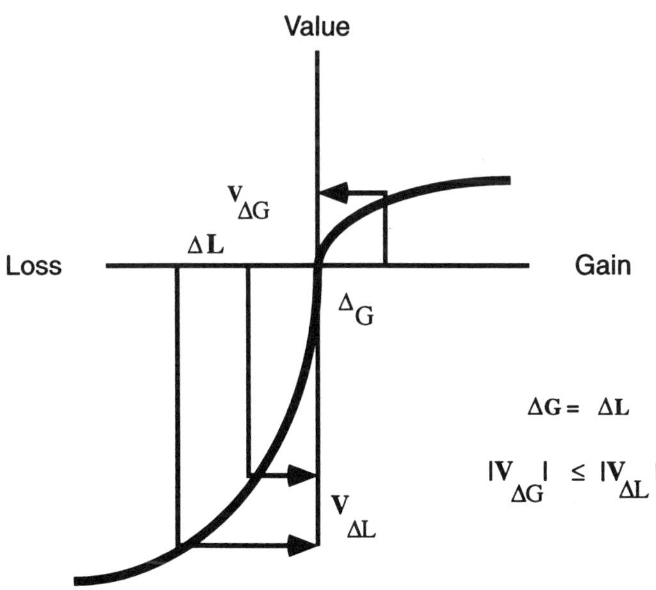

b: Loss Reduction Value Exceeds Compensating Gain Value (source: Zeiss 1991)

Figure 2. Residents' value curves for facility impacts and benefits (source: Zeiss 1991)

The location of the reference point is a crucial factor for value judgments about impacts from hazardous waste facilities. The status quo is a likely candidate but may be modified or rivaled by the community's expectations or social comparisons. While Kahneman, Knetsch, and Thaler (1991) state that the determination of reference points and the influences of status quo entitlements and expectations are well established, these methods are not referenced and may not be so easily developed. Gregory et al. (1991) show that the evaluation of a proposed (environmental) change depends significantly on its framing. Presenting a proposed change as the recovery of a prior loss increased the desirability of the change over that of achieving a gain. Thus, the affected decision makers' knowledge of prior states (conditions), a host community's comparison of its own situation with other communities, or its expectations for the future may cause reference points to shift to a higher level than the status quo. These shifts may not be easy to track but may result in significantly different framing of a proposed facility.

The need to consider changes as sets of gains and losses from a reference point requires that the decision-maker first "frame" the facility. Framing determines whether facility effects are considered gains, losses, or loss reductions from the community's reference point. The setting of the reference point thus determines on which side of the value curve the changes will fall. Framing is important because it may determine the outcome of value judgments.

Frames may be suggested or influenced by the way decisions are posed and can lead to methodological difficulties. Suggested frames, however, may not always be accepted by decision-makers, particularly if the decision-makers are firmly anchored on a reference point and on decision attributes that are strongly suggested by social norms, cultural biases, comparisons, or expectations (Vaughan and Seifert 1992). Further, frames can be rejected because community attributes are kept in different mental accounts (Thaler 1985). Where there is a salient community need, i.e., a perception that the community is below its reference point on the left-hand "loss" side of the value curve of an attribute (e.g., air quality), gains in this attribute may not be intuitively traded off with other kinds of facility impacts (e.g., on water quality). If, however, the facility can be framed, i.e., designed and presented, to address existing discrepancies (where the community status quo is below the reference point), then the value of the facility to the community is likely to be much larger than the value of any new gains as benefits or incentives. In past cases, this has resulted, on the one hand, in better community acceptance (Zeiss 1991) and, on the other, in lower incentives demands (Nieves et al. 1992).

Defining the facility need is synonymous with, first, determining that the status quo is unacceptable (that there is a "need to do something") and, second, selecting a site and designing the facility to improve on the status quo (Zeiss and Paddon 1992; Kunreuther and Susskind 1993). Zeiss and Paddon (1992) and Carlson (1991) mention and illustrate need identification as an element for the host communities of waste facilities and prisons. Jaffe (1991) concludes that need definition should occur before negotiating other package elements. Kemp (1992) and Carnes et al. (1983), however, extend the requirement to justify the need for each compensation and incentive measure. All compensation and incentive measures, it is suggested, should be justified in terms of local need (Kemp 1992). Their purpose, structure, implementation, and consequences should be carefully designed to fit specific community needs. This is proposed principally to avoid the impression of directing attention away from key issues with bribes (Carnes et al. 1983).

Need definition and status quo unacceptability should be established from the host community's perspective. The selection of an optimal facility and site combination to fulfill the needs should be demonstrated and discussed early for the facility and, later, for each compensation and incentive measure.

Impact Management

In order to be able to accurately address impacts with engineered impact management methods, the structure and interaction of facility effects and residents' beliefs and attitudes about waste facility impacts has to be understood. Previous studies revealed that while people generally mention physical impacts and nuisances (odor, noise, traffic, view) as their concerns, specific concerns about impacts at a specific site comprise elements in all physical and nonphysical categories, including the economic impacts on property values; social impacts on community image; and political impacts through loss of control, loss of confidence, and unfairness (Zeiss and Atwater 1987). For example, at a landfill, nuisances (noise, odor, view, litter, traffic, and animals), health risks (air and water emissions), environmental change (air and water quality), property values, community image, and control are perceived to be affected by the facility (see fig. 3). At an incinerator, residents' negative beliefs concern odor and view, air emissions and quality, impacts on property values, community image, control, fairness, and confidence in government, facility owner, and environmental agency.

Furthermore, residents' exposure to physical impacts is statistically associated with the host community responses about their beliefs and, thus, confirms the linkage of physical impacts with beliefs, and

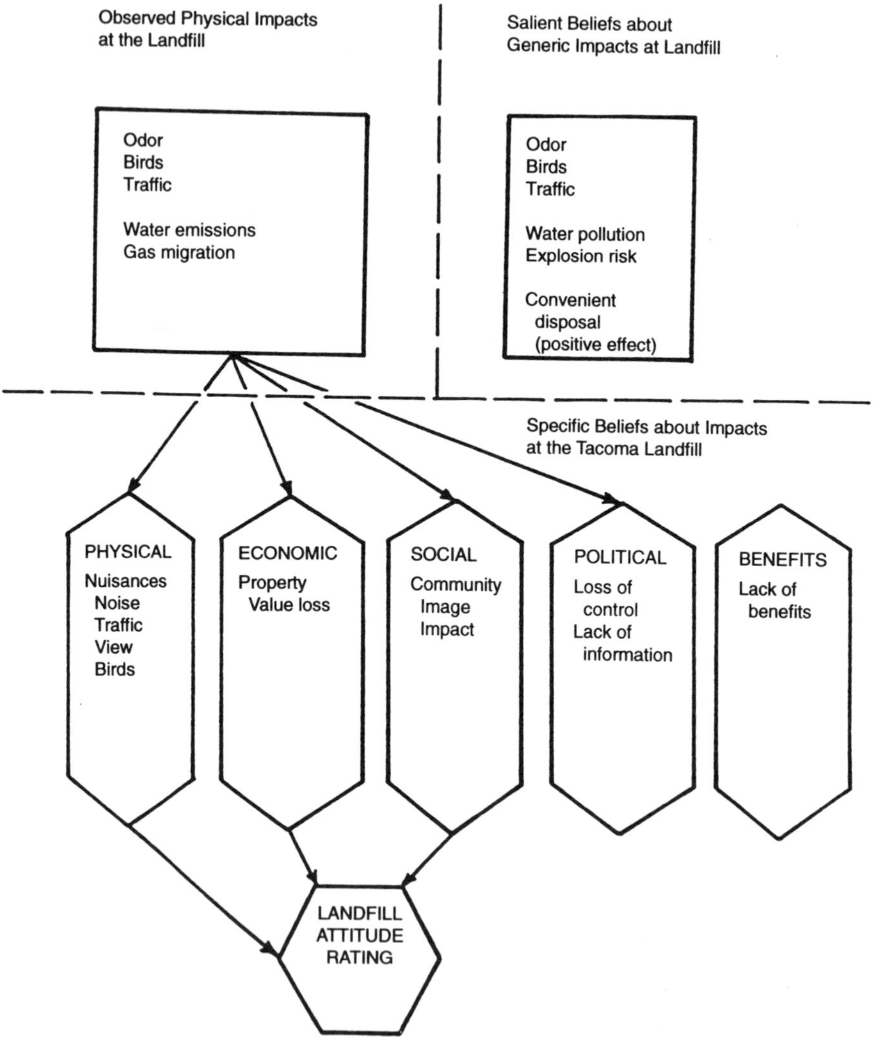

Figure 3. Facility impacts and host community beliefs and attitude (source: Zeiss and Atwater 1991)

indicates connections ("triggers") between physical impacts and beliefs about nonphysical impacts (Zeiss and Atwater 1991).

In summary, residents' spontaneous beliefs about generic facility impacts comprise a narrow range of obvious physical impacts, i.e., nuisances and health risks from water and air emissions. Detailed beliefs about a specific facility, however, include the entire set of physical and nonphysical impacts, including economic, social, and political impacts. The empirical relationship found here supports the hypothesis that residents shift and flatten their concerns about impacts to express only common and intuitive physical impacts about waste facilities in response to simple, straightforward questions. Underlying the obvious and well-documented impacts (odor, noise, air, and water impacts), however, is a larger, broader set of negative beliefs (concerns) about waste facility impacts (economic, social, and political effects). In voicing their concerns, residents may shift their statements to express common physical impacts, rather than the full range of impacts they are concerned about.

The perception of nonphysical impacts is influenced by exposure to obvious physical impacts; that is, property value, community image, control, and fairness concerns may be aggravated by view, noise, and nuisances from a waste facility. This finding represents, in essence, the reverse of "out of sight, out of mind." Apparent but minor physical impacts may trigger among exposed residents much stronger negative beliefs about nonphysical effects than the negative beliefs they cause about directly related physical impacts. Residents appear to interpret the perceived physical impacts in terms of their economic (property value), social-community (image), and political (control, fairness) meanings to themselves.

The research findings result in several essential changes to the planners' and engineers' role in the siting process:

1. Waste management officials must be aware of the interaction of waste facility impacts with the host community's beliefs and attitudes about waste facilities and about itself. Waste facilities are seen to be the cause of obvious physical impacts, such as health risks and nuisances; in addition, they trigger important nonphysical impacts and risks to community image, control, fairness, and confidence. Waste managers must be sensitive to the underlying significance of the changes to the host community.
2. Physical impacts must first be carefully predicted. Then nonphysical impacts must be identified by contrasting the facility impacts with host community characteristics and values. This can be achieved by the expert judgment of, for example, the planner or engineer who knows the community; or for a more detailed analysis, community beliefs about the proposed waste facility, together with host community values, can be elicited by a written survey of residents, personal interviews with formal and informal community leaders, or stakeholders' discussions. The list of possible issues should consist of physical, economic, community image, and political impacts.

During the process of tracing impact-belief-attitude pathways, the waste manager must attempt to acknowledge expressed concerns and beliefs and explore residents' underlying rationale through questions rather than deny or argue seemingly irrational beliefs. First, by acknowledging the residents' concerns, the planner or engineer does not admit that the (physical) impact occurs, or occurs to an exaggerated level; rather, she or he simply validates her or his understanding of the residents' concern. This validation in itself has been shown to reduce residents' stress and fear. Further, the waste manager is interested in mapping the pathway from facility impact to resident attitude by reverse deduction. By listening to the residents' immediate concern and asking questions to determine underlying trigger beliefs, the waste management official can outline the rationale that leads to the basic cause of opposition.

For example, a local farmer, an influential informal leader, expresses during a personal interview the concern about air pollution from an incinerator stack (see the statement in Figure 4). The waste manager acknowledges that, in the farmer's mind, air emissions occur, and could ask further whether, to address the farmer's concern, the proponent should be required to install better pollution control equipment. The farmer, who is well informed, says he understands that the exhaust filtration system will remove over 99% of contaminants

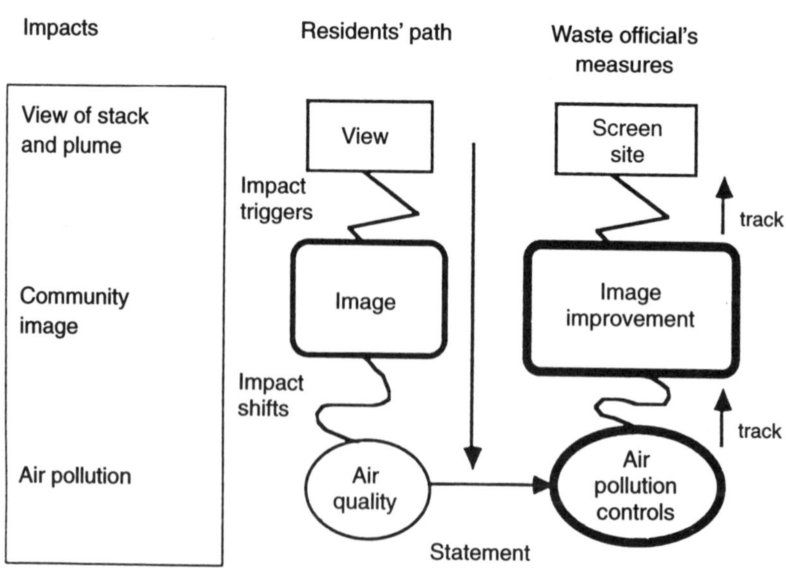

Figure 4. Tracing residents' beliefs through shifts to trigger impacts (source: Zeiss and Atwater 1991)

and that he knows his crops won't be hurt, but expresses his anger at having an industrial facility such as the incinerator in an agricultural area (see shift of impacts in Figure 4). The waste manager again acknowledges the concern without saying whether he or she agrees with the farmer, and asks what feature of the plant most strongly creates the impression of an industrial facility. The farmer says he can see the stack, and often a plume, from his farm (see trigger connection in Figure 4). Through his conversation the waste manager can trace the farmer's rationale from initial physical concern (air pollution) to concern about community image due to the industrial image of stack and plume (see tracking in Figure 4). As a result, the waste official can generate impact management measures to address the real concerns. Parallel to tracing the farmer's concern from the salient "air pollution" to the underlying concern about "community image" to the physical trigger, the view of stack and plume, appropriate impact management measures can be derived to address the farmer's level of concerns. Clearly, the costs and effects of these measures differ. Therefore, the proper choice and the ensuing costs depend on the correct identification of underlying concerns. Then the most effective measure can be chosen to address the underlying concern and enhance residents' attitude. Here, eliminating the steam plume and providing good screening and landscaping may alleviate the farmer's concern about the industrial image of the incinerator. This is, of course, not an alternative to adequate air pollution controls to meet health, environmental quality, and nuisance standards. Rather, this measure may be very effective to address any remaining concerns, such as this farmer's issue with community image impact. These measures may include design engineering changes, siting process changes, or landscaping, screening, or traffic management measures. By identifying and addressing underlying trigger issues, costly but ineffective measures can be avoided and effective measures can be selected instead.

Once impacts have been predicted and host community concerns identified, the proponent can design effective impact management methods.

Four methods to reduce salient impacts and risks to health and the environment are found to be effective in allaying community concerns:

- *Prevention* is defined as a measure that affects the cause prior to the emission source—i.e., before the siting process begins or before the waste stream reaches the facility—and results in a reduction of facility impacts.
- *Controls* are measures that reduce impacts through design or operating measures to reduce emissions, e.g., air pollution con-

trols, landfill liners, etc. In other words, control measures are elements of engineering design or operation that take effect only if and when the facility is operating.

- *Mitigation* comprises all measures to reduce impacts downstream from the facility and after the facility is built. Basically, this category includes measures outside the compliance boundary, i.e., those that reduce impacts at the receptor rather than at the source.
- *Compensation* measures are alternative benefits that are offered instead of impact reduction. Compensation can be in kind, e.g., as the provision of an alternative water source for contaminated groundwater, or in an alternative good, e.g., as a different intangible good or as money.

The prospect theory's asymmetrical value curve furthermore provides engineering priorities for addressing specific concerns about impacts and risks. Because losses weigh more heavily than gains, impact management measures that reduce losses will more effectively improve the prospect of a waste facility than offering new gains as additional compensatory benefits. As a result, engineering efforts to address community concerns should, by this logic, focus on impact and risk reduction rather than on concocting new perks (as fire trucks, community halls, etc.) for putting up with impacts. By using the three categories of impact reduction and a compensation option (see Figure 5), Zeiss (1991) measured host community residents' preferences to test this prediction.

Four general approaches to fit the definitions of impact management measures (see table 3) were submitted to cost estimates and presented to sixty residents in five pairs of host and control communities of existing waste facilities in 1987 and 1988. Despite higher costs, residents clearly preferred prevention (with 85% in first choice) over control and mitigation. All three reduction measures are preferred over compensation, which is placed last by over 90% of respondents (Figure 6; Zeiss 1991).

Three specific engineering impact reductions and a compensation measure to address air emissions from a waste incinerator (table 4) were tested: (1) preventive waste contaminant separation, most effective at 20% reduction but costliest at $21.65 per person per year; (2) exhaust gas emission controls, less effective at 10% but must less costly at $4.10 per person per year and, hence, more cost-effective than prevention on the basis of unit cost for reductions; (3) mitigation such as monitoring and volume reduction, equally effective and in the same cost range as control; and 4) compensation for incinerator impacts, based on payment of a percentage of property values.

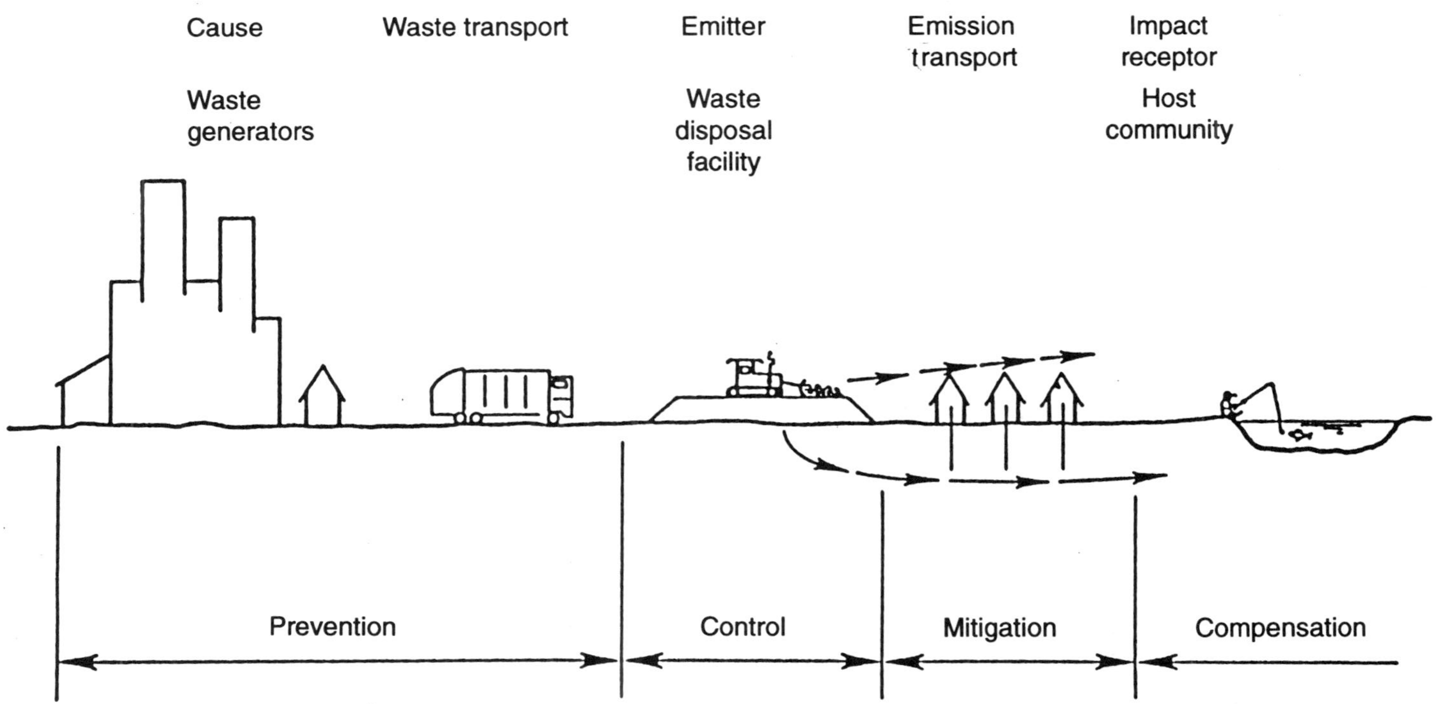

Figure 5. Three categories of impact reduction measures

TABLE 3. General Impact Management Approaches for Waste Disposal Facilities

	Cost per Year per Urban Resident ($)
Prevention	
Develop technical methods and policies to prevent the impacts by eliminating the cause of the problem before it occurs	15–25
Control	
Develop and apply technical emission control methods in the facility	10–15
Mitigation	
Develop remedial design methods to clean up impacts and/or replace damages that actually occur	5–10
Compensation	
Make no changes to the system, but compensate impacts by paying for damages	3–5

Tests in incinerator host and control communities showed that the prevention, control, and mitigation measures for incinerator exhaust emissions are equally effective in improving acceptance, with significantly large proportions of responses indicating that these measures improve acceptability (90, 86, and 96%, respectively; see Figure 7). In contrast, 67% consider compensation to make no difference and constitute a significantly ($n = 21$, $p = 0.005$) larger fraction than those who indicate a positive effect. Impact reduction measures enhance acceptance while compensation does not. Among the specific measures, respondents favor prevention as the first choice (57%, $n = 21$, $p = 0.07$, marginally significant; see Figure 7), but control and mitigation are not clearly distinguished in second and third place ($n = 21$, $p = 0.4$), presumably because of the similarity of effect on the impact level (10–20% reduction) and similar costs ($3 to $5 per person per year). Compensation is clearly ranked in last place (95%, $n = 21$, $p = 0.0001$, significant).

Apart from some blurred preferences (due to effectiveness and cost trade-offs), most residents prefer prevention, despite higher costs, and least prefer compensation with lower costs. These results confirm the theoretical predictions that residents prefer impact reduction over compensation and, furthermore, differentiate between impact reduction measures.

Waste facilities appear to residents as sources of many different physical, economic, social, and political impacts. Residents seem often

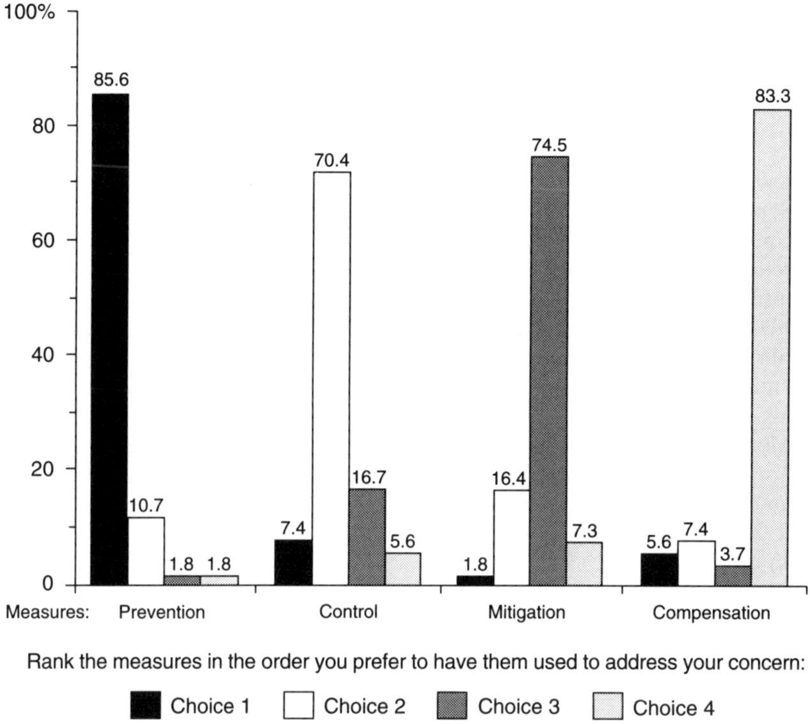

Figure 6. Community preferences for impact management (percentage of responses)

unwilling to accept or value benefits of employment, investment, taxes, etc. in exchange for bearing impacts. Individuals appear to account separately for impact losses and gains in different compartments (Thaler 1985; Kahneman and Tversky 1979) and exhibit reluctance to trade off gains and losses between different compartments, although outcomes would be more favorable were trade-offs to occur. This evidence supports separate management of impacts, first by reduction (prevention, control, and mitigation) and last by compensation.

Finally, stakeholders are sensitive to the fairness of how impacts and benefits are distributed. In contrast to the equal distribution of benefits and financial costs, the facility impacts most strongly affect the host community and least affect distant residents and waste generators (Zeiss and Atwater 1987). As a result, host community residents bear heavy impact losses but enjoy no larger benefits than anyone else in the facility service area. Thus, the ratio of benefits to losses (impacts) is unfair from the perspective of host community residents. The benefit/impact ratio from the host community is likely to decrease with an

TABLE 4. Incinerator Exhaust Gas Emission Impact Management Methods

	Effect in Reducing Air Emissions	*Cost in Taxes per Person per Year*
Prevention		
Separate the garbage and remove contaminants before burning to reduce the emissions to air	20% reduction	21.65
Control		
Install the best available air filtration equipment to reduce air emissions in the exhaust air	10% reduction	4.10
Mitigation		
Monitor air emissions in the air around the plant and reduce the amount of garbage that is burned if the measured emissions are significant	10–20% if proven necessary	3.20—5.90
Compensation		
Compensate homeowners with the dollar equivalent of 10% of property value	No risk to property prices	2.50

increasing number of residents in the service area, because the impacts will generally increase with increasing waste stream and facility size.

The selection of a technology (and a site) that is considered the best option is important for achieving tolerance to move from the status quo in a host community (Easterling 1992). The reduction of impacts through site and technology selection means the politically most amenable site should not always be preferred. The demonstration of technical optimality indicates to host communities that they are not being strategically chosen because they may offer the least political resistance (Kraft and Clary 1991). One site is not always clearly optimal, so a set of "suitable" or "better" sites might also serve the purpose of demonstrating the achievement of a better situation with the proposed facility (McQuaid-Cook and Simons 1989).

Limitations on waste type, quantity, toxicity, sources, and source areas are occasionally identified as relevant components for siting agreements (Nieves et al. 1992; Centaur Associates 1979; Zeiss and Atwater 1987; Zeiss and Paddon 1992). While not always possible because of the facility scope, the exclusion or monitoring of specific wastes might be considered. Waste stream limitations technically prevent impacts, but also contribute to fairness and community control.

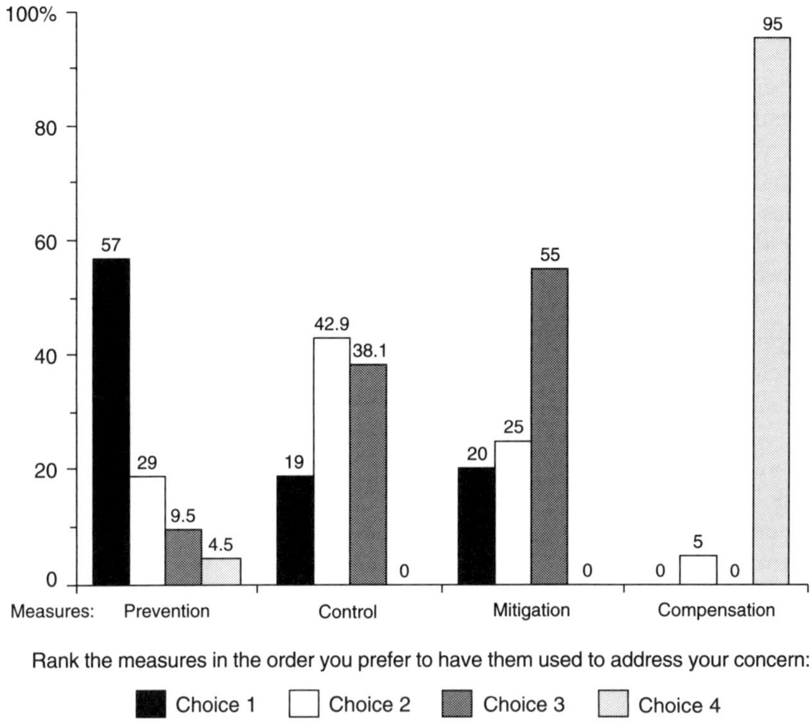

Figure 7. Community preferences for incinerator impact reduction

Benefits, Compensation, and Incentives

Ancillary Benefits Benefits, or ancillary positive effects, of the facility usually consist of local employment during construction and operation, improvements to roads and infrastructure, additional taxes (or payment in lieu of taxes), economic diversification, and attraction of new business. For some facilities and communities, these benefits are significant. Waste facility benefits (particularly local employment and economic diversification) are alone seldom sufficient to make the community equally well or better off (Bord, Ponzurick, and Witzig 1985; Zeiss and Atwater 1987; Bacow and Milkey 1982; Duberg, Frankel, and Niemszewski 1980; Vari, Kemp, and Mumpower 1991; Mank 1991).

Compensation In order to protect the community from additional negative impacts that cannot be eliminated or are too costly or complex to reduce, compensation measures are added to directly replace, offset, or pay the value of impact damages. Compensation is, in this definition, tied to specific impacts connected to the changes caused by the facility. Property value guarantees, improved infrastructure, property tax payments either to the community or to landowners, insurance costs, and closure and cleanup funds are examples of monetary compensation methods, while free disposal services, replacement of affected habitat, water supplies, and provision of emergency vehicles are in-kind compensation tools (Zeiss and Paddon 1992; Gregory et al. 1991; Camerer and Kunreuther 1989; Carnes et al. 1983; Gregory and Kunreuther 1990; Zeiss 1991).

Compensation can be structured as 1) advance—to address fairness, 2) interim—to protect against tangible losses, and 3) ex post facto—to defray losses from future events or losses (accidents, closure, cleanup; Gregory et al. 1991; Smail 1985).

Failure to provide compensation in addition to ancillary facility benefits has been shown to contribute to siting failure (Bord, Ponzurick, and Witzig 1985). In contrast, many studies have found that too much reliance on compensation measures without prior attention to impact management and reduction has also often led to failed siting agreements (Hadden 1991; Gregory et al. 1991; Zeiss and Atwater 1987; Duberg, Frankel, and Niemszewski 1980; McClelland, Schulze, and Hurd 1990; Davis 1984). Some evidence suggests that in-kind compensation is more effective (Gregory and Kunreuther 1990; Camerer and Kunreuther 1989; Davis 1984; Carnes et al. 1983), while many results support considering compensation last, i.e., after impact reduction measures (Bacow and Milkey 1982; Portney 1991; Gregory et al. 1991; Hadden 1991; and others) and identify the principle of carefully connecting each compensation measure to a recognized need (Carnes et al. 1983) in order to avoid the connotation of "bribes."

Compensation measures are considered less effective than impact reduction control and monitoring measures (Portney 1991; Zeiss 1991) and should be negotiated only if local control of the disbursement is effective, and then only if residual health and environmental risks and trigger impacts are below acceptable limits.

Other Incentives Incentives, or equity benefits, consist of all additional payments or in-kind benefits offered to leave the host community better off. These are often presented as payments for inequity (Bacow and Milkey 1982; Carnes et al. 1983; Gregory et al. 1991; Hadden 1991), protection from community stigma (Kunreuther et al. 1990), payments for inconvenience and "someone else's waste" (Himmelberger, Ratick, and White 1991; *Wall Street Journal* 1991; Portney 1991; Kunreuther et al. 1990), incentives for risk assumption (Inhaber 1991), or outright payments to increase facility benefits or enhance the local economy (McQuaid-Cook and Simons 1989).

Incentives, too, are best negotiated only after impact reduction and compensation measures have been exhausted and the perceived risk is reduced below a certain level (Bacow and Milkey 1982; Kunreuther et al. 1990; Mank 1991; Davis 1984) to avoid the perception of bribery. Most frequently reported incentive measures are royalties or tipping fees of $.50 to $7/ton, contributions to local charities, direct rebates to persons, or payments to communities (Smail 1985). While some report good siting success with outright payments (Himmelberger, Ratick, and White 1991), others show moderate to low effect on the attitudes of community residents toward waste facilities (Portney 1991; Zeiss 1991; Carnes et al. 1983; Kunreuther et al. 1990).

Although considered less effective, and hence, best negotiated last in the siting process, benefits, compensation, and other incentives were often designed to address specific facility-related needs and concerns (as suggested by the principle described previously; Carnes et al. 1983) that cannot be further prevented, controlled, or mitigated. Most of the compiled measures focused on specific issues and related the level of compensation to the level of facility impact.

Incentives that are considered extra benefits are best offered as separate items, rather than as a package, in order to improve their separate value (see prospect theory). This approach is predicted to be more effective because the separate value of each incentive is then counted on the steeper initial slope of the gain shown on the right-hand side of the value curve. All additional benefits, compensation, and incentives must be carefully focused on, and justified by, specific impacts.

SUMMARY OF SITING INTERVENTION PRINCIPLES

Prevention, engineering controls, mitigation, and monitoring (fig. 5) are preferred over compensation with alternative benefits (Portney 1991; Zeiss 1991; Johnson 1987). Once stringent measures are effectively used to meet strict standards of physical risk, other incentive measures can be discussed (Zeiss and Paddon 1992; Kemp 1992; Portney 1991; Gregory et al. 1991; Kunreuther et al. 1990; Bacow and Milkey 1982; Duberg, Frankel, and Niemszewski 1980; Mank 1991; and others). Practically, the need to establish values for intangible facility impacts can be avoided by using engineering costs for impact reduction methods. From a planning approach, impact reduction appears to be a more predictable and more effective approach than compensation through alternative goods. Essentially, optimal technology, site choice, and waste stream controls are prevention measures to reduce facility impacts (See fig. 4; Zeiss 1991).

Apparent nuisance impacts (odor, noise, view) can trigger underlying concerns about community image (Zeiss and Atwater 1991).

The trigger effects are easily screened and reduced with buffers and landscaping (McClelland, Schulze, and Hurd 1990). These measures can effectively reduce compensation because they address the community image concern and thereby reduce one of the most important influences on the level of compensation and length of negotiations as reported by Nieves et al. (1992) for twenty-one solid waste sites in Wisconsin. Finally, a focus by the community on infrastructure requirements and issues was determined to reduce the compensation required (Nieves et al. 1992).

Local control, liaison, inspection, monitoring, power to shut down the facility, access to information, and establishment of emergency plans have been identified by several researchers as important components of impact management (Carnes et al. 1983; Elliott 1984; Davis 1984; Johnson 1987; Nieves et al. 1992). Often the element of local control is integrated into the choice of impact management measures and of monitoring and remediation measures (Elliott 1984; Zeiss and Paddon 1992).

Engineering impact management therefore draws on quantitative risk assessment and prospect theory. Impact assessment and reduction are effective engineering siting intervention mechanisms. Prevention, control, mitigation, monitoring, and remediation measures should be applied before facility benefits, compensation, or incentives are offered. Community choice of impact management measures improves the effectiveness of impact management measures.

The next section investigates the effect of applied siting interventions in an analysis of case studies.

Siting Intervention Applications

The use of need justification, impact management, and benefits and compensation as siting interventions was tested by analyzing siting cases. Sixty-seven siting cases were categorized by the use of siting intervention measures and by siting outcome. Four siting interventions were distinguished: (1) need justification, (2) impact management, (3) benefits, compensation, and incentives, and (4) process management. All possible combinations were identified.

In order to test the use of these key principles and interventions in a wide range of actual conditions, sixty-seven siting cases, reported in the literature since 1983, were collated and analyzed to determine applied siting interventions. Three siting cases were selected and analyzed in detail.

Three siting outcomes were used: (a) not sited as a failure, (b) unknown, indeterminate, or siting in process, and (c) sited as a positive outcome.

The results show that most siting outcomes fall into two categories: single-measure siting approaches, all without need identification, lead to unsuccessful outcomes (eleven cases), whereas approaches that combine all measures into an integrated package are clearly more successful with a total of forty-two positive outcomes. While some single-measure cases succeeded (four facilities sited with measure 3, compensation), the community was still opposed. In contrast, some communities rejected integrated siting offers (four sites). The set of siting studies is possibly biased because fewer failed outcomes may have been reported. The combined use, however, of need identification, impact reduction, and compensation, is supported by theory and successful siting attempts.

The detailed comparison of four case studies represents (1) failed single-measure attempts (Bord, Ponzurick, and Witzig 1985), (2) combined-measure attempts that are still undecided (Kemp 1992), and (3) successful combined-measure siting attempts (McQuaid-Cook and Simons 1989; Zeiss and Paddon 1992).

Case Study 1—Pennsylvania Low-Level Nuclear Waste (Bord, Ponzurick, and Witzig 1985)

To fulfill federal regulations to manage Pennsylvania's radioactive waste, a locally known nuclear industrial company proposed to build a low-level treatment facility in the southwestern part of the state. An education program and preliminary survey were completed. Public relations were viewed unnecessary, as the proponent had a good record of waste management. The local need for the facility seemed obvious, as the community required additional industry. Immediately following the public announcement of the facility proposal, a local protest group formed. The group proceeded to dominate public meetings and the town sentiments. The facility was defeated. A follow-up survey showed that the need definition, impact reduction, and benefits were insufficient. Lack of trust in the proponent and the regulator and the request for additional information suggest that the community should have been included in the decision process.

Case Study 2—Central Midwest Interstate Compact (Kemp 1992)

The Central Midwest Interstate Compact was established between Illinois and Kentucky to site a low-level radioactive waste facility in Illinois to deal with the states' wastes. A technical advisory committee was formed to assist a citizens' advisory group. The group consisted of representatives from waste generators, local citizens, and environmental interest groups, with two facilitators to examine technical and siting issues. The entire state was characterized, and the siting process was voluntary. Eight sites were found to be technically feasible, and additional characterization was focused on one site. The facility is an engineered facility for containment of incinerated, compacted, and packaged waste with an overdesigned safety system. A range of compensation and incentive measures is under consideration, each measure being justified in terms of local need with up to $2 million available to the host community. The draft siting plan was publicized and discussed at public meetings. The county is now giving strong opposition, and the siting attempt is still in progress.

Case Study 3—Hazardous Waste Treatment Facility in Alberta (McQuaid-Cook and Simons 1989)

To site a hazardous waste treatment center in Alberta, a comprehensive approach was taken to address environmental and social issues. Strong need definition was completed to make the entire province aware of the need for the facility. All information was made available from the outset, with complete public participation in a voluntary process. The entire province was considered with consistent siting criteria to exclude the technically unfeasible sites. Strict safety criteria and a philosophy of complete containment minimized negative impacts. The facility carried with it significant ancillary benefits to the town and to the province in general. The host community voted in favor of the facility with a majority of 75%.

Case Study 4—Transfer Station Siting Program in Alberta (Zeiss and Paddon 1992)

The failure of the facility proponent in the marooned-attempt siting (at the beginning of this chapter) to acquire a development permit in Morinville led to a detailed analysis of the key elements in the siting decision.

From the observation of previous siting process and outcome in Morinville, it became clear that several key community values and principles had been violated by the linear, approval-oriented approach driven by economic benefits. In order to avoid yet another failure and larger stigma, a different approach had to be developed. A new approach, stepping back and looking at key community values, was derived and implemented within two months—a community study program. Specific engineering siting interventions were developed from the theoretical principles. These interventions were incorporated in the program, which was carried out in fourteen communities throughout Alberta. The practical application and the results of the study program are illustrated in this case study.

The one-way flow of information to a targeted audience may solicit support for general concepts ("It's good to be dealing with hazardous waste; Swan Hills is good for Alberta"), but it is a poor means of soliciting support for actions that are perceived to have a real effect on people's lives ("Not in my backyard, you don't").

While the economic benefits of the transfer stations were recognized, the residents' distrust of the process, the lack of perception of need for the project, concerns over health risks, uncertainties, and the negative stigma for the community image created strong negative attitudes among many residents and stakeholders. The benefits were presented as economic gains but were outweighed by the physical risks and intangible impacts that the hazardous waste transfer station may have generated. This is a familiar cause for similar outcomes of siting attempts. As a result, the transfer station siting program was revised to integrate three siting intervention measures: need justification, impact management, and benefits, compensation, and incentives.

Need Justification The analysis revealed that the need for transfer stations had not been justified at all from the local communities' perspective. In order to obtain the community's interest and commitment at least to negotiate siting conditions, the default option, i.e., the outcome without the facility, must be seen as undesirable. In other words, the disadvantages of not siting the facility must appear clearly to the community. This is simply another way of saying that the community must recognize the need for a hazardous waste transfer station from their perspective.

The principles of prospect theory and of quantitative risk assessment (QRA) were applied to "framing" the transfer station proposals. When in early workshops the concerns about transfer station risks and impacts were voiced, participants were asked to identify the current disposal methods for hazardous wastes from their communities. Very quickly, knowledgeable participants would provide the answer: "To the sewer and to the local landfill, or, worse, into the ditch." Then data could be presented to show the quantity of manifested (that is, accounted-for) hazardous wastes that were disposed of into landfills. On the basis of this information, most workshop participants realized that, to some extent, the proposed transfer stations were a method of improving the community's local quality, gaining control, and taking responsibility for the problem. This insight effectively reframed the proposed facility from a loss (due to its impacts) into a salient reduction of loss in environmental quality under the status quo disposal to landfill, sewer, and ditches.

In the transfer station siting process, stakeholders initially appeared to believe (1) that the problem of hazardous wastes was created by urban and industrial waste generators and hence was not a community problem and (2) that these wastes would not affect them if they resisted the transfer station. As a result, while the majority ascribed to the moral norm and favored proper waste treatment, they were unwilling to take responsible action and host a transfer station. In response, the study program material illustrated for stakeholders the problems and the consequences of maintaining current hazardous waste disposal. It pointed out the transport of contaminants in ground and surface water as well as the consequences to environmental quality, in order to foster awareness of the problems and consequences of the current situation without the transfer stations. Then common business and residential activities were identified to illustrate a community's use and disposal of hazardous products through dry cleaning, painting, household cleaning, and other common activities. These points illustrated and encouraged community responsibility for hazardous waste generation.

Community recognition of the problem, the consequences, and its responsibility were then reinforced by pointing out that the workshop series was intended to develop options for the community to reach a consensus and take action to become part of the solution to the problems and thus foster community pride.

The specific siting interventions applied the concepts of QRA to demonstrate that the status quo in most communities was less than desirable and had negative effects on the local environmental quality. Further, the community was made aware of its responsibility and the consequences of not taking action and was offered the opportunity to consider possible actions through the detailed study of the transfer station proposal. The need justification used risk assessment and prospect theory to reframe the facility as a method to improve current waste disposal and applied altruistic belief-attitude to focus on awareness and responsibility as incentives to take action.

Impact Management Prospect theory's asymmetrical value curve furthermore provides priorities for addressing specific concerns about impacts and risks. Because losses weigh more heavily than gains, impact management measures that reduce losses will more effectively improve the prospect of the transfer station than offering new gains as additional compensatory benefits. As a result, efforts to address community concerns should, by this logic, focus on impact and risk reduction, rather than on concocting new perks (as fire trucks, community halls, etc.) for putting up with impacts.

Further evidence supports separate treatment of impacts. Waste facilities appear to residents as sources of many different physical, economic, social, and political impacts. Residents seem often unwilling to accept or value benefits of employment, investment, taxes, etc., in exchange for bearing impacts. Waste facility impacts create additional barriers because they are frequently intangible (community image) or measured in different units (odor dilutions, noise decibels, air quality–pollutant concentrations, etc.). Intuitive balancing of impact losses and benefits (gains) therefore becomes more difficult. As a consequence of the emphasis on reduction of separate impacts, the majority of measures that were adopted in community workshops to manage impacts focus on impact reduction rather than on compensation. The requirement for fair distribution does not hinge on absolute impact levels and, therefore, will not necessarily be fulfilled if impact levels do not exceed health risk or regulatory guidelines. Rather, fairness is based on a comparison of benefit/impact ratios and on an equal distribution among affected and unaffected residents and waste generators. Increased fairness can result either from increasing benefits or from decreasing or redistributing impacts (losses). When the community is allowed an open field to define a suitable size, catchment area, waste type, monitoring requirements, etc., the residents have the opportunity of balancing benefits and impacts to achieve a fair distribution.

Benefits, Compensation, and Incentives Compensation and incentives offer alternative benefits to a host community for accepting the facility's impacts. In the transfer station siting program, no additional benefits, compensation, or incentives were offered or requested. Compensation appears to be significant only if it is used to redress possible specific losses. Used thus, compensation is a loss reduction measure. The theoretical danger exists of proponents using large incentives to buy off host community concerns about signifcant impacts. The findings here (and elsewhere, Carnes et al. 1983; Zeiss 1991) show, however, that compensation and incentives are least preferred and are effective only when linked to the specific impacts.

More effective, however, was the opportunity to affect tangible changes by allowing stakeholders to choose among different outcomes. In particular, their involvement in the process was voluntary; some communities did not want to participate. Then, during the process, stakeholders had the options of (1) simply completing the study process and choosing a continuing commitment to further education, (2) undertaking a toxic roundup of household hazardous wastes in their community, (3) implementing changes to municipal purchasing and enforcement policies, or (4) negotiating with a proponent to site and operate a collection site or a transfer station. These options represented a sequence of choices that require increasing levels of commitment toward action to participate in hazardous waste management activities in the community. As a result, these choices allow the communities to decide on an action level that best meets the study groups' need for waste management, based on their own level of awareness and responsibility. Each community was given the choice to match their demand for hazardous waste services with the "cost" as effort to cope with risks and impacts.

Community Decision Outcomes By applying key principles to siting transfer stations, the process focused more on offering an appropriate action for the community than on siting a facility. As a result, "success" can hardly be defined solely as an outcome that leads to the siting, construction, and operation of a transfer station. Since the original objective of the study process was to achieve this goal, however, the success rate is defined here as the percentage of participating communities that wished to enter into negotiations with proponents to finalize development agreements for transfer stations. Ironically, the opportunity for choosing other outcomes probably improved the success rate of actually siting transfer stations.

Twenty communities in Alberta were offered the opportunity to participate in the transfer station study process. The set of communities comprised a range of population sizes from 500 to over 60,000, located from 11 to 460 km from one of the two large urban centers of Calgary and Edmonton, with average incomes (in 1986) from a low of $12,775 to a high of $27,146. Education levels, too, varied from a low of 1.9% of the population over fifteen with a university degree to a high of 12.7%. Between 42% and 97% of residential units in the study communities were single-family dwellings, with 47% to 89% of residential units owner-occupied. These values reflect the provincewide averages (see Table 5). Communities that did not agree to negotiate contained (a) the highest average income, (b) the highest percentage of university graduates, and (c) the largest fraction of owner-occupied residential units and single-family dwellings. In all cases the communities that were willing to negotiate contained the next highest values. In contrast, the strongly opposed community of Morinville is also shown on the bottom line of Table 5. The comparison shows that the ten communities that reached consensus to negotiate reflect the full range of demographic and socioeconomic characteristics and do not significantly differ from the province of Alberta or from communities that did not reach consensus agreements. This comparison shows that other, extraneous factors do not appear to explain the success of this program.

Most communities made recommendations to improve education and participate in toxic roundups. Fifteen percent also recommended policies to reduce waste generation through purchasing policies and enforcement procedures. Table 6 summarizes the outcomes of all

TABLE 5. Summary of Community Profiles by Result of Siting Study Processes

						Education		Residential Dwellings	
	Sample Number	*%*	*Population*	*Distance from Large City over 100,000*	*Average Individual Income*	*% Less Than Grade 9*	*% University Graduates*	*% Owned*	*% Single Family*
Province of Alberta									
			2,500,000	n.a.	$14,250	10.5	10.7	62	65
Communities Participating in the Transfer Station Study Process									
Min	20	100	500	11 km	$12,775	5.3	1.9	47	42
Max			60,610	460 km	$27,146	29.0	12.7	89	97
Communities Willing to Negotiate with Proponents over Transfer Station Siting									
Min	10	50	500	11 km	$12,775	5.3	3.2	55	55
Max			60,610	460 km	$21,590	24.1	9.8	80	91
Communities Not Willing to Participate or Negotiate									
Min	10	50	2,553	25 km	$16,870	5.4	3.5	47	42
Max			34,940	435 km	$27,146	29.0	12.7	89	97
Communities Dead Opposed (Morinville)									
	1		5,364	25 km	$18,254	7.2	5.5	80	83

Source: Zeiss and Paddon 1992.

TABLE 6. Community Siting Process Outcomes

Participation	*No. of Communities*	*Percentage (%)*
Study process		
Approached—total	20	100
Not initiated	6	30
Initiated	14	70
Outcome	*No. of Communities*	*Percentage (%)*
Resulting actions		
Study process only	3	15
Education and studies	5	25
Toxic roundups	16	75
New policies for purchasing and enforcement	3	15
Pursue siting negotiations	10	50
Projects built (by October 1995)	2	100

groups and itemizes the recommended actions. Of the twenty communities that were contacted between the fall of 1989 and the spring of 1991, six did not participate in the process, because they either were not interested, lacked time, or were not in a prime location. Eleven groups completed the process by issuing recommendations to their councils for further action. Of particular interest is the number of communities that recommended or completed negotiations with transfer station proponents. Ten communities, over 70% of those that began the process, recommended action to negotiate the siting of transfer stations.

Summary of Case Study Findings

In the first case study, Bord, Ponzurick, and Witzig (1985) report that the low-level radioactive waste incinerator entailed identified health, property value, and traffic effects. Although impact reduction measures were requested by the community to address the other effects, the proponents only reported ancillary benefits (jobs) as an offsetting component of the siting package. As predicted, the siting attempt failed.

In the second study, a group of ambiguous cases, two combined relevant impact reduction measures with substantial compensation, such as property value guarantees, and locally justified the need for compensation. Effort was invested to identify the problems with current hazardous waste disposal and storage and to compare available technical options through intensive community involvement efforts before focusing on several sites. Local need identification led to two host communities, which both identified the facility as a way to maintain their community and lifestyle. Impact management, ancillary benefits, and incentives were then sufficient for successful siting. The facility, therefore, was seen as a way to reduce the potential loss. In addition, significant efforts were made to ensure host community control, monitoring, oversight, and equity. Some success was noted, with one facility sited but opposed and one siting still in process. Thus, substantially improved outcomes were achieved by using, in the prescribed sequence, community involvement with targeted impact reduction and compensation.

The third study concerned the successful case of a hazardous waste treatment facility (McQuaid-Cook and Simons 1989) and the subsequent expansion of this same facility in 1992, resulting from a downturn in the oil economy. The facility was framed to meet a local need in a community accustomed to handling oil and gas products. The fourth case, of siting hazardous waste transfer stations, showed that siting interventions used qualitatively within a workshop study process to show need and manage impacts were effective in most participating communities.

CONCLUSIONS—DESIGN AND USE OF ENGINEERING SITING INTERVENTION MEASURES

In this chapter, the role and effect of engineered siting intervention were identified within the engineering design and the community decision processes (see Fig. 8). The principles of relevant theory were analyzed, and an analysis of siting intervention measures in case studies was presented. The results show that the combined use of (1) need justification, (2) impact management, and (3) benefits, compensation, and incentives coincides in most cases with a successful siting outcome. In this concluding section, a brief synthesis of theory, siting interventions, and examples is provided. This will provide a summary of the key elements and findings.

Need Justification

QRA and prospect theory can be combined to generate estimates of risks, impacts, and performance of current (status quo) facilities and conditions. This frames the proposed facility as an alternative to reduce a current deficiency in the status quo, i.e., to address a need. Quantitative estimates of impact (or risk) can serve as a method to demonstrate the impact reduction effect of the proposed facilities.

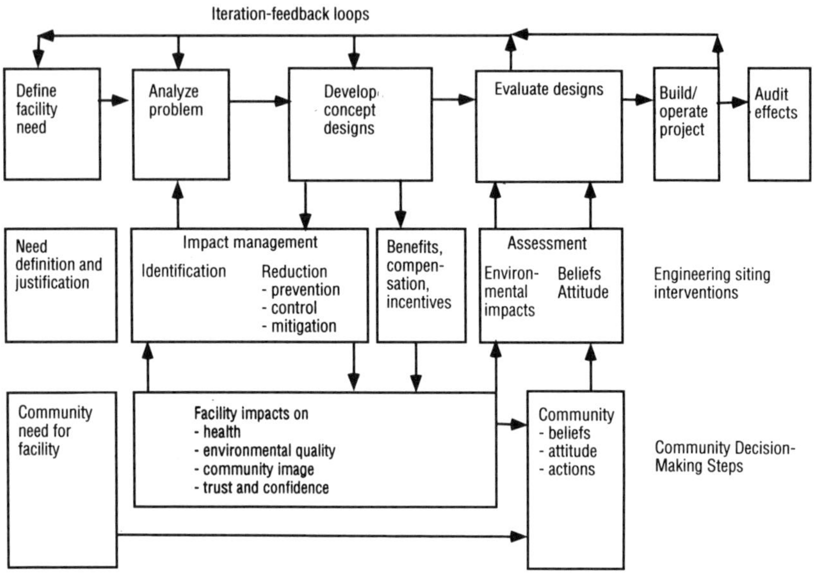

Figure 8. Integrated engineering design process with siting intervention measures

Creating an awareness of current impacts and responsibility are effective messages to persuade people of facility need, as was shown in the Alberta hazardous waste transfer station case study. Specific community needs, too, might be demonstrated to overcome the status quo bias. Different groups might hold predictably different perspectives on need.

Impact Management

Impact management consists of identifying and reducing undesirable facility effects. QRA provides an effective tool to predict facility impacts. Prospect theory then requires that the impacts be effectively reduced through prevention, control, or mitigation measures. Tracing and reducing facility impacts that trigger strong community beliefs and attitudes about the facility are particularly effective. Impacts should be addressed individually, rather than be subjected to horse-trading with offsets, such as additional facility benefits or compensation.

Benefits, Compensation, and Incentives

Generally, compensation of negative effects by means of alternative goods is not desirable, except when impacts have been reduced and the residuals are very small. Voluntary auctions, for example, provide the host community with the opportunity of determining a compensatory value for bearing the facility impacts. Compensation measures, however, that are provided by the proponent must be carefully justified and tailored to address the specific impact or concern. Thus, loose incentive payments have been shown to be ineffective.

The engineering design process consists of six generic steps: define facility need, analyze the limitations and requirements of the facility, develop concept designs, evaluate and select designs, build and operate the facility, and, finally, monitor the facility performance. The design process is iterative and uses the results of later steps to modify the results of earlier steps (see Figure 8). Host communities, too, identify the need for the proposed facility early in the process. Subsequently, communities also identify and assess impacts and form their attitude about the desirability of the facility. This attitude will affect the community's response to the facility. The steps in the community decision process can coincide with the engineering design process if the two processes are scheduled and coordinated by the proponent or the siting agency. In a coordinated siting process, three engineering siting intervention methods have been shown to improve the effectiveness of the siting attempt if they are used in this order: (1) need definition and justification from the proponent's and the community's perspective, (2) impact identification and management to reduce impacts, and (3) benefits, compensation, and incentives. The assessment of environmental impacts and of the community attitude are best integrated into the evaluation and selection of facility design. These intervention measures can be accommodated in the engineering design process and used in the iteration process to improve the resulting facility design. Finally, the monitoring of environmental impacts and community response should also be included in the auditing of facility performance during construction and operation of the facility.

REFERENCES

Armour, A. M. 1991. The Siting of Locally Unwanted Land Uses: Towards a Cooperative Approach. *Progress in Planning* 35, no. 1:1–73.

Bacow, L. S., and J. R. Milkey. 1982. Overcoming Local Opposition to Hazardous Waste Facilities: The Massachusetts Approach. *Harvard Environmental Law Review* 6:265–305.

Bord, R. J., P. J. Ponzurick, and W. F. Witzig. 1985. Community Response to Low-Level Radioactive Waste: A Case Study of an Attempt to Establish a Waste Reduction and Incineration Facility. *IEEE Transactions on Nuclear Science* N.S. 32, no. 6:4466–72.

Camerer, C. F., and H. Kunreuther. 1989. Decision Processes for Low Probability Events: Policy Implications. *Policy Sciences* 8, no. 4:565.

Carlson, K. A. 1991. What Happens and What Counts: Resident Assessments of Prison Impacts on Their Communities. *Humbolt Journal of Social Relations* 17, nos. 1–2:211.

Carnes, S. A., E. D. Copenhaver, E. J. Soderstrom, J. Sorensen, E. Peelle, J. H. Reed, and D. J. Bjornstad. 1983. *Incentives and the Siting of Radioactive Waste Facilities.* Regional and Urban Studies Section, Energy Division, Oak Ridge National Laboratory.

Centaur Associates, Inc. 1979. *Siting of Hazardous Waste Management Facilities and Public Opposition.* Report SW–809. U.S. Environmental Protection Agency, Office of Solid Waste, Washington, D.C.

Davis, C. 1984. Substance and Procedure in Hazardous Waste Facility Siting. *Journal of Environmental Systems* 14, no. 1:51–62.

Duberg, J. A., M. L. Frankel, and C. M. Niemszewski. 1980. Siting of Hazardous Waste Management Facilities and Public Opposition. *Environmental Impact Assessment Review* 1, no. 1:84–89.

Easterling, D. 1992. Fair Rules for Siting a High-Level Nuclear Waste Repository. *Journal of Policy Analysis and Management* 11, no. 3:442.

Elliot, M. L. P. 1984. Improving Community Acceptance of Hazardous Waste Facilities through Alternative Systems for Mitigating and Managing Risk. *Hazardous Waste* 1, no. 3:397–410.

Fishbein, M., and I. Ajzen. 1975. *Belief, Attitude, Intention, and Behavior.* Reading, Mass.: Addison Wesley.

French, M. 1971. *Engineering Design: The Conceptual Stage.* London: Heineman.

Gregory, R., and H. Kunreuther. 1990. Successful Siting Incentives. *Civil Engineering,* April, 73–75.

Gregory, R., H. Kunreuther, D. Easterling, and K. Richards. 1991. Incentives Policies to Site Hazardous Waste Facilities. *Risk Analysis* 11, no. 4:667.

Gregory, R., S. Lichtenstein, and D. MacGregor. 1993. The Role of Past States in Determining Reference Points for Policy Decision. *Organizational Behavior and Human Decision Processes,* 55:195–206.

Hadden, S. G. 1991. Public Perception of Hazardous Waste. *Risk Analysis* 11, no. 1:47–57.

Himmelberger, J. J., S. J. Ratick, and A. L. White. 1991. Compensation for Risks: Host Community Benefits in Siting Locally Unwanted Facilities. *Environmental Management* 15, no. 5:647–58.

Inhaber, H. 1991. A Market-Based Solution to the Problem of Nuclear and Toxic Waste Disposal. *Journal of the Air and Waste Management Association* 41, no. 6:808–16.

Jaffe, M. 1991. Comments on "Public Perception of Hazardous Waste." *Risk Analysis* 11, no. 1:59–60.

Johnson, B. B. 1987. Public Concerns and the Public Role in Siting Nuclear and Chemical Waste Facilities. *Environmental Management* 11, no. 5:571–86.

Kahneman, D., J. Knetsch, and R. Thaler. 1991. Anomalies —the Endowment Effect, Loss Aversion, and Status Quo Bias. *American Economic Review* 5, no. 1 (Winter): 1–14.

Kahneman, D., and A. Tversky. 1979. Prospect Theory: An Analysis of Decision under Risk. *Econometrica* 47 (March): 263–91.

———. 1984. Choices, Values, and Frames. *American Psychologist* April: 341–50.

Kemp, R. 1992. *The Politics of Radioactive Waste Disposal.* Manchester and New York: Manchester University Press.

Knetsch, J. 1983. *Property Rights and Compensation.* Toronto: Butterworths.

Kraft, M. E., and B. B. Clary. 1991. Citizen Participation and the NIMBY Syndrome: Public Response to Radioactive Waste Disposal. *Western Political Quarterly* 44:299–328.

Kunreuther, H., D. Easterling, W. Desvousges, and P. Slovic. 1990. Public Attitudes towards Siting a High-Level Nuclear Waste Respository in Nevada. *Risk Analysis* 18, no. 4:469–84.

Kunreuther, H., and P. Portney. 1991. Wheel of Fortune: A Lottery/Auction Mechanism for Siting of Noxious Facilities. *Journal of Energy Engineering* 117, no. 3:125–32.

Kunreuther, H., and L. Susskind. 1993. *Facility Siting Credo.* Philadelphia: University of Pennsylvania.

Love, S. 1986. *Planning and Creating Successful Engineering Designs.* Los Angeles: Advanced Professional Development.

Mank, B. C. 1991. The Two-Headed Dragon of Siting and Cleaning Up Hazardous Waste Dumps: Can Economic Incentives or Mediation Slay the Monster? *Environmental Affairs* 19:239–85.

McClelland, G. H., W. D. Schulze, and B. Hurd. 1990. The Effect of Risk Beliefs on Property Values: A Case Study of a Hazardous Waste Site. *Risk Analysis* 10, no. 4:485–97.

McQuaid-Cook, J., and C. S. Simons. 1989. Development and Operation of a Waste Management System in Alberta, Canada. *Waste Management and Research* 7:219.

Mitchell, R., and R. Carson. 1989. *Using Surveys to Value Public Goods: The Contingent Valuation Method.* Washington, D.C.: Resources for the Future.

Nieves, L. A., J. J. Himmelberger, S. J. Patrick, and A. L. White. 1992. Negotiated Compensation for Solid-Waste Disposal Facility Siting: An Analysis of the Wisconsin Experience. *Risk Analysis* 12, no. 4:505–11.

Portney, K. E. 1991. *Siting Hazardous Waste Treatment Facilities—the NIMBY Syndrome.* New York: Auburn House.

Slovic, P. 1987. Perception of Risk. *Science* 236, no. 17 (April): 280–85.

Smail, H. E. 1985. Solid Waste Facility Siting—Community Aspects and Incentives in Management of Toxic and Hazardous Waste. In *Management of Toxic and Hazardous Wastes,* edited by H. Bhatt, R. Sykes, and T. Sweeney, 45–63. Chelsea, Mich.: Lewis.

Statistics Canada. 1987. *Profiles—Alberta: Part 1.* Ottawa: Minister of Supply and Services.

———. 1988. *Profiles—Alberta: Part 2.* Ottawa: Minister of Supply and Services.

Thaler, R. 1985. Mental Accounting and Consumer Choice. *Marketing Science* 4, no. 3 (Summer): 199–214.

Tribe, L. 1976. Ways Not to Think about Plastic Trees. In *When Values Conflict,* edited by C. Schilling and J. Voss, 61–91. Cambridge, Mass.: Ballinger.

Ullman, D. 1982. *The Mechanical Design Process*. New York: McGraw-Hill.

Vari, A., R. Kemp, and J. L. Mumpower. 1991. Public Concerns about LLRW Facility Siting: A Comparative Study. *Journal of Cross-Cultural Psychology* 27, no. 1:83–102.

Vaughan, E., and M. Seifert. 1992. Variability in the Framing of Risk Issues. *Journal of Social Issues* 48, no. 4:119–35.

Wall Street Journal. 1991. The Economics of Trash. December 3, 1991, A1, A9.

Williams, E. A., and E. Moffa. 1983. *Siting of Major Facilities: A Practical Approach*. New York: McGraw-Hill.

Zeiss, C. 1991. Community Decision-Making and Impact Management Priorities for Siting Waste Facilities. *Environmental Impact Assessment Review* 11, no. 3 (September): 231–55.

Zeiss, C., and J. Atwater. 1987. Waste Facilities in Residential Communities: Impacts and Acceptance. *American Society of Civil Engineers, Journal of Urban Planning and Development* 113, no. 1 (May): 19–34.

———. 1991. Waste Disposal Facilities and Community Response. *Canadian Journal of Civil Engineering* 18:83–96.

———. 1993. A Case Study for Municipal Waste Landfill Planning. *Environmental Technology* 14:1101–15.

———. 1995. Analytical Framework for Waste Facility Siting. *American Society of Civil Engineers, Journal of Urban Planning and Development* 121, no. 4:1–31.

Zeiss, C., and L. Lefsrud. 1995. Siting Development Host Community for waste Facilities Agreements. *EIA Review*.

Zeiss, C., and B. Paddon. 1992. Management Principles for Negotiating Siting Agreements: A Case Study of Siting Hazardous Waste Transfer Stations in Alberta. *Journal of the Air and Waste Management Association* 42, no. 10 (October): 1296–304.

Contributors

Geoffrey Castle was previously a research associate at the Sustainable Development Research Institute at the University of British Columbia working on waste management strategies. He is currently undertaking graduate studies in business administration at the University of Western Ontario, Canada.

Susan Elliott is an Associate Professor in the Department of Geography, McMaster University. Dr. Elliott has conducted numerous research projects relating to psychosocial impacts of exposure to solid waste facilities in both Ontario and B.C. These entail a number of individual case studies, as well as ongoing longitudinal and comparative analyses. A principal focus of her recent research is the psychosocial effects of the siting process.

Clyde Hertzman is a Professor in the Department of Health Care and Epidemiology at the University of British Columbia, and a Fellow of both the Population Health and Human Development Programs of the Canadian Institute for Advanced Research. In addition to his research on factors that determine the acceptability of siting waste management facilities, his other research activities include a series of ongoing investigations into the determinants of health in Central and Eastern Europe and the Philippines, a historical cohort study of the cancer, mortality, and reproductive experience of sawmill workers exposed to dioxin-contaminated cholorphenol wood preservatives; and the development of a BC Linked Data Base of health services and health outcome files for the whole population.

Michael E. Kraft is Professor of Political Science and Public Affairs and Herbert Fisk Johnson Professor of Environmental Studies at the University of Wisconsin, Green Bay. Among other works, he is author of *Environmental Policy and Politics: Toward the 21st Century* and coeditor and contributing author of *Environmental Policy in the 1990s; Public Reactions to Nuclear Waste: Citizens' Views of Repository Siting;* and *Technology and Politics.* His current research focuses on the politics of nuclear waste disposal in the United States as a case in democratic approaches to the assessment and management of risky technologies.

Howard Kunreuther is Cecelia Yen Koo Professor of Decision Sciences and Public Policy and Co-Director of the Risk Management and Decision Processes Center at the Wharton School of the University of Pennsylvania. He is coauthor of *Risk Analysis and Decision Processes: The Siting of LNG Facilities in Four Countries,* among other works.

David Morell is a consultant with ERM-West, Inc. in California. He is coauthor of *Beyond Superfailure: America's Toxics Policy for the 1990s* (with Daniel Mazmanian) and *Siting Hazardous Waste Facilities* (with Christopher Magorian).

Don Munton is Professor and Chair of International Studies at the University of Northern British Columbia. He is the editor of *Measuring International Behavior,* and coeditor of *Canadian Foreign Policy: Selected Cases* and of *Rethinking National Security: The Public Dimension.* He has written extensively about international environmental issues and is completing a book entitled *Acid Rain and Basic Politics,* dealing with Canadian-American conflict and cooperation over the problem of acid rain.

Aleck Ostry is a research associate in the Department of Health Care and Epidemiology at the University of British Columbia, with a background in health services planning, epidemiology, and history. Besides research on siting waste facilities, he has conducted research in occupational epidemiology and public-health history. He is currently engaged in research on work stress and its impacts on health.

Alun Richards is the former Manager of External Affairs for the Manitoba Hazardous Waste Management Corporation, based in Winnipeg. He is currently a private consultant in community and public relations.

Robert Seeliger received his MA from the University of British Columbia in 1992. He currently is Assistant to Management of Sanne, Kruse & Pape, a German recycling company. He is also completing a PhD at the Institute for Political Science at the University of Tübingen, Germany.

Martin Taylor is Professor and Chair of Geography at McMaster University and Co-Director of the Institute of Environment and Health, a joint initiative of McMaster University and the University of Toronto. He holds degrees from Bristol University (B.A.) and the University of British Columbia (M.A. and Ph.D.). Since coming to McMaster in 1974, a principal focus of his research has been on the effects of environmental factors on human health and well-being, including the impacts of transportation and noise and, more recently, the psychosocial effects of environmental contaminants.

Christopher Zeiss is Associate Professor in the Department of Civil Engineering at the University of Alberta, Canada. He has written extensively on siting issues.

Index

Africa, 234
Alabama, USA, 11, 16, 142
Alberta, Canada, 13, 79, 85–86, 93, 98, 103, 219, 230, 346, 382, 386–87, 390
Alberta, siting in, 21–24, 28, 57, 59–64, 66–67, 74–80, 90–91, 96–97, 325, 338, 344, 348, 360, 382, 388
Alberta Special Waste Corporation, and Special Waste Treatment Centre, 62, 68, 79–80, 98, 219, 356, 360, 382. *See also* Swan Hills, Alberta
American Nuclear Energy Council, 122
Armour, Audrey, 2, 10, 18, 23–24, 28–30, 36, 38, 77, 82, 210, 212, 213, 220–22, 391
Ashcroft, Canada, siting in, 63–66, 275
attitude theory, 27, 291, 294
auctions, siting by 22–23, 28, 338, 391. *See also* lottery

Baden-Württemberg, Germany, 231, 249–53, 255, 257
Basel Convention, 219, 233
Bavaria, Germany, 219, 230–31, 233, 238, 240, 264
Beaudette, Flo, 69–70, 81
Belgium, 215, 220, 244
benzene, 159, 214
Bhopal, India, 290
Biebesheim, Germany, 230–31, 241–45, 252, 254
bioremediation, 34, 322, 328
Boyd County, Nebraska, 103–05
bribes, in siting, 344, 367, 378–379. *See also* compensation
BC Special Waste Company, 63–66
British Columbia, Canada, 26, 28, 62, 64, 79–80, 274, 275, 289
  Ministry of Environment, Lands and Parks, ix, 28, 62–63, 65–66, 81, 262
  Special Waste Advisory Committee, 62–65, 80, 82
British Columbia, siting in, 57–59, 62–66, 74, 76, 78, 275
Browning Ferris Industries, 347–48
Brunsbüttel, Schleswig-Holstein, Germany, 231, 245–47, 249, 252, 254
Burnaby, Canada, 274–76, 281–82, 284–85, 288
Butte, Nebraska, USA 103–5

Cache Creek, Canada, siting in, 63–66, 274–76, 278, 280–81, 284–85, 291
California, USA 18, 25, 31, 33, 35, 142–80, 201, 214, 220–21
  Appeals Board, 144–50, 154–55, 157, 168, 172, 174, 177
  CAP-IT, 161, 162
  Contra Costa County, 153, 156–57, 159–63, 166–69, 177, 179–80
  county hazardous waste management plans, 144–60, 166–69, 175–77, 179
  Department of Health Services, 147, 149–62, 165, 168–69, 179
  Department of Toxic Substances Control, 147, 151, 155–57, 161–62, 165, 173, 179
  environmental impact reports, 158–59, 163–66, 170–74
  Environmental Quality Act, 158, 162–63, 165–66
  EPA, 162, 165, 179
  Hazardous Waste Management Act (state law AB 2948), 25, 143–59, 162–79
  Hazardous Waste Management Council, 143, 144
  local assessment committees, 148–49, 157, 159, 163–67, 171–72, 175, 178, 180
  Southern California Hazardous Waste Management Project, 154–55

Canada, 11, 13, 21–22, 24, 35, 37, 57–82, 89–98, 134, 219–21, 273–88, 338, 345, 350, 352–53, 356, 364
Atomic Energy Control Board, 88–90, 93, 98
Atomic Energy of Canada Limited, 67, 88, 93, 352
community liaison groups, 91–95, 106, 351
Siting Task Forces, for radioactive waste, 90–96, 98
Canada, siting in, 24, 57, 58, 74–79, 85, 96, 101, 257, 274, 321, 325, 335, 338. *See also* Alberta, British Columbia, Manitoba and Ontario
Castle, Geoffrey, ix, 23–24, 79, 81–82, 199, 210, 212, 219
Chalk River, Canada, 88, 92–94
chemical plants, 163, 242, 246
Chemical Waste Management, 143, 170–74, 176
Chernobyl, Ukraine, 70, 290
citizen groups, 1, 135, 143, 174, 201, 251, 296, 312. *See also* environmental groups; citizens' movements
in Germany, 243, 250, 255, 259–61
Citizens' Clearinghouse on Hazardous Waste, 28, 216
citizens' movements, Japan, 187, 190–95, 207, 216–17, 395
collective (and public) goods, 2, 209–10, 341
Colorado, USA, siting in, 16
co-management, 18, 27, 72–75, 79, 97–98, 207, 209–10, 221–22, 313, 321–37
compacts, inter-county, 168
compacts, inter-state, 87–89, 97–102, 105, 381
compensation, 16–17, 21–22, 28, 35–36, 199, 220, 326, 339, 344–56, 364, 367, 372–91
in Europe, 16, 221
in Germany, 240, 254
in Canada, 16, 60, 63, 72–73, 91, 93, 95, 97–98, 221, 329, 344, 382,
in Japan, 16, 26, 190–91, 203–9, 220–21
in USA, 16, 99, 103–4, 110, 117, 133, 135, 167, 205, 220, 344, 382
compensation, nature of, 340, 344, 378
Connecticut, USA, 89, 100–01, 105, 245
constraint mapping, 59
contaminated sites, 5, 7, 33, 89, 94–95, 134, 142–43, 185–86, 189, 197, 215–16, 239, 242, 303
co-operative management. *See* co-management

Darmstadt, Germany, 241, 242
'Decide, Announce, Defend', 10–13, 19, 20–22, 34, 77, 85, 89, 92, 100, 114, 209, 231, 255, 290, 338, 350
Deep River, Canada, siting in, 92–96, 351, 352
Denmark, 215–16, 219–20
dioxins, 8, 34, 159, 240, 247, 249
Dow Chemical Company, 6, 31–32, 143, 157–62, 179–80
dumps, 7, 233, 247, 257. *See also* landfills

Ebenhausen, Bavaria, Germany, 231, 238–41, 254
Edmonton, Canada, 61, 361
Eldorado Resources Ltd, 89, 98
electro-magnetic radiation, 290
engineering design process, 28, 358–60, 391
engineering siting intervention measures, 28, 364, 379–83, 389–91
Envirochem Limited, 63
environmental auditing, 32
environmental ethics, 233
environmental groups, 4, 11, 102, 217, 264, 277, 286, 306, 332, 353, 381
in California, 143, 148, 159, 160–63, 165–66, 171, 173–74, 177
in Germany, 230–65, 234, 236–37, 240–43, 247–55, 258
environmental impact assessment, 89, 158, 163, 238–39, 253, 257, 290, 327–35, 380
environmental monitoring, 72–73
environmental racism, 10, 25, 161, 171, 173
environmental standards, 6, 8, 17, 27, 234, 338–48, 353–54
in Germany, 235, 246

environmental standards (*continued*)
in Japan, 185, 193, 196, 198, 200, 206–7, 211
environmental stress and coping, 27, 292, 296, 298, 314, 315
equity, in siting, 3, 10, 24–25, 57, 77–79, 85, 143–46, 149–55, 169, 177, 326, 345–49, 358, 362, 367, 375, 378
Europe, 7–8, 28, 30–32, 35, 135, 181, 213–14, 219–21, 230, 234, 249, 251
Europe, siting in, 16, 25–26, 97, 208–9, 230–65
European Community, 234, 263, 266
externalities, 4, 341, 345

facilities, emissions and monitoring of, 34, 92, 163, 239–40, 244, 248, 303, 312, 328, 330, 339, 342, 348, 353, 359, 390–91. *See also* pollution effects
emission controls, 17, 239, 248, 370–72, 374
impacts of, 35–36, 62, 191, 290, 339–40, 348, 352, 359–61, 363, 367–68, 371, 374, 379, 383–85, 390–91. *See also* pollution, effects of
costs of, 142, 162, 248, 358
economies of scale, 151, 155
facilities, benefits of. *See* siting, benefits of
facilities, for hazardous waste. *See* hazardous waste facilities
facilities, market needs analysis, 163, 167
facilities, need for, and shortage of, 6–9, 84, 290, 358, 361, 363–67, 381, 383, 389–90
in California, 143, 145, 153, 160, 162, 164, 173, 176, 178
in Canada, 62–63
in Germany, 230, 232–33, 237, 249–52, 259
in Japan, 195, 196, 198
in United States, 6, 25, 31
facility fires, 7, 164, 166, 169, 290
facility siting credo, 15, 133, 136, 345–46, 393
facility standards, 338, 339
'fair share siting', concept of, 145, 149–57, 168–70, 175–77
Fishbein-Ajzen theory of reasoned action, 295, 315–16
France, 134–35, 137, 215, 244, 249–50
Frankfurt, Germany, 241–42
furans, 34, 240

Germany, 18, 34, 183, 201, 209, 215, 219–20, 230–65
Corporation for the Disposal of Special Wastes in Bavaria (GSB), 238–41
Environment Ministry, 237–39, 242–44, 247, 249–50, 259
federal government-Länder relations, 235, 252
Hesse Industrial Waste Corporation (HIM), 242–45, 254
local governments, 245–46, 252–53, 255–56, 258–59, 261
political parties, 245, 248, 251
state (Länder) governments, 242, 252–61
Waste Disposal Act, 235–36, 239, 248, 252
Germany, siting in, 26, 230–65
Gerrard, Michael 29, 33–34, 36–37
Glanbrook Landfill, Canada, 300–6, 310–13
Great Lakes, 95
Greenpeace, 34, 160, 171–73, 234, 263

Hamilton, Canada, 300, 315, 318
Hamilton Solid Waste Reduction Unit, Canada, 300–2, 305–6, 310–12
hazardous waste, 3–9, 13, 25–27, 58–59, 84, 86, 211, 262–63, 385
in/from households, 13, 33, 232, 246
hazardous waste facilities, 1–5, 9, 11–14, 16, 18–19, 22–23, 29–31, 76–79, 84–89, 93, 108, 126, 134, 209–10, 278–81, 299–300, 313, 315, 338, 340–46, 356, 360, 366
in Alberta, 23, 59, 61–62, 76, 79, 85, 96–98, 101, 103, 338, 348, 382, 389

hazardous waste facilities (*continued*)
in Canada, 57, 59, 62–66, 74, 77, 79, 93, 103, 105, 274–75, 280–81, 284, 291, 300, 323
in Europe, 30, 230–265
in Japan, 34, 181–222
in Manitoba, 23, 66–67, 69–71, 79, 96–98, 101, 103, 321, 322–37
in USA, 23, 33, 36, 57, 84, 133, 142–178, 209
hazardous waste, volume of, 5, 142, 146–47, 153, 160, 167, 214
in Japan, 26, 184, 207
in Germany, 231–45, 249, 262–63
in United States, 184
Hesse, Germany, 219, 230–1, 233, 241–45

illegal dumping, 5, 7, 18, 31, 33, 195–98, 201, 204, 208, 215, 232, 242, 304
impact management, 17, 220, 358–91
incentives, for siting, 16, 21, 28, 379. *See also* compensation
incineration
of chemical weapons, 29, 395
of hazardous waste, 31, 158–73, 176, 230–49, 252, 260–61
of medical waste, 30, 274, 277–79
of municipal and solid waste, 37, 232, 263, 274, 299, 300
on North Sea, 8
incinerators, 5, 7–9, 17, 29, 32–34, 79–80, 274, 289, 312, 318, 340–41, 353, 358, 367, 370–72, 376, 388
in California, 25, 143, 153, 155–74
in Canada, 26–27, 62–63, 65–66, 274–288, 294, 300–2, 305–6, 310–12, 318
in Europe, 8, 10, 26, 230–52, 260–61
in Japan, 8, 185, 195, 200, 202–3, 206–7, 210
in United States, 8, 10, 164, 343
Indian bands. *See* native Americans; native Canadians
international cooperation, 34, 342
International Joint Commission, 95
itai-itai disease, 205
Italy, 33
Iwate, Japan, 200, 204

Japan, 34, 135, 181–222, 262
'Big Four' environmental court cases, 205
Clean Japan Center, 195, 212, 219, 224
culture of, 25, 182, 187, 188, 189, 190, 208, 209
Environment Agency, 195, 197, 212, 214–17, 219, 224
Environmental Disputes Coordination Commission, 189
factional politics model, 186
industrialization, 183, 191
Keidanren (Federation of Economic Organizations), 188
Liberal Democratic Party, 193
local governments, 193–94, 207
Ministry of Health and Welfare, 196–97, 201, 207, 212, 218–20, 224
Ministry of International Trade and Industry (MITI), 212, 221
pollution prevention agreements, 193, 203, 206–7
Pollution-related Health Damage Compensation Act, 206
Waste Disposal Law, 185, 197–201, 203, 206, 211, 213, 218
'Japan, Inc.' model, 186, 208
Jonan Island facility, Japan, 194

Kanegafuchi Chemical Company, 202, 207, 210
Kasperson, Roger, 18–20, 28, 34, 37–38, 44, 119, 122, 127, 132, 135–36, 138, 220, 224, 354
Kehl, Germany, 250, 251
Kettleman Hills facility, California, 170–73, 176
Kings County, California, 170–73, 176
Kobe City, Japan, 195, 203
Kötö ward, Tokyo, Japan, 196
Kraft, Michael, 23, 25, 28, 35, 45, 108–141, 213, 225
Kunreuther, Howard, 15–16, 22–23, 27–28, 36, 45, 119, 127, 132–33, 136, 138–39, 199, 225, 338–57, 392–93
Kupchanko, Eugene, 22, 45, 59–60, 79, 80, 82
Kyushu, Japan, 185, 194, 200

Laidlaw Inc., 176
land use decisions, 342
landfills, 5, 17, 28, 161, 346, 356, 358, 367, 372, 383
  hazardous waste in, 33, 220, 263
  in California, 32, 142, 144, 151, 155, 158, 161, 165, 168, 170
  in Canada, 26–27, 62–63, 90, 185, 274–75, 278, 284–86, 297, 299–313, 329
  in Germany, 230, 232–34, 238, 242–43, 247, 257
  in Japan, 185, 194–96, 200
  in USA, 7, 11, 33, 142, 170, 185, 215, 343–45, 347–48
  radioactive waste in, 88
local control, 339, 367, 369, 380, 383
local rights, 258
local veto, 12–14, 22, 58, 75, 143, 144, 258
longitudinal attitude studies, 301. *See also* public attitudes
loss aversion, 350, 357. *See also* prospect theory
lotteries, for siting, 38, 338–39, 348–50, 352–53. *See also* auctions
Louisiana, USA, 142, 158, 162
Love Canal, USA, 30, 70, 185, 220, 290, 297
LULUs (locally unwanted land use), 1, 28, 57, 74–76

Magorian, Christopher, 11–12, 14, 21, 28, 32, 34–37, 47, 61, 78, 82, 213, 226
Manitoba, Canada, 66–73, 78–79, 85, 93, 96–98, 103, 219, 230, 321–23, 325, 327–38
Manitoba, Canada
  community advisory committees, 325, 328, 334
  Hazardous Waste Management Corporation, 81, 82
  Manitoba Hazardous Waste Management Corporation (MHWMC), 27, 66–73, 322–32, 334, 337
Manitoba, siting in, 23–24, 27, 57, 59, 66–78, 96, 321–38
Martinez, California, 162–69, 177–79
Massachusetts, USA, 15, 16, 137
Mazmanian, Daniel, 6–7, 9, 19, 28, 34–35, 47, 121, 133, 140, 155, 180–81, 199, 201, 213, 215, 225, 292, 317
McKean, Margaret, 186–93, 196, 205–7, 209, 212, 216–19, 221, 225–26
mediation, 15, 26, 134, 173, 178, 230–31, 257–60, 265
Michigan, USA, 89, 99–101, 128
Mie Prefecture, Japan, 191, 206
Milton Landfill, Canada, 300–11, 313
Minamata disease (mercury poisoning), 187, 189, 205
Mishima, Japan, 191, 193
MMM (3M) Company, 6
monitored retrievable storage (MRS) facility, 116, 125, 127, 347
Montcalm, Canada, 68–73, 78–79, 81, 322–29, 335, 337
Montcalm, siting facility in, 68–69, 71, 73–74
Morell, David, 6–7, 9, 11–12, 14, 19, 21, 23, 25, 28, 32–33, 35–37, 47, 61, 78, 82, 133, 140, 142–78, 155, 180–81, 199, 201, 213, 215, 225, 259, 292, 317
Morinville, Canada, 361–62, 382, 386, 387

native Americans, 113, 125–27, 347
native Canadians, 65, 80, 351, 353
Nebraska, USA, siting in, 24, 85, 101–5
negotiation, 15–17, 20, 22–23, 35, 74–76, 135, 265, 331, 349, 353, 367, 380, 383, 385–88
  in Germany, 26, 230–31, 253, 257
  in United States, 15, 168, 173, 177–78
  in Canada, 57–58, 79, 326–29, 351
  in Japan, 193, 203–4, 206–7, 209
Netherlands, 35, 183, 215, 220
Nevada, USA, 99, 117–19, 137
Nevada Initiative, 122–23, 138
New Jersey, USA, 11, 30, 35, 100, 220
New York, USA, 31, 35, 89, 347–48
NIMBY, 1–2, 4, 7, 10, 12–13, 28–30, 34, 38, 57, 83–84, 106, 132, 137, 180, 199, 205, 209, 264, 290, 338,

NIMBY (*continued*)
354–56, 383, 395. *See also* public opposition to facilities
in Asia Pacific, 10
in California, 1, 25, 142–44, 150, 153, 161, 165, 170, 175, 177
in Canada, 9–11, 13, 24–26, 64, 77, 84, 86, 90, 92, 291, 346
in Europe, 25, 208, 209
in Germany, 9, 26, 209, 243, 247, 250, 251, 252, 256
in Japan, 10, 25, 181–82, 186, 190, 194–96, 208, 210
in USA, 1, 9, 25, 61, 84, 98, 100–1, 104, 108–11, 142–45, 148, 150–52, 157, 167–69, 174–78, 216, 395
NIMBY, contextual variables, 110, 128–32
NIMBY, nature and roots of, 2–3, 21, 258, 395
NKK Corporation, 182, 210, 212–13
non-hazardous waste facilities, 1, 290–315
Norway, 349, 354
nuclear power plants, 1, 29, 86–87, 88, 100, 111, 127, 191–92, 242, 246–47, 265, 299, 343
nuclear waste. *See* radioactive waste

OECD, 183, 212–13
Ohio, USA, 35–36, 217
oil refineries, 29, 191, 246
Öita, Japan, 200, 206, 220
on-site treatment. *See* waste treatment, on-site
Ontario, Canada, 24, 88, 97, 294, 300, 312, 314–18
Ontario Waste Management Corporation (OWMC) , 4, 11, 30, 81, 291, 230
Siting Task Force, 351–353
Ontario, Canada, siting in, 4, 30, 77, 85–86, 89–90, 93–95, 101, 103, 105, 291–92, 300, 339, 350–53
open houses, on siting, 81, 324, 326, 332
Osaka, Japan, 183, 193
override, 14–15, 22, 35, 147–48, 152, 168, 199
PCBs, 7, 9, 66, 79, 185, 202–3, 207, 210, 215, 239, 290, 304, 329
petro-chemical plants, 163–64, 183, 191, 192, 206
plebiscite, 61. *See also* referenda
Poland, 33
polluter pays principle, 198
pollution
air, 159, 163, 189, 191, 193, 206, 208, 211, 302, 311, 316, 342, 346, 370–71
water, 181, 191, 206, 303, 311, 384
pollution prevention, 4, 31–32, 191, 193, 206, 207, 215
pollution prevention agreements. *See* Japan
pollution, agricultural effects of, 305, 311
pollution, health effects of, 2, 7, 17, 27, 30, 108, 150, 159, 165, 171, 189–91, 205–6, 234, 242–44, 247, 290–95, 297, 299, 301–11, 314–17, 327, 332, 340, 367–69, 385, 390
pollution, psychosocial effects, 292–315
definition, 292
pollution, solution through dilution, 183
Port Hope, Canada, 89–90, 93–97
Portney, Kent, 3, 14, 16, 18, 36, 48, 79, 82–83, 110, 132, 137, 140, 182, 199, 202, 205, 209, 211, 219–20, 226, 292, 300, 317, 348, 356
preemption, 12–15, 19, 22, 35, 143–46, 149, 151–53, 199
property taxes, 340, 344–45, 378
property values, siting impact on, 16, 36, 171, 290–91, 294–95, 303, 306, 309–10, 328, 340, 344, 367–69, 372, 378, 388, 393
prospect theory, 28, 33, 350, 364–66, 375, 379–80, 383–84, 389–90, 393
public consultation, 338. *See also* public participation, public involvement
public education, 13–14, 35, 85, 96, 159, 161, 199, 331, 363, 381, 385–86, 388
public facility location theory, 27, 291, 294, 315

public (and collective) goods, 2, 209, 210
public health, 148, 150–51, 154, 167
public hearings, 264, 290–91, 326
in Canada, 59, 278–79, 323
in Germany, 236–237, 240, 244, 248–49, 254, 256
in USA, 101, 132, 135, 159, 161–62, 165, 170–71
public information, 174, 257, 274, 288, 312, 329, 353
in Germany, 241, 243, 259
public involvement, 105, 108, 148, 256, 257, 277, 312–13, 382, 389
in Germany, 237, 254, 257, 259, 261
in Japan. *See* citizens' movements
public opinion, 28–29, 123, 129–31, 190, 194, 217, 271–88, 290–315, 342–45
public opposition to facilities, 1–2, 4–5, 7, 9, 11–12, 14, 17, 25, 28, 30, 57, 77–80, 84, 108, 111, 127, 131, 151, 161, 164, 167, 170, 172, 178, 205, 220, 222, 255, 290–92, 294, 333, 338–41, 346, 358, 369, 381. *See also* NIMBY
Canada, 11–12, 60, 62, 64–67, 70–71, 76, 90, 94, 96, 274, 278–79, 302–5, 308, 362
Germany, 209, 230, 231, 234, 241, 243, 245, 247, 251, 253, 255–56
Japan, 10, 181–82, 186–96, 199–200, 208, 210
USA, 12, 13, 25, 96, 100, 104, 109–10, 114–23, 127, 129–35, 140, 143–78, 209
public participation, 2, 14, 22, 35, 37, 345, 347, 350, 382
in Germany, 241, 257, 258, 260
in Canada, , 89, 96, 98, 313, 321
in Japan, 26, 208. *See also* citizens' movements
in USA, 25, 100, 102–3, 108–9, 114–16, 128–36, 133, 144, 175, 200
public perception of risks, 108–11, 273–88, 290–315. *See also* NIMBY
public trust, 93, 98, 103, 109, 122, 124–25, 130–33, 142, 200–1, 234, 241, 285–86, 290, 303, 312, 326, 328, 331–32, 336, 351, 354, 362–63, 381
public utilities, for waste management, 199, 208. *See also* Ontario, OWMC

Rabe, Barry, 23–24, 28, 49, 79, 83, 84–106, 107, 133, 140, 180, 199, 210, 213, 257, 268, 292, 313, 317, 338, 344–45, 348, 356
radioactive waste, 24, 25, 28, 86–88, 92, 95, 111–12, 322
high-level, 86–87, 108–37, 213, 339, 343, 347
low-level, 84–106, 111, 213, 338, 339, 347, 350–52, 381, 388
radioactive waste facilities, 84–106, 108–37
recycling, 31, 34, 84–85, 212, 263, 312
in California, 146–47, 155, 160, 162, 164, 175
in Germany, 233–35, 244, 247, 250–51
in Japan, 198, 212, 214
referenda, 20, 22, 37, 75, 79–80, 339–40, 343, 350–52
in Canada, 57–58, 60–61, 63–67, 69, 71, 76, 94, 177, 322, 325–26, 329, 333–34, 382
in USA, 101, 103, 347–48
Reid-Crowther, Ltd, 58, 64, 79, 83
Rhone-Poulenc Basic Chemical Company, 143, 162–70, 179
risk assessment, 28, 243, 314, 383, 384, 389, 390
risk mitigation, 18, 22, 36, 72, 75, 199, 207–9, 211, 222, 312, 338, 340–43, 353, 372, 374
risk perception, 75, 273, 289, 342
risk reduction, 18, 20, 97, 167, 190, 199, 211, 371, 372. *See also* risk mitigation
risk sharing. *See* YIMBY
risk substitution, 18, 30, 199, 202, 208, 210, 211, 253
risks, transboundary, 339–42, 346, 348
risks, voluntary acceptance of, 75

San Francisco and San Francisco Bay area, 156, 164–65, 167–70, 175

Sanyu Plant Service, Japan, 195, 218
Schleswig Holstein, Germany, 231, 245, 246
siting, framing of issues, 256, 264, 366, 383, 389
siting "fair share" concept, 145, 149–51, 153, 155, 177
siting authority, public, 339, 341–42, 346, 348, 354
siting criteria, 35, 147, 149, 150–51, 153, 159, 163, 249–50, 255, 325, 333, 346–47, 358, 382
siting, equity of. *See* equity
siting facilities, benefits of, 344, 338
  economic benefits of, 21, 24, 60, 69, 70–72, 76–77, 97, 127, 145, 149, 173, 211, 240–41, 243, 326–28, 330, 339, 344, 348, 350, 361–65, 367, 374–75, 377, 379, 382–83, 388–89
siting, types of facilities
  airports, 1, 29, 105, 395
  amusement theme parks, 1, 395
  group residences, 1
  highways, 1, 29, 250, 395
  hospitals, 100, 105
  mental health facilities, 295
  non-nuclear power plants, 1, 183, 191–92
  nuclear power plants, 1, 29, 191, 265, 395
  oil refineries and chemical plants, 29, 192, 395
  prisons, 1, 367, 395
  public housing, 1, 105, 395
  research laboratories, 1, 395
siting facilities, legitimacy of, 258
siting facilities by voluntary processes. *See* voluntary choice
siting referenda. *See* referenda
siting, strategies for, 10–19, 22, 26, 97, 105, 145, 176, 198–211, 219, 231, 236, 242, 255–62, 338
siting, success of, 106, 134, 381. *See also* Alberta; Canada; Germany; Japan; Manitoba; Nebraska; Ontario; Swan Hills; Montcalm; and USA, siting in
social costs, 339–41
social networks, 292, 297, 299–300, 302, 304, 307–8
South Carolina, USA, 88, 99–101
SPEC (Society for the Prevention of Environmental Pollution), Canada, 277
St. Basile-le-Grande, 7, 290
'St. Florians', in Germany, 250, 256, 264
Strasbourg, France, 250–51
surveys. *See* public attitudes
Swan Hills, Alberta, siting in, 13, 22–23, 61–62, 66, 68, 78–80, 291, 325, 334, 344–45, 360–62, 383
Sweden, 30, 134, 135, 220
Switzerland, 347, 356

Texas, USA, 142, 164
thermal power plants, 183, 191–92
Three Mile Island, USA, 290, 297, 299, 316
Tokyo, Japan, 183, 186, 194, 195–97, 203, 212, 214, 218
Toronto, Canada, 294, 316, 318
toxic roundups, 385–88
transfer stations, 382–84, 386–87, 388–90
transparency, in siting, 75

United Kingdom, 36, 137, 215
United States, 6–7, 11, 13, 15, 24, 28, 31, 33, 35–37, 68, 84, 87, 181–84, 209, 214–16, 218, 220, 222, 265, 338, 345, 347, 356, 364
  Bush administration, 122, 125
  Central Midwest Interstate Compact, 381
  Carter administration, 112–13, 137
  Clinton administration, 35
  Congress, 29, 98–99, 113, 116–17, 120–21, 125–27, 129, 137, 139, 141, 220
  Department of Energy, 25, 111–32, 134, 136–37
  Energy Policy Act, 1992, 120
  Environmental Protection Agency, 8, 11, 28, 31–33, 35, 51, 113, 118, 120, 126, 141, 150, 165, 213–15, 227, 228
  General Accounting Office, 51, 112, 118–19, 122, 141, 213, 228
  Low-Level Radioactive Waste Policy Act, 1980, 98–101

United States (*continued*)
National Academy of Sciences, 115, 120–21, 124
Nuclear Regulatory Commission, 113, 115, 118–21, 128
nuclear waste negotiator, 125–28, 133, 347
Nuclear Waste Policy Act, and Amendments, 110, 113–21, 125, 128, 131, 136
Office of Technology Assessment, 28, 31, 36, 51, 136, 141, 213, 228
Reagan administration, 35, 116, 131
Resource Conservation and Recovery Act, 161, 218
Superfund, 32–33, 143, 220
United States, siting in, 8–9, 12, 16, 22, 24, 77, 85–87, 91, 96–106, 108–37, 161, 347–48
US Ecology Inc, 101–5
University of British Columbia, ix, 212, 274–81, 284–85, 287

Vancouver, Canada, 63–65, 274–75, 278, 288–89
voluntary choice siting, 19–28, 37, 74–75, 85, 103–5, 199, 210, 331, 333, 335, 339–41, 345–47, 350–54, 356, 382, 385–87, 391
in Alberta, Canada, 22–23, 24, 28, 85–86, 91
in Canada, 21–22, 24–25, 27–28, 57–58, 60, 62, 74–79, 86, 90–93, 96, 177, 291, 351
in Germany, 26, 260
in Japan, 203
in Manitoba, Canada, 24, 85, 322, 324, 325, 326, 327
in United States, 21–25, 85, 98, 102–5, 133, 330
voluntary choice, nature of, and link to auction, 22–23
volunteer communities, 1, 20, 37, 57, 59–61, 63–64, 66, 74–76, 85, 91, 93–94, 96, 102–5, 328, 346, 348, 351, 386

Washington, USA, 88, 99
waste facilities, 238, 276–77, 286, 290, 303, 313. *See also* facilities
waste management corporations (public-private), 233, 242
waste minimization, 4–6, 9, 18, 31, 175, 182. *See also* waste reduction
waste reduction, 4–6, 9, 18, 31–32, 84–85, 157–58, 160, 165, 168, 170, 175, 212–13, 234, 244, 312, 372, 386. *See also* waste minimization
in Canada, 87
in Germany, 233, 235, 244, 247, 250–51, 254
in Japan, 182, 185, 198, 202, 211
in USA, 6, 87, 147
link to facility siting, 4, 9, 97
waste storage, on site, 84, 86, 88
waste transfer stations, 28, 230, 233, 238, 360–61, 362
waste transport, 85, 93, 158, 159, 163, 165, 244, 263, 294, 322, 346, 352, 373
waste treatment
incineration. *See* incinerators
landfills. *See* landfills
on-site, 4–5, 18, 34, 36, 158, 176, 185, 210–11, 222
physical-chemical, 29, 62–63, 66, 232, 238, 242–43, 247, 322, 358
wastes (general), 4–5, 8, 19, 58, 84, 87–88, 91, 93–96, 99, 144–45, 151, 153, 163, 303, 311–13
biomedical, 84, 86, 274, 278, 279
hazardous. *See* hazardous waste
household, 13, 215, 385
radioactive. *See* radioactive waste
industrial, 182–84, 214, 220
solid and municipal, 26, 31, 33, 37, 84, 86–87, 211, 214–15, 220, 265, 274–75, 278, 289, 299, 313, 340, 343, 347, 380
wastes, volume of, 30–31. *See also* hazardous wastes, volume of
wastes, export of, 6, 142, 162, 196, 232–34, 263–64, 294
wastes, imports of, 18, 62, 72, 99, 210, 245, 286–87, 329
Winnipeg, Canada, siting in, 67–69, 71, 77–78, 83, 322, 331, 337
Wisconsin, USA, 115, 128, 220, 380

Yakuza, in Japan, 198
YIMBY, 18, 25, 199, 201, 259, 354
in Canada, 96, 98

YIMBY (*continued*)
in California, 18–19, 25, 154–55. *See also* 'fair share' concept
in Japan, 208
Yokkaichi, Japan, 191, 206
Yokohama, Japan, 182, 193, 195, 206, 210
Yucca Mountain, Nevada, 114, 116–20, 122–25, 128–30, 137, 140–41, 339

Zeiss, Christopher, 23, 28, 36, 53, 180, 199, 215, 228–29, 276, 289, 294, 338, 357, 358–391, 394